This book presents a detailed and mostly elementary exposition of the generalized Riemann integrals discovered by Henstock, Kurzweil, and McShane. Along with the classical results, it contains some recent developments connected with lipeomorphic change of variables, higher-dimensional multipliers, and the divergence theorem for discontinuously differentiable vector fields.

Defining the Lebesgue integral in Euclidean spaces from McShane's point of view has a clear pedagogical advantage, since the initial stages of development are both conceptually and technically simpler. The McShane integral evolves directly from the basic ideas about integration taught in elementary calculus. The difficult transition from subdividing the domain to subdividing the range, intrinsic to the Lebesgue definition, is completely bypassed. The unintuitive Carathéodory concept of measurability is also made more palatable by means of locally fine partitions.

The mathematical significance of the generalized Riemann integrals emerges when the Henstock–Kurzweil approach is used to define the Denjoy–Perron integral. While there is little similarity between the classical definitions of Lebesgue, Denjoy, and Perron, the Lebesgue and Denjoy–Perron integrals are naturally connected via the McShane and Henstock–Kurzweil definitions. This is used to obtain a coordinate free multidimensional integral which provides an unrestricted Gauss–Green theorem for vector fields with large sets of singularities.

Although written as a monograph, the book can be used as a graduate text, and certain portions of it can be presented even to advanced undergraduate students with a working knowledge of limits, continuity, and differentiation on the real line.

CAMBRIDGE TRACTS IN
MATHEMATICS

General Editors
B. BOLLOBAS, P. SARNAK, C.T.C. WALL

109   The Riemann approach to integration

Washek F. Pfeffer

Department of Mathematics
University of California, Davis

The Riemann approach to integration

Local geometric theory

CAMBRIDGE
UNIVERSITY PRESS

Published by the Press Syndicate of the University of Cambridge
The Pitt Building, Trumpington Street, Cambridge CB2 1RP
40 West 20th Street, New York, NY 10011, USA
10 Stamford Road, Oakleigh, Melbourne 3166, Australia

First published 1993

Printed in the United States of America

Library of Congress Cataloging-in-Publication Data
Pfeffer, Washek F.
The Riemann approach to integration / Washek F. Pfeffer.
   p.   cm. – (Cambridge tracts in mathematics; 109)
Includes bibliographical references and indexes.
ISBN 0-521-44035-1
1. Riemann integral.   I. Title.   II. Series.
QA311.P44   1993
515′.43 – dc20   93–18565
                                 CIP

A catalog record for this book is available from the British Library.

ISBN 0-521-44035-1 hardback

To Lida

Contents

Preface   *page* **xi**

Acknowledgments   **xv**

I   One-dimensional integration

1   Preliminaries   **3**
   1.1   Lengths   4
   1.2   Partitions   5
   1.3   Stieltjes sums   7

2   The McShane integral   **8**
   2.1   The integral   8
   2.2   Absolute integrability   12
   2.3   Convergence theorems   17
   2.4   Connections with derivatives   24
   2.5   Gap functions   30
   2.6   Integration by parts   32

3   Measure and measurability   **37**
   3.1   Extended real numbers   37
   3.2   Measures   38
   3.3   Measurable sets   44
   3.4   Calculating measures   52
   3.5   Negligible sets   55
   3.6   Measurable functions   56
   3.7   The α_A-measure   60

4   Integrable functions   **64**
   4.1   Integral and measure   64
   4.2   Semicontinuous functions   69
   4.3   The Perron test   71
   4.4   Approximations   77

5   Descriptive definition   **81**
   5.1   AC functions   81
   5.2   Covering theorems   86

| | 5.3 | Differentiation | 91 |
| | 5.4 | Singular functions | 98 |

6	**The Henstock–Kurzweil integral**	**102**
	6.1	The P-integral	102
	6.2	Integration by parts	108
	6.3	Connections with measures	111
	6.4	AC$_*$ functions	115
	6.5	Densities	118
	6.6	Almost differentiable functions	120
	6.7	Gages and calibers	124

II Multidimensional integration

7	**Preliminaries**	**133**
	7.1	Intervals	134
	7.2	Volumes	136
	7.3	Partitions	139
	7.4	Stieltjes sums	141

8	**The McShane integral**	**143**
	8.1	The integral	143
	8.2	Dirac volumes	146
	8.3	The divergence theorem	148
	8.4	Measures and measurability	153
	8.5	The Perron test	158
	8.6	The Fubini theorem	161

9	**Descriptive definition**	**168**
	9.1	AC functions	168
	9.2	Covering theorems	170
	9.3	Derivability	174

10	**Change of variables**	**180**
	10.1	Integrating over a set	180
	10.2	Luzin maps	185
	10.3	Lipschitz maps	188
	10.4	The Rademacher theorem	193
	10.5	The main formula	198
	10.6	Almost differentiable maps	203

11   The gage integral   **207**
    11.1   A motivating example   207
    11.2   Continuous additive functions   211
    11.3   Gages and calibers   214
    11.4   The g-integral   218
    11.5   Improper integrals   225
    11.6   Connections with the McShane integral   228
    11.7   Almost derivable functions   230

12   The $\mathcal{F}$-integral   **239**
    12.1   Shape and regularity   239
    12.2   The $\mathcal{F}$-integral   242
    12.3   Derivability relative to $\mathcal{F}$ 246
    12.4   Integration by parts   251
    12.5   The quasi-Hausdorff measure   253
    12.6   Solids   258
    12.7   Change of variables   261
    12.8   Multipliers   265

13   Recent developments   **273**
    13.1   The $\mathcal{S}$-integral   273
    13.2   The perimeter   278
    13.3   The flux   285
    13.4   The $\mathcal{BV}$-integral   287

Bibliography   **293**

List of symbols   **296**

Index   **299**

Preface

> If we think of the Lebesgue integral as God sent, then differentiable functions whose derivatives are not Lebesgue integrable may appear evil. The challenge is to resolve the conflict, particularly in higher dimensions.

This book presents a detailed and mostly elementary exposition of the generalized Riemann–Stieltjes integrals discovered by Henstock, Kurzweil, and McShane more than thirty years ago. Aside from the classical results, it contains some recent developments connected with lipeomorphic change of variables, higher-dimensional multipliers, and the divergence theorem for discontinuously differentiable vector fields.

Roughly speaking, the generalized Riemann integrals differ from the classical Riemann integral in that *uniformly fine* partitions of the integration domain are replaced by *locally fine* partitions. This idea is perhaps best explained by looking at the numerical evaluation of integrals by means of rectangular approximations. For instance, suppose we want to evaluate the integral

$$\int_{10^{-3}}^{10^3} t^{-1} \sin t^{-1}\, dt\,.$$

The behavior of the integrand in the intervals $[10^{-3}, 1]$ and $[1, 10^3]$ is very different: it oscillates rapidly in the first interval and decreases steadily to zero in the second. From this observation, it is easy to conclude that employing uniformly fine (e.g., equidistant) partitions would be wasteful. The most efficient evaluation of the integral is provided by a rectangular approximation based on a partition of $[10^{-3}, 10^3]$ that is *fine* in $[10^{-3}, 1]$ and *coarse* in $[1, 10^3]$. Such a partition is an example of a locally fine partition of the interval $[10^{-3}, 10^3]$. In general, the mesh of a locally fine partition varies from point to point. The full impact of this idea is well illustrated by Example 2.2.9 below.

The ingenious passage from uniformly fine to locally fine partitions has profound consequences: instead of the classical Riemann integral we obtain the Lebesgue and Denjoy–Perron integrals, depending upon whether the Mc-Shane or Henstock–Kurzweil approach is used.

While in the final analysis there is no substitute for the standard definition of the Lebesgue integral in an abstract measure space, defining the Lebesgue integral in Euclidean spaces from McShane's point of view has a clear pedagogical advantage: the initial stages of development are appreciably simpler, both conceptually and technically. The McShane integral evolves naturally from the initial ideas about integration we learn in the first courses of calculus. The traumatic transition from subdividing the domain to subdividing the range, intrinsic to the Lebesgue definition, is completely bypassed. The unintuitive Carathéodory concept of measurability is also made more palatable by means of locally fine partitions.

What has been said about the Lebesgue integral is true to an even larger degree when the Henstock–Kurzweil approach is used to define the Denjoy–Perron integral. While there is little similarity between the classical definitions of Lebesgue, Denjoy, and Perron, introducing the Lebsgue integral by McShane's definition provides a natural path to the more delicate integral of Henstock and Kurzweil. In my opinion, the successful multidimensional generalizations of the Denjoy–Perron integral obtained during the past decade are due mainly to the simplicity of the Henstock–Kurzweil definition.

Beyond advanced calculus, the prerequisites for understanding this book amount to little more than mathematical sophistication. My primary goal has been to make the material understandable to beginners whose background does not exceed the first year of graduate school. In fact, a large portion should be accessible to any undergraduate student who has a working knowledge of limits, continuity, and differentiation on the real line. On the other hand, it is only fair to admit that some familiarity with the Lebesgue integral will help the reader to appreciate the subject. While certain aspects of Perron and descriptive Denjoy integration enter implicitly into the exposition, the constructive Denjoy integral is never used.

Parts I and II of the book deal with one-dimensional and multidimensional integration, respectively. Although formally the one-dimensional McShane integral is a special case of the m-dimensional one, I believe that integration in higher dimensions is conceptually more complicated and often difficult to understand by those who are not properly initiated on the real line. This is particularly true for the conditionally convergent integrals of Henstock–Kurzweil type: here even the formal similarity breaks down when passing to dimensions greater than one. The unavoidable consequence of the two-level presentation is a certain amount of repetition, which I have strived to minimize. It is limited to those cases when repeating an argument from a slightly different point of view enhances understanding. Trivial repetitions are invariably left to the reader.

The book is organized into thirteen chapters. The first two are devoted to a completely elementary and self-contained development of the McShane integral on the real line, including the convergence theorems and integration by parts. These chapters, which elaborate my paper [39], are inspired by the original work of McShane (see [28] and [29]). They are fully accessible to more advanced undergraduate students; I have covered them successfully in several undergraduate courses.

In Chapter 3, locally fine partitions are used to define measures on the real line. This is conceptually more difficult but should present few problems to the readers who absorbed Chapters 1 and 2. All the basic properties of measures and measurable sets are carefully derived, and measurable functions are introduced. We reconcile the additivity of generalized length on nonoverlapping intervals with that of the induced measure on disjoint measurable sets.

The relationship between measure and the McShane integral is investigated in Chapter 4. We show that the measurability of a bounded set is equivalent to the integrability of its characteristic function, and that the value of the integral equals the measure of the set. Using sets of measure zero, we extend the definition of the integral to functions which have infinite values and are defined only almost everywhere. The main result is the Vitali–Carathéodory theorem, which is used to show that the McShane and Lebesgue integrals are equivalent.

Chapter 5 moves to the more demanding field of differentiation. It contains the covering theorems of Vitali and Besicovitch, the Radon–Nikodym theorem, and the Lebesgue decomposition of increasing functions.

The Henstock–Kurzweil integral is introduced in Chapter 6. We establish its main properties, including the relationship to the McShane integral. The central topic is the evolution of the fundamental theorem of calculus that leads to the multidimensional generalization presented in Chapter 11. Here the Stepanoff theorem plays an important role. We prove it in the real line only, but in such a way that the generalization to any dimension is obvious.

Chapters 7, 8, and 9, which are the beginning chapters of Part II, contain results concerning the multidimensional McShane integral. The divergence theorem is obtained independently of Fubini's theorem. Its proof is attuned to the spirit of generalized Riemann integration and points to further generalizations discussed in Chapters 11, 12, and 13.

In Chapter 10 we show that the McShane integral is invariant with respect to changes of coordinates that are more general than Lipschitz. This result is derived from the equivalence between the McShane and Lebesgue integrals, and does not have a direct relationship to McShane's definition.

Chapters 11, 12, and 13 deal with multidimensional generalizations of the Henstock–Kurzweil integral. They are motivated by the desire to recover the flux of any differentiable vector field by integrating its divergence, which may not be Lebesgue integrable. The topics are relatively new and will introduce the reader to current research in the area. For various reasons given in Chapter 13, it appears that the most useful generalization is the $\mathcal{BV}$-integral utilizing sets of finite perimeter. We sketch its definition and hint at some of its properties without going into details. Many interesting results concerning this integral have been obtained in recent years, but a definitive treatment awaits future development. The area remains a subject of vigorous research and, in my view, is not ready for presentation in book form. The interested reader is referred to papers [19], [20], [21], [24], [37], [36], [25], and [38].

Numerous exercises are scattered throughout the text, the majority of them containing easily provable results that form an integral part of the exposition. Thus a diligent effort should be made to work them out.

For completeness, proofs of several well-known theorems have been included. No originality is claimed: these proofs are simple adaptations of those found in standard texts such as [11], [43], [44], and [49].

I consciously avoided abstractions such as Henstock's division spaces, generalized limits, or functions with values in Banach spaces. The readers who wish to pursue a more abstract approach to Riemann integration may consult the books of Henstock ([14], [15], [16]), Kurzweil ([23]), and McShane ([28], [30]).

The present book contains no historical comments.

Davis, California   W. F. P.

Acknowledgments

It is a pleasure to acknowledge the contributions of my teachers, colleagues, and friends towards the completion of this work.

I was introduced to nonabsolutely convergent integration by J. Mařík, and learned about the Riemann approach to it from the works of R. Henstock, J. Kurzweil, and the late E. J. McShane; many of my ideas grew out of personal contacts with McShane. The present book may not have been written, however, if it were not for the encouragement and moral support I received from P. S. Bullen and R. D. Mouldin during the time I worked on the divergence theorems for discontinuously differentiable vector fields.

A major portion of the book was thought out during my visits to the University of Palermo in Italy, the Royal Institute of Technology in Sweden, and the Catholic University of Louvain in Belgium. I am indebted to B. Bongiorno, M. Giertz, and J. Mawhin for inviting me to their respective institutions and for their willingness to collaborate on problems related to the theme of this book. The invitations by A. Volčič to lecture on the generalized Riemann integral at the School of Measure Theory and Real Analysis in Grado, Italy, helped me greatly to organize the presented material.

During the preparation of the manuscript I benefited from discussions with N. L. Burkett, G. D. Chakerian, M. Chlebík, A. Fialowski, R. J. Gardner, C. Gorez, M. A. Jodeit, W. B. Jurkat, J. Král, J. G. Kupka, P. Y. Lee, P. Mattila, M. Miranda, D. J. F. Nonnenmacher, M. J. Paris, K. Prikry, P. Pucci, A. Salvadori, I. Tamanini, and R. Výborný. A thoughtful criticism given by B. S. Thomson substantially improved the exposition; in particular, he suggested the treatment of measures adopted in Chapter 3. Z. Buczolich constructed some essential examples; J. W. Mortensen and A. Novikov coauthored the main results of Sections 12.8 and 12.7, respectively.

Several improvements are due to R. J. Battig and A. S. Jiang, who read various segments of the manuscript. E. J. Howard and J. W. Mortensen kindly agreed to undertake the tedious task of proofreading the final version and weeded out a multitude of misprints and errors; those which remain are the sole responsibility of the author. Throughout this writing V. H. DuBose and N. R. Staargaard selflessly provided on-line help with the intricacies of English grammar. In this regard I am also obliged to the Cambridge University Press, in particular to L. C. Gruendel and R. S. Wells, for their editorial help.

W. F. P.

Part I

One-dimensional integration

Chapter 1

Preliminaries

If E is a set and Π is a property of the elements of E, we denote by

$$\{x \in E : \Pi(x)\}$$

the set of all elements of E that have the property Π. A *countable set* is either finite or countably infinite. A *map* $f : X \to Y$ from a set X into a set Y is a set $f \subset X \times Y$ such that for each $x \in X$ there is a unique $y \in Y$ with $(x, y) \in f$. As usual, we write $y = f(x)$ instead of $(x, y) \in f$. We note that if $X = \emptyset$ (the empty set), then for any set Y there is a unique map $f : X \to Y$, namely the *empty map* $f = X \times Y = \emptyset$. An *enumeration* of a countable set C is a one-to-one map $n \mapsto c_n$ from a finite or infinite set $\{p, p + 1, \ldots\}$ of integers onto C; in most cases $p = 1$, but this is not required.

The set of all real numbers is denoted by $\mathbf{R}$. An *interval* is a set

$$[a, b] = \{x \in \mathbf{R} : a \le x \le b\}$$

where $a, b \in \mathbf{R}$. We say that an interval $[a, b]$ is *degenerate* if $a \ge b$. A nondegenerate interval is called a *cell*. Thus an interval is a cell if and only if its interior is nonempty. The intersection of two intervals is again an interval; however, the intersection of two cells need not be a cell. A collection of intervals is called *nonoverlapping* if their interiors are disjoint. Note that a degenerate interval overlaps no interval (including itself). Aside from intervals, we shall consider the *segments*

$$(a, b) = \{x \in \mathbf{R} : a < x < b\},$$
$$[a, b) = \{x \in \mathbf{R} : a \le x < b\},$$
$$(a, b] = \{x \in \mathbf{R} : a < x \le b\}.$$

For an $x \in \mathbf{R}$ and a real number $r > 0$, we let

$$U(x,r) = (x - r, x + r) \qquad \text{and} \qquad U[x,r] = [x - r, x + r].$$

The *closure*, *interior*, and *boundary* of a set $E \subset \mathbf{R}$ are denoted by E^-, E°, and ∂E, respectively.

A *function* on a set E is a map $f : E \to \mathbf{R}$. Thus, in the absence of qualifying attributes, functions are assumed to be *real-valued*. When no confusion is possible, we use the same symbol to denote a function on a set E as well as its restrictions to various subsets of E (cf. Propositions 2.1.9 and 2.1.10). The algebraic operations and order among functions on the same set are defined *pointwise*. For instance, if f and g are functions defined on a set E, then

$$(f + g)(x) = f(x) + g(x)$$

for all $x \in E$, and $f \le g$ means that $f(x) \le g(x)$ for each $x \in E$.

Let f be a function on a set $E \subset \mathbf{R}$. We say that f is, respectively, *increasing* or *decreasing* whenever

$$f(x) \le f(y) \qquad \text{or} \qquad f(x) \ge f(y)$$

for all $x, y \in E$ with $x < y$; if the above inequalities are strict, we say that f is *strictly increasing* or *strictly decreasing*, respectively. The function f is called *monotone* if it is either increasing or decreasing, and *strictly monotone* if it is either strictly increasing or strictly decreasing.

1.1 Lengths

Let F be a function on a set $S \subset \mathbf{R}$. For an interval $[a, b] \subset S$, we let

$$F([a, b]) = \begin{cases} F(b) - F(a) & \text{if } a \le b, \\ 0 & \text{if } a > b. \end{cases}$$

Thus the function F of points of S determines uniquely a function of subintervals of S, still denoted by F and called the *associated interval function*. It is easy to verify that the associated interval function is *additive* in the following sense: if an interval $B \subset S$ is the union of nonoverlapping intervals C and D, then

$$F(B) = F(C) + F(D).$$

On the other hand, let G be a function defined on all subintervals of an interval $A = [a, b]$ that is additive in the above sense. If D is a degenerate subinterval of A, then

$$G(D) = G(D \cup D) = G(D) + G(D)$$

and consequently, $G(D) = 0$. Choose a $t \in \mathbf{R}$, and define a function G_t on A by setting

$$G_t(x) = G([a, x]) + t$$

for every $x \in A$. When $[c, d]$ is a subinterval of A with $c \leq d$, then

$$G_t([c, d]) = G_t(d) - G_t(c) = G([a, d]) - G([a, c]) = G([c, d]) \,.$$

It follows that G is associated with G_t for each t.

Let α be an *increasing* function on a set $S \subset \mathbf{R}$. Then the associated interval function α is nonnegative, and we call it a *length* in S. If B is a subinterval of S, the nonnegative number $\alpha(B)$ is called the α-*length* of B.

By λ we denote the *identity function* on $\mathbf{R}$, i.e., $\lambda(x) = x$ for each $x \in \mathbf{R}$. The length in $\mathbf{R}$ associated with λ is called the *Lebesgue length*. If $A = [a, b]$ is an interval with $a \leq b$, then $\lambda(A) = b - a$ and we see that the λ-length of A is the *usual length* of this interval.

Convention. Throughout Part I of this book, we always tacitly assume that an *increasing* function α is defined on each cell in question.

1.2 Partitions

Definition 1.2.1 A *Lebesgue partition*, or simply a *partition*, is a collection

$$P = \{(A_1, x_1), \ldots, (A_p, x_p)\}$$

where $A_1, \ldots, A_p$ are nonoverlapping cells, and $x_1, \ldots, x_p$ are points of $\mathbf{R}$. If $x_i \in A_i$ for $i = 1, \ldots, p$, we call P a *Perron partition*, abbreviated as P-partition.

Note. In a partition $P = \{(A_1, x_1), \ldots, (A_p, x_p)\}$, either Lebesgue or Perron, the points $x_1, \ldots, x_p$ need not be distinct. When P is a Perron partition, the map $i \mapsto x_i$ is at most two-to-one. On the other hand, if P is a Lebesgue partition, then for no $i = 1, \ldots, p$ is it necessary that $x_i \in A_i$, and all points x_i may coalesce.

Definition 1.2.2 Let δ be a positive function defined on a set $E \subset \mathbf{R}$. A partition $\{(A_1, x_1), \ldots, (A_p, x_p)\}$ with $\{x_1, \ldots, x_p\} \subset E$ is called δ-*fine* whenever $A_i \subset U(x_i, \delta(x_i))$ for $i = 1, \ldots, p$.

Remark 1.2.3 It is important to keep in mind that the empty collection is also a partition, called the *empty partition*. Trivially, the empty partition $P = \emptyset$ is a Perron partition that is δ-fine for any positive function δ defined on any set $E \subset \mathbf{R}$. We remind the reader that the empty function $\delta = \emptyset$ on the empty set $E = \emptyset$ is trivially positive. These facts will be used, often tacitly, in Chapter 3.

Let $P = \{(A_1, x_1), \ldots, (A_p, x_p)\}$ be a partition and let A be a cell. If $\{x_1, \ldots, x_p\}$ and $\bigcup_{i=1}^p A_i$ are subsets of A, we say that P is a partition *in* A; if, in addition, $A = \bigcup_{i=1}^p A_i$, we say that P is a partition *of* A.

The following proposition, often referred to as *Cousin's lemma*, is pivotal for our further exposition.

Proposition 1.2.4 *For each positive function δ on a cell A, there is a δ-fine Perron partition P of A.*

PROOF. Let $A = [a, b]$ and let $c \in (a, b)$. If P_a and P_b are δ-fine P-partitions of the cells $[a, c]$ and $[c, b]$, respectively, then $P = P_a \cup P_b$ is a δ-fine P-partition of A. Using this observation, we proceed by contradiction. Assuming that the proposition is false, we construct inductively cells

$$A = A_0 \supset A_1 \supset \ldots$$

such that for $n = 0, 1, \ldots$, no δ-fine P-partition of A_n exists and $\lambda(A_n) = (b - a)/2^n$. The nested intervals theorem ([42, Theorem 2.38, p. 38]) implies that

$$\bigcap_{n=0}^{\infty} A_n = \{z\}$$

for a point $z \in A$. Since $\delta(z) > 0$, we can find an integer $k \geq 0$ so that $\lambda(A_k) < \delta(z)$. Thus $\{(A_k, z)\}$ is a δ-fine P-partition of A_k, contrary to our assumption. ∎

Corollary 1.2.5 *Let δ be a positive function on a cell A. Each δ-fine partition P in A is a subset of a δ-fine partition Q of A, which is Perron whenever P is Perron.*

PROOF. Let $P = \{(A_1, x_1), \ldots, (A_p, x_p)\}$ and let $B_1, \ldots, B_k$ be cells such that $\{A_1, \ldots, A_p, B_1, \ldots, B_k\}$ is a nonoverlapping family with

$$\left(\bigcup_{i=1}^p A_i \right) \cup \left(\bigcup_{j=1}^k B_j \right) = A.$$

By Cousin's lemma, there are δ-fine P-partitions P_j of B_j for $j = 1, \ldots, k$. Thus $Q = P \cup (\bigcup_{j=1}^{k} P_k)$ is the desired δ-fine partition of A. $\blacksquare$

Note. Notwithstanding its simplicity, Cousin's lemma is a nontrivial result that is equivalent to the compactness of a cell A. Indeed, let $\mathcal{U}$ be an open cover of A. For each $x \in A$ there is a positive number $\delta(x)$ such that $U(x, \delta(x))$ is contained in some $U \in \mathcal{U}$. By Cousin's lemma, there is a δ-fine partition $\{(A_1, x_1), \ldots, (A_p, x_p)\}$ of A. Clearly, each A_i is contained in some $U_i \in \mathcal{U}$, and consequently $\{U_1, \ldots, U_p\} \subset \mathcal{U}$ is a finite cover of A. This establishes the compactness of A.

1.3 Stieltjes sums

Let α be an increasing function on a cell A, and let $P = \{(A_1, x_1), \ldots, (A_p, x_p)\}$ be a partition in A. For any function f on $\{x_1, \ldots, x_p\}$, we set

$$\sigma(f, P; \alpha) = \sum_{i=1}^{p} f(x_i)\alpha(A_i)$$

and call this number the α-*Stieltjes sum*, or simply the *Stieltjes sum*, of f associated with P. Often we write $\sigma(f, P)$ or $\sigma(P)$ instead of $\sigma(f, P; \alpha)$. Since the empty sum is by definition equal to zero, we see that $\sigma(f, P; \alpha) = 0$ whenever any of the following cases occurs: $f = 0$, $P = \emptyset$, or α is a constant.

If f is a function defined on a cell A, then an *integral* of f over A with respect to α is a real number that can be approximated by α-Stieltjes sums of f. Depending on the method of approximation, we obtain different kinds of integrals.

Chapter 2
The McShane integral

In this chapter we define the generalized Riemann integral of McShane and develop it into a powerful analytic tool by establishing most of its classical properties.

2.1 The integral

Definition 2.1.1 Let α be an increasing function on a cell A. A function f on A is called *integrable* over A with respect to α if there is a real number I having the following property: given $\varepsilon > 0$, we can find a positive function δ on A such that

$$|\sigma(f, P; \alpha) - I| < \varepsilon$$

for each δ-fine partition P of A.

Recall that $\sigma(f, P; \alpha)$ denotes the Stieltjes sum defined in Section 1.3.

Remark 2.1.2 Cousin's lemma implies that the number I from Definition 2.1.1 is determined uniquely by the integrable function f. Indeed, if a number J different from I also satisfies Definition 2.1.1, let $\varepsilon = |I - J|/2$, and find positive functions δ_I and δ_J on A so that $|\sigma(f, P; \alpha) - I| < \varepsilon$ for each δ_I-fine partition P of A, and $|\sigma(f, P; \alpha) - J| < \varepsilon$ for each δ_J-fine partition P of A. Now let $\delta = \min\{\delta_I, \delta_J\}$, and use Cousin's lemma to find a δ-fine partition P of A. Then

$$|I - J| \leq |I - \sigma(f, P; \alpha)| + |\sigma(f, P; \alpha) - J| < 2\varepsilon = |I - J|,$$

which is a contradiction.

In view of Remark 2.1.2, we call the number I from Definition 2.1.1 the *integral* of f over A with respect to α, denoted by $\int_A f \, d\alpha$ or $\int_a^b f \, d\alpha$ if $A = [a, b]$. When no confusion is possible, we use also the symbols $\int_A f$ or $\int_a^b f$. By $\mathcal{R}(A, \alpha)$ we denote the family of all functions integrable over A with respect to α.

Convention. For technical reasons, it is useful to let $\int_A f \, d\alpha = 0$ whenever $A = [a, b]$ is a degenerate interval; in particular, $\int_a^a f \, d\alpha = 0$.

Note. In the literature, the integral $\int_A f \, d\alpha$ is usually called the *generalized Riemann integral of McShane* (cf. [27, Section S8.3, p. 237]), or simply the *McShane integral*. Clearly, Definition 2.1.1 remains meaningful if we use only P-partitions. This modification leads to the integral of Henstock and Kurzweil, which is quite different from that of McShane; we shall discuss it in Chapter 6. In the present chapter, P-partitions will be used only as an auxiliary tool for obtaining certain properties of the McShane integral (see Theorems 2.4.8 and 2.6.1).

Proposition 2.1.3 *Let A be a cell, let f and g be in $\mathcal{R}(A, \alpha)$, and let c be a real number. Then $f + g$ and cf belong to $\mathcal{R}(A, \alpha)$, and*

$$\int_A (f + g) \, d\alpha = \int_A f \, d\alpha + \int_A g \, d\alpha \quad \text{and} \quad \int_A (cf) \, d\alpha = c \int_A f \, d\alpha \, .$$

If, in addition, $f \le g$, then $\int_A f \, d\alpha \le \int_A g \, d\alpha$.

PROOF. Observe that for each partition P of A, we have

$$\sigma(f + g, P; \alpha) = \sigma(f, P; \alpha) + \sigma(g, P; \alpha) \, ,$$
$$\sigma(cf, P; \alpha) = c\sigma(f, P; \alpha) \, .$$

Moreover, if $f \le g$ then $\sigma(f, P; \alpha) \le \sigma(g, P; \alpha)$. The proposition follows. ▮

Note. The statement of Proposition 2.1.3 is usually abbreviated as follows: *The set $\mathcal{R}(A, \alpha)$ is a linear space of real-valued functions, and the map $f \mapsto \int_A f \, d\alpha$ is a nonnegative linear functional on $\mathcal{R}(A, \alpha)$.*

Corollary 2.1.4 *If $f \in \mathcal{R}(A, \alpha)$ and $|f| \le c$ for some $c \in R$, then*

$$\left| \int_A f \, d\alpha \right| \le c\alpha(A) \, .$$

PROOF. Observe that $\int_A c \, d\alpha = c\alpha(A)$, and apply Proposition 2.1.3 to $\int_A f \, d\alpha$ and $\int_A (-f) \, d\alpha$. ▮

Proposition 2.1.5   *Let α and β be increasing functions on a cell A, let $c \geq 0$, and let $f \in \mathcal{R}(A, \alpha) \cap \mathcal{R}(A, \beta)$. Then $f \in \mathcal{R}(A, \alpha + \beta) \cap \mathcal{R}(A, c\alpha)$ and*

$$\int_A f \, d(\alpha + \beta) = \int_A f \, d\alpha + \int_A f \, d\beta \qquad and \qquad \int_A f \, d(c\alpha) = c \int_A f \, d\alpha \, .$$

If $\beta = \alpha + c$, then $\mathcal{R}(A, \alpha) = \mathcal{R}(A, \beta)$ and $\int_A f \, d\alpha = \int_A f \, d\beta$.

PROOF.   Observe that for each partition P of A, we have

$$\sigma(f, P; \alpha + \beta) = \sigma(f, P; \alpha) + \sigma(f, P; \beta) \, ,$$

$$\sigma(f, P; c\alpha) = c\sigma(f, P; \alpha) \, .$$

Moreover, if $\beta = \alpha + c$, then $\beta(B) = \alpha(B)$ for each interval $B \subset A$. ∎

Corollary 2.1.6   *If α is a constant function on a cell A, then each function f on A belongs to $\mathcal{R}(A, \alpha)$ and $\int_A f \, d\alpha = 0$.* ∎

Remark 2.1.7   It follows from the last statement of Proposition 2.1.5 that the integral $\int_A f \, d\alpha$ depends on the length associated with α, rather than on the increasing function α itself. There are compelling reasons, however, why we prefer to start with an increasing function.

- Tradition, which stems from the simple relationship between the two concepts described in Section 1.1.
- Conceptual simplicity, which is important from the pedagogical point of view.
- The simplicity of the general formula for integration by parts (Theorem 2.6.1).

All these reasons disappear in higher dimensions, and so accordingly the starting point of Part II will be an additive function of cells.

Our next goal is to show that a function f integrable over a cell A is also integrable over each cell $B \subset A$. As we are not able to guess the value of $\int_B f \, d\alpha$, Definition 2.1.1 cannot be used directly. Instead we need a tool referred to as *Cauchy's test* for integrability.

Proposition 2.1.8   *A function f on a cell A belongs to $\mathcal{R}(A, \alpha)$ if and only if for each $\varepsilon > 0$ there is a positive function δ on A such that*

$$|\sigma(f, P; \alpha) - \sigma(f, Q; \alpha)| < \varepsilon$$

for all δ-fine partitions P and Q of A.

PROOF. For $n = 1, 2, \ldots$, choose a positive function δ_n on A so that the condition of the proposition is satisfied for $\varepsilon = 1/n$. Replacing δ_n by $\min\{\delta_1, \ldots, \delta_n\}$, we may assume that $\delta_1 \geq \delta_2 \geq \ldots$. If P_n is a δ_n-fine partition of A, then it is easy to verify that the sequence $\{\sigma(P_n)\}$ is Cauchy and hence convergent ([42, Theorem 3.11(c), p. 53]). Let $I = \lim \sigma(P_n)$, and choose an $\varepsilon > 0$. There is an integer $k > 2/\varepsilon$ such that $|\sigma(P_k) - I| < \varepsilon/2$, and we set $\delta = \delta_k$. Now if P is a δ-fine partition of A, then

$$|\sigma(P) - I| \leq |\sigma(P) - \sigma(P_k)| + |\sigma(P_k) - I| < \frac{1}{k} + \frac{\varepsilon}{2} < \varepsilon.$$

Consequently, $I = \int_A f$ and $f \in \mathcal{R}(A, \alpha)$. The converse is obvious. $\blacksquare$

Proposition 2.1.9 *If A is a cell and $f \in \mathcal{R}(A, \alpha)$, then $f \in \mathcal{R}(B, \alpha)$ for each cell $B \subset A$.*

PROOF. Choose an $\varepsilon > 0$ and find a positive function δ on A so that $|\sigma(P) - \sigma(Q)| < \varepsilon$ for each pair of δ-fine partitions P and Q of A. If B is a proper subcell of A, then either A is the union of nonoverlapping cells B and C, or A is the union of nonoverlapping cells B, C, and D. As both cases are similar, we consider only the latter. By Cousin's lemma, there are δ-fine partitions P_C and P_D of C and D, respectively. Now if P_B and Q_B are δ-fine partitions of B, then $P = P_B \cup P_C \cup P_D$ and $Q = Q_B \cup P_C \cup P_D$ are δ-fine partitions of A. As

$$\sigma(P) = \sigma(P_B) + \sigma(P_C) + \sigma(P_D),$$
$$\sigma(Q) = \sigma(Q_B) + \sigma(P_C) + \sigma(P_D),$$

we have

$$\varepsilon > |\sigma(P) - \sigma(Q)| = |\sigma(P_B) - \sigma(Q_B)|.$$

An application of Cauchy's test completes the argument. $\blacksquare$

Proposition 2.1.10 *Let f be a function on a cell $[a, b]$, and let $c \in (a, b)$. If $f \in \mathcal{R}([a, c], \alpha) \cap \mathcal{R}([c, b], \alpha)$, then $f \in \mathcal{R}([a, b], \alpha)$ and*

$$\int_a^b f \, d\alpha = \int_a^c f \, d\alpha + \int_c^b f \, d\alpha.$$

PROOF. Set $I = \int_a^c f + \int_c^b f$, and choose an $\varepsilon > 0$. There are positive functions δ_a and δ_b on the cells $[a, c]$ and $[c, b]$, respectively, such that

$$\left| \sigma(P_a) - \int_a^c f \right| < \frac{\varepsilon}{2} \qquad \text{and} \qquad \left| \sigma(P_b) - \int_c^b f \right| < \frac{\varepsilon}{2}$$

for each δ_a-fine partition P_a of $[a, c]$ and each δ_b-fine partition P_b of $[c, b]$. Define a positive function δ on $[a, b]$ by setting

$$\delta(x) = \begin{cases} \min\{\delta_a(x), c - x\} & \text{if } x < c, \\ \min\{\delta_b(x), x - c\} & \text{if } x > c, \\ \min\{\delta_a(c), \delta_b(c)\} & \text{if } x = c. \end{cases}$$

Now choose a δ-fine partition $P = \{(A_1, x_1), \ldots, (A_p, x_p)\}$ of $[a, b]$. From the choice of δ, we see that $x_i \in [a, c]$ whenever $A_i \subset [a, c]$, $x_i \in [c, b]$ whenever $A_i \subset [c, b]$, and $x_i = c$ whenever $c \in A_i$. Thus, if each A_i is contained in either $[a, c]$ or $[c, b]$, then $P = P_a \cup P_b$ where P_a is a δ_a-fine partition of $[a, c]$ and P_b is a δ_b-fine partition of $[c, b]$. Since $\sigma(P) = \sigma(P_a) + \sigma(P_b)$, we have

$$|\sigma(P) - I| \le \left|\sigma(P_a) - \int_a^c f\right| + \left|\sigma(P_b) - \int_c^b f\right| < \varepsilon.$$

If there is an A_i contained in neither $[a, c]$ nor $[c, b]$, then applying the above argument to the partition

$$Q = \{(A_1, x_1), \ldots, (A_i \cap [a, c], x_i), (A_i \cap [c, b], x_i), \ldots, (A_p, x_p)\}$$

yields $|\sigma(Q) - I| < \varepsilon$. As $\sigma(Q) = \sigma(P)$, the proposition is established. ∎

2.2 Absolute integrability

Our task is to show that the absolute value of an integrable function is also integrable. We establish this by elaborating on Cauchy's test for integrability. Our arguments will rely heavily on the fact that we are dealing with Lebesgue rather than Perron partitions (cf. the *Note* following Definition 2.1.1).

Lemma 2.2.1 *A function f on a cell A belongs to $\mathcal{R}(A, \alpha)$ if and only if for each $\varepsilon > 0$ there is a positive function δ on A such that*

$$|\sigma(f, R; \alpha) - \sigma(f, S; \alpha)| < \varepsilon$$

for all partitions $R = \{(A_1, x_1), \ldots, (A_n, x_n)\}$ and $S = \{(A_1, y_1), \ldots, (A_n, y_n)\}$ of A that are δ-fine.

PROOF.   Choose an $\varepsilon > 0$, and find a positive function δ on A so that the condition of the lemma is satisfied. Let $P = \{(B_1, u_1), \ldots, (B_p, u_p)\}$ and

$Q = \{(C_1, v_1), \dots, (C_q, v_q)\}$ be δ-fine partitions of A. For $i = 1, \dots, p$ and $j = 1, \dots, q$ set

$$A_{i,j} = B_i \cap C_j, \quad x_{i,j} = u_i, \quad y_{i,j} = v_j,$$

and denote by N the collection of all ordered pairs (i, j) for which $A_{i,j}$ is a cell, i.e., for which $(A_{i,j})^{\circ} \neq \emptyset$. Then

$$R = \{(A_{i,j}, x_{i,j}) : (i,j) \in N\} \quad \text{and} \quad S = \{(A_{i,j}, y_{i,j}) : (i,j) \in N\}$$

are δ-fine partitions of A,

$$
\begin{aligned}
\sigma(R) &= \sum_{(i,j) \in N} f(x_{i,j}) \alpha(A_{i,j}) = \sum_{i=1}^{p} \sum_{j=1}^{q} f(x_{i,j}) \alpha(A_{i,j}) \\
&= \sum_{i=1}^{p} f(u_i) \sum_{j=1}^{q} \alpha(B_i \cap C_j) = \sum_{i=1}^{p} f(u_i) \alpha(B_i) = \sigma(P),
\end{aligned}
$$

and similarly $\sigma(S) = \sigma(Q)$. Thus

$$|\sigma(P) - \sigma(Q)| = |\sigma(R) - \sigma(S)| < \varepsilon,$$

and $f \in \mathcal{R}(A, \alpha)$ by Cauchy's test. The converse is clear. ∎

Lemma 2.2.2 *A function f on a cell A belongs to $\mathcal{R}(A, \alpha)$ if and only if for each $\varepsilon > 0$ there is a positive function δ on A such that*

$$\sum_{i=1}^{n} |f(x_i) - f(y_i)| \alpha(A_i) < \varepsilon$$

for all partitions $\{(A_1, x_1), \dots, (A_n, x_n)\}$ and $\{(A_1, y_1), \dots, (A_n, y_n)\}$ in A that are δ-fine.

PROOF. Let $f \in \mathcal{R}(A, \alpha)$ and $\varepsilon > 0$. Find a positive function δ on A so that the condition of Lemma 2.2.1 is satisfied. In view of Corollary 1.2.5, it suffices to consider δ-fine partitions

$$P = \{(A_1, x_1), \dots, (A_n, x_n)\} \quad \text{and} \quad Q = \{(A_1, y_1), \dots, (A_n, y_n)\}$$

of A. After a suitable reordering, we may assume that there is an integer k with $0 \leq k \leq n$ and such that $f(x_i) \geq f(y_i)$ for $i = 1, \dots, k$, and $f(x_i) < f(y_i)$ for $i = k+1, \dots, n$. Now it is clear that

$$
\begin{aligned}
R &= \{(A_1, x_1), \dots, (A_k, x_k), (A_{k+1}, y_{k+1}), \dots, (A_n, y_n)\}, \\
S &= \{(A_1, y_1), \dots, (A_k, y_k), (A_{k+1}, x_{k+1}), \dots, (A_n, x_n)\},
\end{aligned}
$$

are δ-fine partitions of A. Thus, by the choice of δ, we have

$$\varepsilon > |\sigma(R) - \sigma(S)|$$

$$= \sum_{i=1}^{k}[f(x_i) - f(y_i)]\alpha(A_i) + \sum_{i=k+1}^{n}[f(y_i) - f(x_i)]\alpha(A_i)$$

$$= \sum_{i=1}^{n}|f(x_i) - f(y_i)|\alpha(A_i).$$

The converse follows directly from Lemma 2.2.1. ∎

Proposition 2.2.3   *Let A be a cell. If $f \in \mathcal{R}(A, \alpha)$, then $|f| \in \mathcal{R}(A, \alpha)$ and*

$$\left|\int_A f \, d\alpha\right| \leq \int_A |f| \, d\alpha.$$

PROOF.   As $||f(x)| - |f(y)|| \leq |f(x) - f(y)|$, the integrability of $|f|$ follows from Lemma 2.2.2. The inequality is a consequence of Proposition 2.1.3. ∎

Corollary 2.2.4   *Let A be a cell. If f and g belong to $\mathcal{R}(A, \alpha)$, then so do* $\max\{f, g\}$ *and* $\min\{f, g\}$.

PROOF.   It is easy to verify that

$$\max\{f, g\} = \frac{1}{2}(f + g + |f - g|) \quad \text{and} \quad \min\{f, g\} = \frac{1}{2}(f + g - |f - g|),$$

and so the corollary follows from Propositions 2.1.3 and 2.2.3. ∎

If f is a function, let

$$f^+ = \max\{f, 0\} \quad \text{and} \quad f^- = \max\{-f, 0\},$$

and observe that $f = f^+ - f^-$ and $|f| = f^+ + f^-$.

Corollary 2.2.5   *Let f be a function on a cell A. Then f belongs to $\mathcal{R}(A, \alpha)$ if and only if both f^+ and f^- do.*

PROOF.   The corollary is a direct consequence of Proposition 2.1.3 and Corollary 2.2.4. ∎

Note. Corollary 2.2.5 is important because it reduces many questions about integrable functions to questions about *nonnegative* integrable functions (cf. Theorems 2.3.12 and 2.6.1).

Corollary 2.2.6 *Let A be a cell, and let f and g be bounded functions from $\mathcal{R}(A, \alpha)$. Then fg belongs to $\mathcal{R}(A, \alpha)$.*

PROOF. Choose an $\varepsilon > 0$, and positive constants a and b so that $|f| \le a$ and $|g| \le b$. By Lemma 2.2.2, there are positive functions δ_f and δ_g on A such that

$$\sum_{i=1}^{n} |f(x_i) - f(y_i)| \alpha(A_i) < \frac{\varepsilon}{2b} \quad \text{or} \quad \sum_{i=1}^{n} |g(x_i) - g(y_i)| \alpha(A_i) < \frac{\varepsilon}{2a}$$

whenever the partitions

$$P = \{(A_1, x_1), \dots, (A_n, x_n)\} \quad \text{and} \quad Q = \{(A_1, y_1), \dots, (A_n, y_n)\}$$

in A are δ_f-fine or δ_g-fine, respectively. Now set $\delta = \min\{\delta_f, \delta_g\}$, and assume that P and Q are δ-fine. Then

$$\sum_{i=1}^{n} |f(x_i)g(x_i) - f(y_i)g(y_i)| \alpha(A_i)$$

$$\le \sum_{i=1}^{n} |g(x_i)| \cdot |f(x_i) - f(y_i)| \alpha(A_i) + \sum_{i=1}^{n} |f(y_i)| \cdot |g(x_i) - g(y_i)| \alpha(A_i)$$

$$\le b \sum_{i=1}^{n} |f(x_i) - f(y_i)| \alpha(A_i) + a \sum_{i=1}^{n} |g(x_i) - g(y_i)| \alpha(A_i) < \varepsilon,$$

and it follows from Lemma 2.2.2 that $fg \in \mathcal{R}(A, \alpha)$. ∎

Exercise 2.2.7 A function g on $\mathbf{R}$ is called *Lipschitz* if there is a constant c such that $|g(u) - g(v)| \le c|u - v|$ for all real numbers u and v. If A is a cell and $f \in \mathcal{R}(A, \alpha)$, show that $g \circ f$ belongs to $\mathcal{R}(A, \alpha)$ for each Lipschitz function g on $\mathbf{R}$.

Theorem 2.2.8 *Each continuous function defined on a cell A belongs to $\mathcal{R}(A, \alpha)$.*

PROOF. If f is a continuous function on A, then f is uniformly continuous on A by [42, Theorem 4.19, p. 91]. Hence given $\varepsilon > 0$, there is a $\Delta > 0$ such that

$$|f(x) - f(y)| < \frac{\varepsilon}{\alpha(A) + 1}$$

for each x and y in A with $|x - y| < 2\Delta$. Set $\delta(x) = \Delta$ for every $x \in A$, and let $\{(A_1, x_1), \dots, (A_n, x_n)\}$ and $\{(A_1, y_1), \dots, (A_n, y_n)\}$ be δ-fine partitions in A. Since

$$A_i \subset U(x_i, \Delta) \cap U(y_i, \Delta),$$

we have $|x_i - y_i| < 2\Delta$ for $i = 1, \ldots, n$. Thus

$$\sum_{i=1}^{n} |f(x_i) - f(y_i)| \alpha(A_i) \leq \frac{\varepsilon}{\alpha(A) + 1} \sum_{i=1}^{n} \alpha(A_i) \leq \varepsilon \frac{\alpha(A)}{\alpha(A) + 1} < \varepsilon,$$

and the proposition follows from Lemma 2.2.2. ∎

In the proof of Theorem 2.2.8 we managed to get by with a *constant* function δ. The next example shows that this is not always the case.

Example 2.2.9 For each $x \in [0, 1]$ set

$$f(x) = \begin{cases} 1 & \text{if } x \text{ is rational,} \\ 0 & \text{if } x \text{ is irrational.} \end{cases}$$

According to [42, Chapter 6, Exercise 4, p. 138], the function f is not Riemann integrable in the classical sense of [42, Definition 6.1, p. 120]. We show, however, that f belongs to $\mathcal{R}([0, 1], \lambda)$ and that $\int_0^1 f \, d\lambda = 0$ (recall that the function λ on $\mathbf{R}$ has been defined towards the end of Section 1.1).

To this end choose an $\varepsilon > 0$, and let $\{r_1, r_2, \ldots\}$ be an enumeration of all rational numbers from $[0, 1]$. Define a positive function δ on $[0, 1]$ by setting

$$\delta(x) = \begin{cases} \varepsilon 2^{-n-1} & \text{if } x = r_n \text{ and } n = 1, 2, \ldots, \\ 1 & \text{if } x \text{ is irrational,} \end{cases}$$

and let $P = \{(A_1, x_1), \ldots, (A_p, x_p)\}$ be a δ-fine partition of $[0, 1]$. If the points $x_{i_1}, \ldots, x_{i_k}$ are equal to r_n, then

$$\bigcup_{j=1}^{k} A_{i_j} \subset U(r_n, \delta(r_n)),$$

and we have

$$\sum_{j=1}^{k} f(x_{i_j}) \lambda(A_{i_j}) = \sum_{j=1}^{k} \lambda(A_{i_j}) < \varepsilon 2^{-n}.$$

Since $f(x) = 0$ whenever x is irrational, it follows that

$$0 \leq \sigma(f, P; \lambda) < \sum_{n=1}^{\infty} \varepsilon 2^{-n} = \varepsilon,$$

and our claim is established.

On the other hand, if δ is a positive constant, we can find δ-fine partitions $P = \{(A_1, x_1), \ldots, (A_p, x_p)\}$ and $Q = \{(A_1, y_1), \ldots, (A_p, y_p)\}$ of $[0, 1]$ such that $x_1, \ldots, x_p$ are rational, and $y_1, \ldots, y_p$ are irrational. Thus $\sigma(f, P; \lambda) = 1$ and $\sigma(f, Q; \lambda) = 0$.

Note. The positive function δ of Definition 2.1.1 need not be completely arbitrary. A class of these functions that is sufficient for defining the McShane and Henstock–Kurzweil integrals has been described in [35].

2.3 Convergence theorems

We begin by proving an important result commonly referred to as the *Henstock lemma*.

Lemma 2.3.1 *Let A be a cell and let $f \in \mathcal{R}(A, \alpha)$. Given $\varepsilon > 0$, there is a positive function δ on A such that*

$$\sum_{i=1}^{p} \left| f(x_i)\alpha(A_i) - \int_{A_i} f \, d\alpha \right| < \varepsilon$$

for each δ-fine partition $\{(A_1, x_1), \ldots, (A_p, x_p)\}$ in A.

PROOF. By Definition 2.1.1, given $\varepsilon > 0$, there is a positive function δ on A such that $|\sigma(P) - \int_A f| < \varepsilon/3$ for each δ-fine partition P of A. In view of Corollary 1.2.5, it suffices to consider a δ-fine partition $\{(A_1, x_1), \ldots, (A_p, x_p)\}$ of A. After a suitable reordering, we may assume that there is an integer k with $0 \leq k \leq p$ and such that the number $f(x_i)\alpha(A_i) - \int_{A_i} f$ is nonnegative for $i = 1, \ldots, k$, and negative for $i = k+1, \ldots, p$. Using Cousin's lemma and Proposition 2.1.9, we can find a δ-fine partition P_i of A_i so that $|\sigma(P_i) - \int_{A_i} f| < \varepsilon/3p$ for $i = 1, \ldots, p$. Now

$$P = \{(A_1, x_1), \ldots, (A_k, x_k)\} \cup \bigcup_{i=k+1}^{p} P_i \,,$$

$$Q = \{(A_{k+1}, x_{k+1}), \ldots, (A_p, x_p)\} \cup \bigcup_{i=1}^{k} P_i$$

are δ-fine partitions of A. Thus

$$\frac{\varepsilon}{3} > \left| \sigma(P) - \int_A f \right| \geq \sum_{i=1}^{k} \left[f(x_i)\alpha(A_i) - \int_{A_i} f \right] - \left| \sum_{i=k+1}^{p} \left[\sigma(P_i) - \int_{A_i} f \right] \right|$$

$$\geq \sum_{i=1}^{k} \left| f(x_i)\alpha(A_i) - \int_{A_i} f \right| - (p-k)\frac{\varepsilon}{3p}$$

and similarly

$$\frac{\varepsilon}{3} > \sum_{i=k+1}^{p} \left| f(x_i)\alpha(A_i) - \int_{A_i} f \right| - k\frac{\varepsilon}{3p} \,.$$

The lemma follows by adding the last two inequalities. ∎

Corollary 2.3.2 *Let $f \in \mathcal{R}([a, b], \alpha)$, and set $F(x) = \int_a^x f \, d\alpha$ for each x in the cell $[a, b]$. If α is right continuous at $c \in [a, b)$ then so is F, and if α is left continuous at $c \in (a, b]$ then so is F.*

PROOF. Choose an $\varepsilon > 0$, and use Henstock's lemma to find a positive function δ on $[a, b]$ so that

$$\sum_{i=1}^p \left| f(x_i)\alpha(A_i) - \int_{A_i} f \right| < \frac{\varepsilon}{2}$$

for each δ-fine partition $\{(A_1, x_1), \dots, (A_p, x_p)\}$ in $[a, b]$. Now assume that α is right continuous at $c \in [a, b)$. There is a $\Delta > 0$ such that $|f(c)|\alpha(B) < \varepsilon/2$ for each cell $B \subset [a, b] \cap [c, c + \Delta)$. Set $\eta = \min\{\Delta, \delta(c)\}$, and select a point $x \in (c, c + \eta) \cap [a, b]$. Now $\int_c^x f = F(x) - F(c)$ by Proposition 2.1.10. Since $\{([c, x], c)\}$ is a δ-fine partition in $[a, b]$, we have

$$\left| |f(c)|\alpha([c, x]) - |F(x) - F(c)| \right| \le \left| f(c)\alpha([c, x]) - \int_c^x f \right| < \frac{\varepsilon}{2},$$

and it follows that $|F(x) - F(c)| < \varepsilon$. The rest of the corollary is proved similarly. ∎

Remark 2.3.3 The previous argument yields a more precise result. Indeed, interpreting the continuity at $c \in [a, b]$ as that from the right or left when $c = a$ or $c = b$, respectively, the following equivalence holds: *the function F is continuous at c if and only if α is continuous at c or $f(c) = 0$.*

Unless specified otherwise, convergence of functions is always defined *pointwise*. Thus we say that a sequence of functions $\{f_n\}$ defined on a set E converges to a function f on E, and write $\lim f_n = f$, whenever $\lim f_n(x) = f(x)$ for each $x \in E$.

A sequence $\{f_n\}$ of functions is called, respectively, *increasing* or *decreasing* if $f_n \le f_{n+1}$ or $f_n \ge f_{n+1}$ for $n = 1, 2, \dots$. When an increasing or decreasing sequence $\{f_n\}$ of functions converges to a function f, we write $f_n \nearrow f$ or $f_n \searrow f$, respectively.

Now we are ready to prove the crucial *monotone convergence theorem*.

Theorem 2.3.4 *Let f be a function on a cell A, and let $\{f_n\}$ be a sequence in $\mathcal{R}(A, \alpha)$. If $f_n \nearrow f$ and $\lim \int_A f_n \, d\alpha$ is finite, then $f \in \mathcal{R}(A, \alpha)$ and*

$$\int_A f \, d\alpha = \lim \int_A f_n \, d\alpha.$$

PROOF. Choose an $\varepsilon > 0$ and for $n = 1, 2, \ldots$, use Henstock's lemma to find positive functions δ_n on A so that

$$\sum_{i=1}^{q} \left| f_n(y_i)\alpha(B_i) - \int_{B_i} f_n \right| < \varepsilon 2^{-n}$$

for each δ_n-fine partition $\{(B_1, y_1), \ldots, (B_q, y_q)\}$ in A. Let $I = \lim \int_A f_n$, and select a positive integer r with $\int_A f_r > I - \varepsilon$. Finally, for each $x \in A$ find an integer $n(x) \geq r$ such that $|f_{n(x)}(x) - f(x)| < \varepsilon$. Now define a positive function δ on A by setting $\delta(x) = \delta_{n(x)}(x)$ for every $x \in A$. We prove the theorem by showing that

$$|\sigma(P) - I| < \varepsilon[2 + \alpha(A)]$$

for each δ-fine partition $P = \{(A_1, x_1), \ldots, (A_p, x_p)\}$ of A. To this end, we observe first that

$$\left| \sigma(P) - \sum_{i=1}^{p} f_{n(x_i)}(x_i)\alpha(A_i) \right| \leq \sum_{i=1}^{p} |f(x_i) - f_{n(x_i)}(x_i)|\alpha(A_i)$$

$$\leq \varepsilon \sum_{i=1}^{p} \alpha(A_i) = \varepsilon\alpha(A).$$

The integers $n(x_i)$ need not be distinct. If

$$\{n(x_1), \ldots, n(x_p)\} = \{k_1, \ldots, k_s\}$$

where $k_1 < \cdots < k_s$, then $\{1, \ldots, p\}$ is the disjoint union of the sets $T_j = \{i : n(x_i) = k_j\}$ with $j = 1, \ldots, s$. Since for each $i \in T_j$ we have

$$A_i \subset U(x_i, \delta(x_i)) = U(x_i, \delta_{n(x_i)}(x_i)) = U(x_i, \delta_{k_j}(x_i)),$$

the collection $\{(A_i, x_i) : i \in T_j\}$ is a δ_{k_j}-fine partition in A. Therefore

$$\left| \sum_{i=1}^{p} f_{n(x_i)}(x_i)\alpha(A_i) - \sum_{i=1}^{p} \int_{A_i} f_{n(x_i)} \right| \leq \sum_{j=1}^{s} \sum_{i \in T_j} \left| f_{k_j}(x_i)\alpha(A_i) - \int_{A_i} f_{k_j} \right|$$

$$\leq \sum_{j=1}^{s} \varepsilon 2^{-k_j} < \sum_{k=1}^{\infty} \varepsilon 2^{-k} = \varepsilon.$$

As $r \leq n(x_i) \leq k_s$ for $i = 1, \ldots, p$, Propositions 2.1.3 and 2.1.10 yield

$$I - \varepsilon < \int_A f_r = \sum_{i=1}^{p} \int_{A_i} f_r \leq \sum_{i=1}^{p} \int_{A_i} f_{n(x_i)}$$

$$\leq \sum_{i=1}^{p} \int_{A_i} f_{k_s} = \int_A f_{k_s} \leq I < I + \varepsilon,$$

and consequently

$$\left| \sum_{i=1}^{p} \int_{A_i} f_{n(x_i)} - I \right| < \varepsilon .$$

The desired inequality follows. ∎

Lemma 2.3.5 *Let A be a cell, and let $f_n, g \in \mathcal{R}(A, \alpha)$ be such that $f_n \geq g$ for $n = 1, 2, \ldots$. Then $\inf f_n$ belongs to $\mathcal{R}(A, \alpha)$.*

PROOF. If $g_n = \min\{f_1, \ldots, f_n\}$ for $n = 1, 2, \ldots$, then $g_n \in \mathcal{R}(A, \alpha)$ by Corollary 2.2.4, and $g_n \searrow \inf f_n$. Since $g \leq g_n$ for all n, we have

$$\int_A g \leq \lim \int_A g_n \leq \int_A g_1 .$$

Applying the monotone convergence theorem to the sequence $\{-g_n\}$ completes the proof. ∎

The next proposition, called the *Fatou lemma*, is useful for dealing with arbitrary sequences of functions.

Proposition 2.3.6 *Let f be a function on a cell A, and let $f_n, g \in \mathcal{R}(A, \alpha)$ be such that $f_n \geq g$ for $n = 1, 2, \ldots$. If $f = \liminf f_n$ and $\liminf \int_A f_n \, d\alpha$ is finite, then $f \in \mathcal{R}(A, \alpha)$ and*

$$\int_A f \, d\alpha \leq \liminf \int_A f_n \, d\alpha .$$

PROOF. If $g_n = \inf_{k \geq n} f_k$ for $n = 1, 2 \ldots$, then $g_n \in \mathcal{R}(A, \alpha)$ by Lemma 2.3.5, and $g_n \nearrow f$. Since $g_n \leq f_n$ for all n, we have

$$\int_A g_1 \leq \lim \int_A g_n \leq \liminf \int_A f_n .$$

Now the monotone convergence theorem implies that $f \in \mathcal{R}(A, \alpha)$ and

$$\int_A f = \lim \int_A g_n . \quad ∎$$

An easy corollary of Fatou's lemma is the *Lebesgue dominated convergence theorem*.

Corollary 2.3.7 *Let A be a cell, and suppose that $f_n, g \in \mathcal{R}(A, \alpha)$ are such that $|f_n| \leq g$ for $n = 1, 2, \ldots$. If $\lim f_n = f$, then $f \in \mathcal{R}(A, \alpha)$ and*

$$\int_A f \, d\alpha = \lim \int_A f_n \, d\alpha .$$

PROOF. We see immediately from Fatou's lemma that $f \in \mathcal{R}(A, \alpha)$. Applying Fatou's lemma to the sequences $\{f_n\}$ and $\{-f_n\}$, we obtain

$$\int_A f = \int_A \liminf f_n \le \liminf \int_A f_n$$
$$\le \limsup \int_A f_n \le \int_A \limsup f_n = \int_A f,$$

and the corollary follows. ∎

Exercise 2.3.8 Let A be a cell, and let $\{f_n\}$ be a sequence in $\mathcal{R}(A, \alpha)$ which converges *uniformly* to a function f on A (see [42, Definition 7.7, p. 147]). Show that $f \in \mathcal{R}(A, \alpha)$ and $\int_A f \, d\alpha = \lim \int_A f_n \, d\alpha$.

Exercise 2.3.9 Let A be a cell, and suppose that $f \in \mathcal{R}(A, \alpha)$ maps A into a cell B. Show that $g \circ f$ belongs to $\mathcal{R}(A, \alpha)$ for each continuous function g on B. *Hint.* Use Corollary 2.2.6 if g is a polynomial, and then apply the Weierstrass approximation theorem ([42, Theorem 7.26, p. 159]).

Using the Lebesgue dominated convergence theorem, we shall elaborate on Theorem 2.2.8.

Theorem 2.3.10 *Let f be a bounded function on a cell A. If f has only finitely many discontinuities, then $f \in \mathcal{R}(A, \alpha)$.*

PROOF. Let $A = [a, b]$, and assume that f has a discontinuity only at b. The case when f has a discontinuity only at a is similar, and the general case follows from Proposition 2.1.10. For $n = 1, 2, \ldots$, let $b_n = b - (b - a)/n$, and set $f_n(x) = f(x)$ if $x \in [a, b_n]$, and

$$f_n(x) = f(b_n) + \frac{f(b) - f(b_n)}{b - b_n}(x - b_n)$$

if $x \in (b_n, b]$ (draw a picture!). Then each f_n is continuous and bounded by $\sup_{x \in A} |f(x)|$. Since $\lim f_n = f$, Theorem 2.2.8 and the Lebesgue dominated convergence theorem imply that $f \in \mathcal{R}(A, \alpha)$. ∎

Theorem 2.3.11 *Each monotone function f on a cell A belongs to $\mathcal{R}(A, \alpha)$.*

PROOF. Since f is monotone, $f(A)$ is contained in a cell $[a, b]$. Let $c = b - a$, and fix an integer $n \ge 1$. Set $A_0 = \{x \in A : f(x) = a\}$, and for $k = 1, \ldots, n$, let

$$A_k = \left\{ x \in A : a + (k - 1)\frac{c}{n} < f(x) \le a + k\frac{c}{n} \right\}.$$

Then A is the disjoint union of $A_0, \ldots, A_n$, and since f is monotone, each A_k is an interval or a segment (draw a picture!). Now define a function f_n on A by setting

$$f_n(x) = a + (k-1)\frac{c}{n}$$

for $x \in A_k$ and $k = 0, \ldots, n$. Then f_n has at most $n-1$ discontinuities, and $|f - f_n| \leq c/n$. Thus the sequence $\{f_n\}$ converges uniformly to f, and the theorem follows from Theorem 2.3.10 and Exercise 2.3.8. ∎

By means of the monotone convergence theorem, we can substantially improve Corollary 2.2.6.

Theorem 2.3.12 *Let A be a cell, and let f and g be functions in $\mathcal{R}(A, \alpha)$. If f or g is bounded, then fg belongs to $\mathcal{R}(A, \alpha)$.*

PROOF. Since $fg = f^+g^+ + f^-g^- - f^+g^- - f^-g^+$, it suffices to prove the theorem for nonnegative functions f and g (see Proposition 2.1.3 and Corollary 2.2.5). By symmetry, we may assume that g is bounded by a constant c. If $f_n = \min\{f, n\}$ for $n = 1, 2, \ldots$, then $f_n g \in \mathcal{R}(A, \alpha)$ by Corollary 2.2.6, and $\int_A f_n g \leq c \int_A f$. As $f_n g \nearrow fg$, an application of the monotone convergence theorem completes the proof. ∎

If $\{f_n\}$ is a sequence of functions on a cell A, we say that $\sum_{n=1}^{\infty} f_n$ *converges* whenever $\sum_{n=1}^{\infty} f_n(x)$ converges to a *real number* for each $x \in A$.

Theorem 2.3.13 *Let A be a cell, and let $\{f_n\}$ be a sequence of functions in $\mathcal{R}(A, \alpha)$ such that $\sum_{n=1}^{\infty} |f_n|$ and $\sum_{n=1}^{\infty} \int_A |f_n| \, d\alpha$ converge. Then $\sum_{n=1}^{\infty} f_n$ belongs to $\mathcal{R}(A, \alpha)$, and*

$$\int_A \left(\sum_{n=1}^{\infty} f_n \right) d\alpha = \sum_{n=1}^{\infty} \int_A f_n \, d\alpha.$$

PROOF. For $k = 1, 2, \ldots$, set $g_k = \sum_{n=1}^{k} |f_n|$, and let $g = \sum_{n=1}^{\infty} |f_n|$. By Propositions 2.2.3 and 2.1.3, each g_k belongs to $\mathcal{R}(A, \alpha)$. Since $g_k \nearrow g$ and

$$\lim_{k \to \infty} \int_A g_k = \lim_{k \to \infty} \sum_{n=1}^{k} \int_A |f_n| = \sum_{n=1}^{\infty} \int_A |f_n| < +\infty,$$

it follows from the monotone convergence theorem that $g \in \mathcal{R}(A, \alpha)$. As

$$\left| \sum_{n=1}^{k} f_n \right| \leq g_k \leq g \qquad \text{and} \qquad \lim_{k \to \infty} \sum_{n=1}^{k} f_n = \sum_{n=1}^{\infty} f_n,$$

the Lebesgue dominated convergence theorem implies that $\sum_{n=1}^{\infty} f_n$ belongs to $\mathcal{R}(A, \alpha)$ and

$$\int_A \left(\sum_{n=1}^{\infty} f_n \right) = \lim_{k \to \infty} \int_A \left(\sum_{n=1}^{k} f_n \right) = \lim_{k \to \infty} \sum_{n=1}^{k} \int_A f_n = \sum_{n=1}^{\infty} \int_A f_n . \ \blacksquare$$

The last proposition of this section addresses the *integral's dependence on a parameter*.

Proposition 2.3.14 *Let A be a cell, and let f be a continuous function on $A \times U$ where U is an open subset of R. If $F(t) = \int_A f(x, t) \, d\alpha(x)$ for each $t \in U$, then F is a continuous function on U. Moreover, if the partial derivative $\partial f / \partial t$ is continuous in $A \times U$, then F' is continuous in U, and*

$$F'(t) = \int_A \frac{\partial}{\partial t} f(x, t) \, d\alpha(x)$$

for each $t \in U$.

PROOF. Choose a sequence $\{t_n\}$ in U that converges to a $t \in U$, and find a cell $B \subset U$ and a positive integer k so that $t \in B$ and $t_n \in B$ for all $n \geq k$. As f is continuous, $\lim f(\cdot, t_n) = f(\cdot, t)$ and $|f(\cdot, t_n)| \leq a$ for a positive constant a and each $n \geq k$ ([42, Theorem 4.15, p. 89]). It follows from the dominated convergence theorem that $F(t_n) \to F(t)$, and the continuity of F is established.

Next suppose that $\partial f / \partial t$ is continuous in $A \times U$, and that $t_n \neq t$ for $n = 1, 2, \ldots$. Again by [42, Theorem 4.15, p. 89], we can find a positive constant b so that $|\partial f / \partial t| \leq b$ on $A \times B$. For each $x \in A$ and $n = 1, 2, \ldots$, set

$$g_n(x) = \frac{f(x, t_n) - f(x, t)}{t_n - t},$$

and observe that by the mean value theorem ([42, Theorem 5.10, p. 108]), $|g_n| \leq b$ whenever $n \geq k$. It follows from Proposition 2.1.3 and the dominated convergence theorem that

$$F'(t) = \lim \frac{F(t_n) - F(t)}{t_n - t} = \lim \int_A g_n = \int_A \lim g_n = \int_A \frac{\partial}{\partial t} f(x, t) \, d\alpha(x) .$$

Now the continuity of F' in U is a consequence of the first part of the proposition applied to $\partial f / \partial t$. $\blacksquare$

2.4 Connections with derivatives

Under certain conditions, integrating with respect to a general increasing function α can be reduced to integrating with respect to the identity function λ.

Theorem 2.4.1 *Let $A = [a, b]$ be a cell, and let $g \in \mathcal{R}(A, \alpha)$ be a nonnegative function. Setting $\beta(x) = \int_a^x g \, d\alpha$ for each $x \in [a, b]$ defines an increasing function β on A. A function f on A belongs to $\mathcal{R}(A, \beta)$ if and only if fg belongs to $\mathcal{R}(A, \alpha)$, in which case*

$$\int_A f \, d\beta = \int_A fg \, d\alpha \, .$$

PROOF. Assume first that there is a positive integer n such that $|f(x)| \leq n$ for all $x \in A$, and choose an $\varepsilon > 0$. Using Henstock's lemma, find a positive function δ on A so that

$$\sum_{i=1}^p \left| g(x_i) \alpha(A_i) - \int_{A_i} g \, d\alpha \right| < \frac{\varepsilon}{n}$$

for each δ-fine partition $P = \{(A_1, x_1), \ldots, (A_p, x_p)\}$ of A. For such a partition P, we have

$$|\sigma(fg, P; \alpha) - \sigma(f, P; \beta)| \leq n \sum_{i=1}^p \left| g(x_i) \alpha(A_i) - \int_{A_i} g \, d\alpha \right| < \varepsilon \, ,$$

and the theorem follows directly from Definition 2.1.1.

Next assume that $f \geq 0$, and let $f_n = \min\{f, n\}$ for $n = 1, 2, \ldots$. Then $f_n \nearrow f$ and $f_n g \nearrow fg$, and so it suffices to apply the first part of the proof and the monotone convergence theorem.

Finally, if f is an arbitrary function, we apply the second part of the proof to functions f^+ and f^-, and use Corollary 2.2.5. ∎

Theorem 2.4.2 *Let α be an increasing function on a cell B, and let Φ be a continuous strictly monotone map from a cell A onto B. A function f on B belongs to $\mathcal{R}(B, \alpha)$ if and only if $f \circ \Phi$ belongs to $\mathcal{R}(A, \pm \alpha \circ \Phi)$, in which case*

$$\int_A f \circ \Phi \, d(\pm \alpha \circ \Phi) = \int_B f \, d\alpha \, ;$$

here the sign $+$ or $-$ is used according to whether Φ is increasing or decreasing, respectively.

PROOF. Let $f \in \mathcal{R}(B, \alpha)$ and let $\varepsilon > 0$. There is a positive function δ_B on B such that

$$\left| \sigma(f, Q; \alpha) - \int_B f \, d\alpha \right| < \varepsilon$$

for each δ_B-fine partition Q of B. Since Φ is continuous, we can find a positive function δ_A on A so that $|\Phi(t) - \Phi(x)| < \delta_B[\Phi(x)]$ for each $t, x \in A$ with $|t - x| < \delta_A(x)$. Now let $P = \{(A_1, x_1), \ldots, (A_p, x_p)\}$ be a δ_A-fine partition of A. If $B_i = \Phi(A_i)$ and $y_i = \Phi(x_i)$ for $i = 1, \ldots, p$, then it is easy to verify that $Q = \{(B_1, y_1), \ldots, (B_p, y_p)\}$ is a δ_B-fine partition of B. Moreover, $\sigma(f, Q; \alpha) = \sigma(f \circ \Phi, P; \pm\alpha \circ \Phi)$ where the sign $+$ or $-$ is used according to whether Φ is increasing or decreasing, respectively. Consequently

$$\left| \sigma(f \circ \Phi, P; \pm\alpha \circ \Phi) - \int_B f \, d\alpha \right| < \varepsilon,$$

and it follows that $f \circ \Phi$ belongs to $\mathcal{R}(A, \pm\alpha \circ \Phi)$ and

$$\int_A f \circ \Phi \, d(\pm\alpha \circ \Phi) = \int_B f \, d\alpha.$$

If $f \circ \Phi$ belongs to $\mathcal{R}(A, \pm\alpha \circ \Phi)$, then f belongs to $\mathcal{R}(B, \alpha)$ by the first part of the argument applied to the map $\Phi^{-1} : B \to A$. ∎

We say that a function F on a cell $[a, b]$ is *differentiable* on $[a, b]$ whenever F has a *finite* derivative $F'(x)$ for each $x \in (a, b)$ and also *finite* one-sided derivatives $F'_+(a)$ and $F'_-(b)$, which we denote by $F'(a)$ and $F'(b)$, respectively.

Theorem 2.4.3 *Let F be a differentiable function defined on a cell $[a, b]$. If $F' \geq 0$, then $F' \in \mathcal{R}([a, b], \lambda)$.*

PROOF. Define a differentiable function G on $[a, b + 1]$ by setting

$$G(x) = \begin{cases} F(x) & \text{if } x \in [a, b], \\ F(b) + F'(b)(x - b) & \text{if } x \in (b, b + 1]. \end{cases}$$

Then G is increasing and $G' = F'$ in $[a, b]$. If

$$G_n(x) = \frac{G\left(x + \frac{1}{n}\right) - G(x)}{\frac{1}{n}}$$

for $n = 1, 2, \ldots$ and $x \in [a, b]$, then $G_n \geq 0$ and $\lim G_n = F'$. Since G is continuous, each G_n belongs to $\mathcal{R}([a, b], \lambda)$. Applying Theorem 2.4.2 to the function G on $[a + 1/n, b + 1/n]$ and the map $\Phi(x) = x + 1/n$, we obtain

$$\int_a^b G\left(x + \frac{1}{n}\right) d\lambda(x) = \int_{a+\frac{1}{n}}^{b+\frac{1}{n}} G \, d\lambda,$$

because $\lambda \circ \Phi = \lambda + 1/n$ (see Proposition 2.1.5). Thus, by Proposition 2.1.10,

$$\int_a^b G_n \, d\lambda = n\left(\int_{a+\frac{1}{n}}^{b+\frac{1}{n}} G \, d\lambda - \int_a^b G \, d\lambda\right)$$

$$= n\left(\int_b^{b+\frac{1}{n}} G \, d\lambda - \int_a^{a+\frac{1}{n}} G \, d\lambda\right) \leq G\left(b + \frac{1}{n}\right) - G(a)$$

for $n = 1, 2, \ldots$. Consequently $\liminf \int_a^b G_n \, d\lambda < +\infty$, and the theorem follows from Fatou's lemma. $\blacksquare$

Remark 2.4.4 We shall see in Example 2.4.12 below that the assumption $F' \geq 0$ cannot be omitted from Theorem 2.4.3.

Exercise 2.4.5 Let F be a differentiable function on a cell A. Show that if F' is bounded, then $F' \in \mathcal{R}(A, \lambda)$.

Exercise 2.4.6 Let $f \in \mathcal{R}([a, b], \lambda)$, and set $F(x) = \int_a^x f \, d\lambda$ for each x in the cell $[a, b]$. Show that if f is continuous at $c \in (a, b)$, then $F'(c)$ exists and equals $f(c)$.

Lemma 2.4.7 *Let F be a differentiable function on a cell $[a, b]$. Then given $\varepsilon > 0$, there is a positive function δ on $[a, b]$ such that*

$$\left|\sigma(F', P; \lambda) - [F(b) - F(a)]\right| < \varepsilon$$

for each δ-fine Perron partition P of $[a, b]$.

PROOF. Choose an $\varepsilon > 0$, and find a positive function δ on $[a, b]$ so that

$$\left|\frac{F(t) - F(x)}{t - x} - F'(x)\right| < \frac{\varepsilon}{b - a}$$

for each $t, x \in [a, b]$ with $0 < |t - x| < \delta(x)$. If $P = \{([t_0, t_1], x_1), \ldots,$

$([t_{p-1}, t_p], x_p)\}$ is a δ-fine P-partition of $[a, b]$, then $t_{i-1} \le x_i \le t_i$ for $i = 1, \ldots, p$. Thus

$$|F(t_i) - F(t_{i-1}) - F'(x_i)(t_i - t_{i-1})|$$
$$\le |F(t_i) - F(x_i) - F'(x_i)(t_i - x_i)|$$
$$\quad + |F(x_i) - F(t_{i-1}) - F'(x_i)(x_i - t_{i-1})|$$
$$< \frac{\varepsilon}{b-a}[(t_i - x_i) + (x_i - t_{i-1})] = \frac{\varepsilon}{b-a}(t_i - t_{i-1}),$$

and consequently

$$\left| \sigma(F', P) - [F(b) - F(a)] \right|$$
$$= \left| \sum_{i=1}^{p} F'(x_i)(t_i - t_{i-1}) - \sum_{i=1}^{p} [F(t_i) - F(t_{i-1})] \right|$$
$$\le \sum_{i=1}^{p} \left| F'(x_i)(t_i - t_{i-1}) - [F(t_i) - F(t_{i-1})] \right| < \varepsilon. \quad \blacksquare$$

Theorem 2.4.8 *Let $[a, b]$ be a cell, and let $f \in \mathcal{R}([a, b], \lambda)$. If there is a differentiable function F on $[a, b]$ such that $F'(x) = f(x)$ for each $x \in [a, b]$, then*

$$\int_a^b f \, d\lambda = F(b) - F(a).$$

PROOF. Given $\varepsilon > 0$, find a positive function δ on $[a, b]$ so that

$$\left| \sigma(f, Q) - \int_a^b f \right| < \frac{\varepsilon}{2}$$

for each δ-fine partition Q of $[a, b]$. By Cousin's lemma and Lemma 2.4.7, there is a δ-fine P-partition P of A such that

$$\left| \sigma(f, P) - [F(b) - F(a)] \right| < \frac{\varepsilon}{2}.$$

Thus we have

$$\left| \int_a^b f - [F(b) - F(a)] \right|$$
$$\le \left| \int_a^b f - \sigma(f, P) \right| + \left| \sigma(f, P) - [F(b) - F(a)] \right| < \varepsilon,$$

and the theorem follows from the arbitrariness of ε. $\quad \blacksquare$

Theorem 2.4.8 is an important link between two principal concepts of calculus: the derivative and integral. As usual, we call it the *fundamental theorem of calculus.*

Corollary 2.4.9 *Let f be a function on a cell A. If α is an increasing differentiable function on A, then f belongs to $\mathcal{R}(A, \alpha)$ if and only if $f\alpha'$ belongs to $\mathcal{R}(A, \lambda)$, in which case*

$$\int_A f \, d\alpha = \int_A f\alpha' \, d\lambda \, .$$

PROOF. Let $A = [a, b]$. Since $\alpha' \geq 0$, Theorems 2.4.3 and 2.4.8 imply that $\alpha' \in \mathcal{R}(A, \lambda)$ and that

$$\alpha(x) = \alpha(a) + \int_a^x \alpha' \, d\lambda$$

for each $x \in [a, b]$. In view of Proposition 2.1.5, the corollary follows from Theorem 2.4.1. ∎

The next corollary gives the usual *change of variable formula.*

Corollary 2.4.10 *Let f be a function on a cell B, and let Φ be a differentiable strictly monotone map from a cell A onto B. Then f belongs to $\mathcal{R}(B, \lambda)$ if and only if $(f \circ \Phi)|\Phi'|$ belongs to $\mathcal{R}(A, \lambda)$, in which case*

$$\int_A (f \circ \Phi)|\Phi'| \, d\lambda = \int_B f \, d\lambda \, .$$

PROOF. According to whether Φ is increasing or decreasing, $|\Phi'| = \Phi'$ or $|\Phi'| = -\Phi'$, respectively. As $\lambda \circ \Phi = \Phi$, the corollary follows immediately from Theorem 2.4.2 and Corollary 2.4.9. ∎

Example 2.4.11 Choose a real number s, and define a function f on $[0, 1]$ by setting $f(0) = 0$ and $f(x) = x^s$ for $x \in (0, 1]$. According to Theorem 2.4.8, for each $t \in (0, 1)$, we have $f \in \mathcal{R}([t, 1], \lambda)$ and

$$\int_t^1 f \, d\lambda = \begin{cases} \frac{1}{s+1}(1 - t^{s+1}) & \text{if } s \neq -1, \\ -\ln t & \text{if } s = -1. \end{cases}$$

It follows from Corollary 2.3.2 that $f \notin \mathcal{R}([0, 1], \lambda)$ whenever $s \leq -1$.

Now let $s > -1$, and for $n = 1, 2, \ldots$, set

$$f_n(x) = \begin{cases} 0 & \text{if } x \in [0, 1/n), \\ f(x) & \text{if } x \in [1/n, 1]. \end{cases}$$

Theorem 2.3.10 and Corollary 2.3.2 imply that $f_n \in \mathcal{R}([0, 1/n], \lambda)$ and $\int_0^{1/n} f_n \, d\lambda = 0$. Thus $f_n \in \mathcal{R}([0, 1], \lambda)$ by Proposition 2.1.10, and as $f_n \nearrow f$, it follows from the monotone convergence theorem that $f \in \mathcal{R}([0, 1], \lambda)$ and

$$\int_0^1 f \, d\lambda = \frac{1}{s+1}.$$

Example 2.4.12 Set $F(0) = f(0) = 0$, and for $x \neq 0$ let

$$F(x) = x^2 \cos \pi x^{-2} \qquad \text{and} \qquad f(x) = 2x \cos \pi x^{-2} + 2\pi x^{-1} \sin \pi x^{-2}.$$

It is easy to check that $F'(x) = f(x)$ for each $x \in R$; indeed, this is trivial if $x \neq 0$, and it follows from the definition of the derivative if $x = 0$. We claim, however, that $f \notin \mathcal{R}([0, 1], \lambda)$ (cf. Remark 2.4.4).

Proceeding towards a contradiction, assume that $f \in \mathcal{R}([0, 1], \lambda)$. Under this assumption, Proposition 2.2.3 implies that the function g defined by

$$g(x) = \begin{cases} x^{-1} |\sin \pi x^{-2}| & \text{if } x \neq 0, \\ 0 & \text{if } x = 0, \end{cases}$$

belongs to $\mathcal{R}([0, 1], \lambda)$. Applying Corollary 2.4.10 to the map $\Phi(x) = x^{-2}$, and employing the more customary dx and dt instead of $d\lambda(x)$ and $d\lambda(t)$, respectively, we obtain

$$\int_0^1 g \, d\lambda \geq \int_{1/\sqrt{n}}^1 \frac{|\sin \pi x^{-2}|}{x} \, dx = \frac{1}{2} \int_1^n \frac{|\sin \pi t|}{t} \, dt$$

$$= \frac{1}{2} \sum_{k=2}^n \int_{k-1}^k \frac{|\sin \pi t|}{t} \, dt \geq \frac{1}{2} \sum_{k=2}^n \frac{1}{k} \int_{k-1}^k |\sin \pi t| \, dt = \frac{1}{\pi} \sum_{k=2}^n \frac{1}{k}$$

for $n = 1, 2, \ldots$. This is a contradiction, since $\sum_{k=1}^\infty (1/k) = +\infty$ ([42, Theorem 3.28, p. 62]).

Remark 2.4.13 Example 2.4.12 shows that the assumption $f \in \mathcal{R}([a, b], \lambda)$ cannot be omitted from Theorem 2.4.8. On closer examination, this is surprising: irrespective of whether f is integrable or not, $F(b) - F(a)$ is a well-defined real number that depends only on f. Thus we may have a *non-integrable* function and yet know the value of its "integral." This deficiency of the McShane integral is, perhaps, the main motivation for introducing the integral of Henstock and Kurzweil (see Chapter 6; in particular, Theorem 6.1.2).

2.5 Gap functions

Throughout this section, let $A = [a, b]$ be a fixed cell, and let $S = \{s_n\}$ be a sequence of distinct points of (a, b). Furthermore, let $\{t_n\}$ be a sequence in $[0, 1]$, and let $\{c_n\}$ be a sequence of nonnegative real numbers such that the series $\sum_{n=1}^{\infty} c_n$ converges. For $n = 1, 2, \ldots$ and $x \in \mathbf{R}$, set

$$\gamma_n(x) = \begin{cases} 0 & \text{if } x < s_n, \\ t_n & \text{if } x = s_n, \\ 1 & \text{if } x > s_n. \end{cases}$$

Then

$$\gamma = \sum_{n=1}^{\infty} c_n \gamma_n$$

is an increasing function on $\mathbf{R}$, called a *gap function* determined by the triple sequence $\{s_n, t_n, c_n\}$. Since

$$\gamma(x) = \begin{cases} \sum_{s_n < x} c_n & \text{if } x \in \mathbf{R} - S, \\ \sum_{s_n < x} c_n + c_k t_k & \text{if } x = s_k \text{ and } k = 1, 2, \ldots, \end{cases}$$

it is not difficult to verify that γ is continuous at each $x \in \mathbf{R} - S$, and discontinuous at s_k if and only if $c_k > 0$. Moreover,

$$\gamma(x) = \begin{cases} \gamma(a) & \text{if } x < a, \\ \gamma(b) & \text{if } x > b. \end{cases}$$

Lemma 2.5.1   *Let f be a bounded function on A. Then $f \in \mathcal{R}(A, \gamma)$ and*

$$\int_A f \, d\gamma = \sum_{n=1}^{\infty} c_n f(s_n).$$

PROOF.   Since there is a positive constant c such that $|f| \leq c$, the series $\sum_{n=1}^{\infty} c_n f(s_n)$ converges. Choose an $\varepsilon > 0$ and find an integer $k \geq 1$ so that $\sum_{n > k} c_n < \varepsilon/(2c)$. Define a positive function δ on A by setting

$$\delta(x) = \min\{|x - s_n| : 1 \leq n \leq k, \ s_n \neq x\}$$

for each $x \in A$, and let $P = \{(A_1, x_1), \ldots, (A_p, x_p)\}$ be a δ-fine partition of A. By the choice of δ, each cell A_i contains at most one point s_j with $1 \leq j \leq k$, and if such an s_j is in A_i, then $x_i = s_j$. Moreover, we may assume that such an s_j is an interior point of the cell A_i. Indeed, if s_j is a boundary point of two cells A_u and A_v, then $x_u = x_v = s_j$ is the interior point of $A_u \cup A_v$ and

$$Q = \{(A_u \cup A_v, s_j)\} \cup \{(A_i, x_i) : i = 1, \ldots, p;\ i \neq u, v\}$$

is a δ-fine partition of A with $\sigma(f, Q; \gamma) = \sigma(f, P; \gamma)$. Hence, after a suitable reordering, we obtain that $s_i = x_i$ is an interior point of A_i for $i = 1, \ldots, k$, and $A_i \cap \{s_1, \ldots, s_k\} = \emptyset$ for $i = k+1, \ldots, p$. Now if $\beta = \sum_{n>k} c_n \gamma_n$, then $\beta(A) < \varepsilon/(2c)$ and

$$\gamma(A_i) = \begin{cases} c_i + \beta(A_i) & \text{if } i = 1, \ldots, k, \\ \beta(A_i) & \text{if } i = k+1, \ldots, p. \end{cases}$$

Thus we have

$$\left| \sigma(f, P; \gamma) - \sum_{n=1}^{\infty} c_n f(s_n) \right| = \left| \sum_{i=1}^{p} f(x_i)\beta(A_i) - \sum_{n=k+1}^{\infty} c_n f(s_n) \right|$$

$$\leq c \sum_{i=1}^{p} \beta(A_i) + c \sum_{n=k+1}^{\infty} c_n = c\beta(A) + \frac{\varepsilon}{2} < \varepsilon,$$

and the lemma is proved. ∎

Proposition 2.5.2 *A function f on A belongs to $\mathcal{R}(A, \gamma)$ if and only if the series $\sum_{n=1}^{\infty} c_n f(s_n)$ converges absolutely, in which case*

$$\int_A f\, d\gamma = \sum_{n=1}^{\infty} c_n f(s_n).$$

PROOF. It is easy to see that the series $\sum_{n=1}^{\infty} c_n f(s_n)$ converges absolutely if and only if both series $\sum_{n=1}^{\infty} c_n f^+(s_n)$ and $\sum_{n=1}^{\infty} c_n f^-(s_n)$ converge, in which case

$$\sum_{n=1}^{\infty} c_n f(s_n) = \sum_{n=1}^{\infty} c_n f^+(s_n) - \sum_{n=1}^{\infty} c_n f^-(s_n).$$

Thus, in view of Corollary 2.2.5, it suffices to prove the proposition for $f \geq 0$. If $f_k = \min\{f, k\}$ for $k = 1, 2, \ldots$, then $f_k \in \mathcal{R}(A, \gamma)$ and

$$\lim_{k \to \infty} \int_A f_k\, d\gamma = \lim_{k \to \infty} \sum_{n=1}^{\infty} c_n f_k(s_n) = \sum_{n=1}^{\infty} c_n f(s_n)$$

by Lemma 2.5.1. Since $f_n \nearrow f$, the proposition follows from the monotone convergence theorem. ∎

Note. Proposition 2.5.2 shows that neither the integrability of a function f on A nor the value of the integral $\int_A f \, d\gamma$ depends on the sequence $\{t_n\}$. For this reason, we usually set each $t_n = 0$ or each $t_n = 1$, which makes γ, respectively, left or right continuous in A.

2.6 Integration by parts

Throughout this section, $A = [a, b]$ is a fixed cell, α is an increasing function on A, and f is a function in $\mathcal{R}(A, \alpha)$. We define a function F on A by letting $F(x) = \int_a^x f \, d\alpha$ for each $x \in [a, b]$. According to Remark 2.3.3, the function F is continuous at $x \in A$ if and only if α is continuous at x or $f(x) = 0$.

Theorem 2.6.1 *If F is continuous and G is an increasing function on A, then fG belongs to $\mathcal{R}(A, \alpha)$, F belongs to $\mathcal{R}(A, G)$, and*

$$\int_a^b fG \, d\alpha = F(b)G(b) - \int_a^b F \, dG \, .$$

PROOF (*simplified by M.J. Paris*). The integrability of fG with respect to α follows from Theorems 2.3.11 and 2.3.12; the integrability of F with respect to G is guaranteed by Theorem 2.2.8.

Assume first that $f \geq 0$. Choose an $\varepsilon > 0$ and using the uniform continuity of F, find a $\Delta > 0$ so that $|F(x) - F(y)| < \varepsilon$ for all $x, y \in A$ with $|x - y| < 2\Delta$ ([42, Theorem 4.19, p. 91]). By Theorem 2.4.1, the integral $\int_a^b G \, dF$ exists and equals $\int_a^b fG \, d\alpha$. Hence there is a positive function δ on A such that

$$\left| \sigma(G, P; F) - \int_a^b fG \, d\alpha \right| < \varepsilon \quad \text{and} \quad \left| \sigma(F, P; G) - \int_a^b F \, dG \right| < \varepsilon$$

for each δ-fine partition P of A. With no loss of generality, we may assume that $\delta \leq \Delta$. Now Cousin's lemma implies that there is a δ-fine P-partition $P = \{([t_0, t_1], x_1), \ldots, ([t_{p-1}, t_p], x_p)\}$ of A. As $t_{i-1} \leq x_i \leq t_i$ for $i = 1, \ldots, p$, we see that

$$\sum_{i=1}^p [G(x_i) - G(t_{i-1})] \leq \sum_{i=1}^p [G(t_i) - G(t_{i-1})] = G(A) \, .$$

Since

$$\begin{aligned}
F(b)G(b) &= \sum_{i=1}^p [F(t_i)G(t_i) - F(t_{i-1})G(t_{i-1})] \\
&= \sum_{i=1}^p \Big(G(t_{i-1})[F(t_i) - F(t_{i-1})] + F(t_i)[G(t_i) - G(t_{i-1})] \Big),
\end{aligned}$$

we obtain

$$\left| \int_a^b fG \, d\alpha + \int_a^b F \, dG - F(b)G(b) \right| \le \left| \int_a^b fG \, d\alpha - \sigma(G, P; F) \right|$$

$$+ \left| \int_a^b F \, dG - \sigma(F, P; G) \right| + |\sigma(G, P; F) + \sigma(F, P; G) - F(b)G(b)|$$

$$< 2\varepsilon + \left| \sum_{i=1}^p \Big(G(x_i)[F(t_i) - F(t_{i-1})] + F(x_i)[G(t_i) - G(t_{i-1})] \right.$$

$$\left. - G(t_{i-1})[F(t_i) - F(t_{i-1})] - F(t_i)[G(t_i) - G(t_{i-1})] \Big) \right|$$

$$\le 2\varepsilon + \sum_{i=1}^p [G(x_i) - G(t_{i-1})] \cdot |F(t_i) - F(t_{i-1})|$$

$$+ \sum_{i=1}^p |F(x_i) - F(t_i)| \cdot [G(t_i) - G(t_{i-1})] \le 2\varepsilon[1 + G(A)],$$

and the desired equality follows from the arbitrariness of ε.

If f is arbitrary, let $F_\pm(x) = \int_a^x f^\pm \, d\alpha$ for each $x \in [a,b]$. By Corollary 2.2.5, Remark 2.3.3, and the first part of the proof, we have

$$\int_a^b f^\pm G \, d\alpha = F_\pm(b)G(b) - \int_a^b F_\pm \, dG \,.$$

Since $f = f^+ - f^-$ and $F = F_+ - F_-$, the theorem follows from Proposition 2.1.3. ∎

Corollary 2.6.2 *Let $g \in \mathcal{R}(A, \alpha)$, and let G be a function on A such that $G(x) = \int_a^x g \, d\alpha$ for each $x \in [a,b]$. If F is continuous on A, then both fG and Fg belong to $\mathcal{R}(A, \alpha)$ and*

$$\int_a^b fG \, d\alpha = F(b)G(b) - \int_a^b Fg \, d\alpha \,.$$

PROOF. The integrability of Fg with respect to α follows from Theorems 2.2.8 and 2.3.12. If $G_\pm(x) = \int_a^x g^\pm \, d\alpha$ for each $x \in [a,b]$, then $fG_\pm \in \mathcal{R}(A, \alpha)$ and

$$\int_a^b fG_\pm \, d\alpha = F(b)G_\pm(b) - \int_a^b F \, dG_\pm = F(b)G_\pm(b) - \int_a^b Fg^\pm \, d\alpha$$

by Theorems 2.6.1 and 2.4.1. Since $g = g^+ - g^-$ and $G = G_+ - G_-$, an application of Proposition 2.1.3 completes the proof. ∎

Theorem 2.6.1 and Corollary 2.6.2 are typical results referred to as *integration by parts*. They are very important from a theoretical as well as practical point of view. The usual integration by parts formula of elementary calculus is an appreciably simpler proposition.

Proposition 2.6.3 *Let u and v be differentiable functions on A. If both u' and v' belong to $\mathcal{R}(A, \lambda)$, then so do $u'v$ and uv' and*

$$\int_a^b u'v \, d\lambda = u(b)v(b) - u(a)v(a) - \int_a^b uv' \, d\lambda .$$

PROOF.   Since u and v are continuous functions, the integrability of $u'v$ and uv' follows from Theorems 2.2.8 and 2.3.12. Thus $(uv)' = u'v + uv'$ is integrable, and

$$\int_a^b u'v \, d\lambda + \int_a^b uv' \, d\lambda = u(b)v(b) - u(a)v(a)$$

by the fundamental theorem of calculus. ∎

The next example shows that the integrability of *both* derivatives u' and v' is an essential assumption of Proposition 2.6.3 (cf. Proposition 6.2.5).

Example 2.6.4   Set $u(0) = v(0) = 0$, and for $x \neq 0$ let

$$u(x) = x \qquad \text{and} \qquad v(x) = x^2 \cos \pi x^{-3} .$$

Then u and v are differentiable functions on $\mathbf{R}$ with $u'v \in \mathcal{R}([0, 1], \lambda)$. On the other hand,

$$u(x)v'(x) = 2x^2 \cos \pi x^{-3} + 3\pi x^{-1} \sin \pi x^{-3}$$

for each $x \neq 0$. Arguing as in Example 2.4.12, we conclude that uv' does not belong to $\mathcal{R}([0, 1], \lambda)$.

We close this section by presenting the *first* and *second mean value theorems of integral calculus.*

Theorem 2.6.5   *Let f be a continuous function on A. Then there is a $\xi \in A$ such that*

$$\int_A f \, d\alpha = f(\xi)\alpha(A) .$$

PROOF. Let m and M denote, respectively, the minimum and maximum value of f on A ([42, Theorem 4.16, p. 89]). Then

$$m\alpha(A) \le \int_A f \, d\alpha \le M\alpha(A),$$

and so

$$m \le \frac{1}{\alpha(A)} \int_A f \, d\alpha \le M$$

whenever $\alpha(A) > 0$. Thus, if $\alpha(A) > 0$, the existence of ξ follows from the intermediate value theorem for continuous functions ([42, Theorem 4.23, p. 93]). On the other hand, any $\xi \in A$ works if $\alpha(A) = 0$. $\blacksquare$

Theorem 2.6.6 *If F is continuous and G is a monotone function on A, then there is a $\xi \in A$ such that*

$$\int_a^b fG \, d\alpha = G(a) \int_a^\xi f \, d\alpha + G(b) \int_\xi^b f \, d\alpha.$$

PROOF. Replacing G by $-G$ if necessary, we may assume that G is increasing. By Theorem 2.6.1,

$$\int_a^b fG \, d\alpha = F(b)G(b) - \int_a^b F \, dG,$$

and by Theorem 2.6.5, there is a $\xi \in A$ such that

$$\int_a^b F \, dG = F(\xi)G(A) = F(\xi)[G(b) - G(a)].$$

Thus, in view of Proposition 2.1.10,

$$
\begin{aligned}
\int_a^b fG \, d\alpha &= F(b)G(b) - F(\xi)[G(b) - G(a)] \\
&= G(a)F(\xi) + G(b)[F(b) - F(\xi)] \\
&= G(a) \int_a^\xi f \, d\alpha + G(b) \int_\xi^b f \, d\alpha,
\end{aligned}
$$

and the theorem is proved. $\blacksquare$

Example 2.6.7 Let $b > a > 0$ and $s > 0$. Writing dx instead of $d\lambda(x)$, Theorem 2.6.6 yields

$$\left| \int_a^b x^{-s} \cos x \, dx \right| \le a^{-s} \left| \int_a^\xi \cos x \, dx \right| + b^{-s} \left| \int_\xi^b \cos x \, dx \right| \le 2(a^{-s} + b^{-s}).$$

From this we see that

$$\lim_{a,b\to+\infty} \int_a^b x^{-s} \cos x \, dx = 0$$

and consequently a finite

$$\lim_{a\to+\infty} \int_1^a x^{-s} \cos x \, dx$$

exists ([42, Theorem 3.11(c), p. 53]). We shall use this fact in Example 6.1.7.

Chapter 3

Measure and measurability

Throughout this chapter, we assume that α is a fixed increasing function defined on $\mathbf{R}$, and for each $c \in \mathbf{R}$ we let

$$\alpha(c+) = \lim_{x \to c+} \alpha(x) \qquad \text{and} \qquad \alpha(c-) = \lim_{x \to c-} \alpha(x).$$

Our goal is to show that a concept similar to the α-length of an interval can be defined for any subset of $\mathbf{R}$. Among several ways of doing this, the one due to Thomson ([40]) appears particularly suitable: while following the spirit of the McShane integral, it still resembles the traditional approach of Carathéodory, to which it is equivalent. The exposition is formally independent of Chapter 2.

3.1 Extended real numbers

Along with the set $\mathbf{R}$ of all real numbers we introduce the set

$$\overline{\mathbf{R}} = \mathbf{R} \cup \{+\infty, -\infty\},$$

whose elements are called the *extended real numbers*. We say that an $x \in \overline{\mathbf{R}}$ is *finite* or *infinite* according to whether $x \in \mathbf{R}$ or $x = \pm\infty$, respectively.

The order of $\mathbf{R}$ is extended to $\overline{\mathbf{R}}$ by stipulating that $-\infty < x < +\infty$ for each $x \in \mathbf{R}$. As $\overline{\mathbf{R}}$ has the smallest and largest elements, $\inf E$ and $\sup E$ exist in $\overline{\mathbf{R}}$ for every set $E \subset \overline{\mathbf{R}}$; in particular,

$$\inf E = -\sup E = +\infty$$

whenever E is empty. We let

$$\pm\infty + x = x \pm \infty = \pm\infty \pm \infty = \pm\infty$$

for each $x \in \mathbf{R}$, and set

$$-(\pm\infty) = \mp\infty \quad \text{and} \quad |+\infty| = |-\infty| = +\infty.$$

The symbols $+\infty - \infty$ and $-\infty + \infty$, as well as all ratios involving $\pm\infty$, are not defined. For each $x \in \overline{\mathbf{R}}$, we let

$$x(\pm\infty) = (\pm\infty)x = \begin{cases} \pm\infty & \text{if } x > 0, \\ \mp\infty & \text{if } x < 0, \\ 0 & \text{if } x = 0. \end{cases}$$

While the symbol $0/0$ is not defined, for every $x \in \mathbf{R} - \{0\}$, we set

$$\frac{x}{0} = \begin{cases} +\infty & \text{if } x > 0, \\ -\infty & \text{if } x < 0. \end{cases}$$

We also let $x^0 = 1$ for each nonnegative $x \in \mathbf{R}$.

Note. There is no "inner logic" in defining $0 \cdot (\pm\infty) = 0$ or $0^0 = 1$; it is merely a convention that will prove useful in our presentation.

Let E be a set. A map $f : E \to \overline{\mathbf{R}}$ is called an *extended real-valued function* on E. The algebraic operations (when meaningful), order, and convergence among extended real-valued functions on E are defined pointwise. As before, a map $f : E \to \mathbf{R}$ is called a *function* on E; for emphasis, we often add the attributes "real-valued" or "finite."

3.2 Measures

We say that a partition $\{(A_1, x_1), \ldots, (A_p, x_p)\}$ is *anchored* in a set $E \subset \mathbf{R}$ whenever $\{x_1, \ldots, x_p\}$ is a subset of E. If δ is a positive function on a set $E \subset \mathbf{R}$, let

$$\alpha^\delta(E) = \sup \sum_{i=1}^p \alpha(A_i)$$

where the supremum is taken over all partitions $\{(A_1, x_1), \ldots, (A_p, x_p)\}$ anchored in E that are δ-fine. Note that $\alpha^\theta(E) \leq \alpha^\delta(E)$ for each positive function θ on E with $\theta \leq \delta$.

Definition 3.2.1 The α-*measure* of the set $E \subset \mathbf{R}$ is the number

$$\alpha^*(E) = \inf \alpha^\delta(E)$$

where the infimum is taken over all positive functions δ defined on E.

Proposition 3.2.2 *If $A = [a, b]$ is an interval with $a \leq b$, then*

$$\alpha^*(A) = \alpha(b+) - \alpha(a-)\,;$$

in particular, $\alpha(A) \leq \alpha^*(A)$.

PROOF. Let δ be a positive function on A. If $a < b$ then by Proposition 1.2.4, there is a δ-fine partition P of A; if $a = b$ let $P = \emptyset$. Considering the δ-fine partition

$$P \cup \left\{ \Big([a - \delta(a)/2, a], a \Big), \Big([b, b + \delta(b)/2], b \Big) \right\},$$

it is clear that $\alpha^\delta(A) > \alpha(b+) - \alpha(a-)$ and so $\alpha^*(A) \geq \alpha(b+) - \alpha(a-)$. On the other hand, if $\delta = 1/n$ where n is a positive integer, it is easy to verify that

$$\alpha^*(A) \leq \alpha^\delta(A) < \alpha\left(b + \frac{1}{n}\right) - \alpha\left(a - \frac{1}{n}\right).$$

It follows that $\alpha^*(A) \leq \alpha(b+) - \alpha(a-)$, and the proposition is proved. ∎

Exercise 3.2.3 Show that $\alpha^*((a, b)) = \alpha(b-) - \alpha(a+)$ for any segment (a, b) with $a < b$, and that

$$\alpha^*(\mathbf{R}) = \lim_{x \to +\infty} \alpha(x) - \lim_{x \to -\infty} \alpha(x)\,.$$

Proposition 3.2.4 *The following statements are true.*

1. $\alpha^*(\emptyset) = 0$.
2. $\alpha^*(C) \leq \alpha^*(D)$ for each $C \subset D \subset \mathbf{R}$.
3. If $\{E_n\}$ is a sequence of subsets of $\mathbf{R}$, then

$$\alpha^* \left(\bigcup_{n=1}^{\infty} E_n \right) \leq \sum_{n=1}^{\infty} \alpha^*(E_n)\,.$$

PROOF. The first statement is correct because only the empty partition $P = \emptyset$ is used in the definition of $\alpha^*(\emptyset)$ (cf. Remark 1.2.3). If $C \subset D \subset \mathbf{R}$, then

$$\alpha^*(C) \leq \alpha^\delta(C) \leq \alpha^\delta(D)$$

for each positive function δ on D, and the second statement follows.

To prove the third statement, assume first that the sets E_n are disjoint. If $\delta_1, \delta_2, \ldots$ are positive functions on $E_1, E_2, \ldots$, respectively, we define a positive function δ on $E = \bigcup_{n=1}^{\infty} E_n$ by letting $\delta(x) = \delta_n(x)$ whenever $x \in E_n$ and $n = 1, 2, \ldots$. Let $P = \{(A_1, x_1), \ldots, (A_p, x_p)\}$ be a partition anchored in E, and for each positive integer n, let $P_n = \{(A_i, x_i) : x_i \in E_n\}$. If P is δ-fine, then P_n is δ_n-fine and consequently

$$\sum_{i=1}^{p} \alpha(A_i) = \sum_{n=1}^{\infty} \sum_{x_i \in E_n} \alpha(A_i) \leq \sum_{n=1}^{\infty} \alpha^{\delta_n}(E_n).$$

This and the arbitrariness of P imply that

$$\alpha^*(E) \leq \alpha^{\delta}(E) \leq \sum_{n=1}^{\infty} \alpha^{\delta_n}(E_n).$$

As the functions δ_n are arbitrary, the desired inequality follows. Now if the sets E_n are any subsets of $\mathbf{R}$, the previous result in conjunction with the second statement yields

$$\alpha^* \left(\bigcup_{n=1}^{\infty} E_n \right) = \alpha^* \left(\bigcup_{n=1}^{\infty} \left[E_n - \bigcup_{k=1}^{n-1} E_k \right] \right)$$

$$\leq \sum_{n=1}^{\infty} \alpha^* \left(E_n - \bigcup_{k=1}^{n-1} E_k \right) \leq \sum_{n=1}^{\infty} \alpha^*(E_n),$$

and the proposition is established. ∎

Note. The properties 1, 2, and 3 stated in Proposition 3.2.4 are often referred to as the *normalization*, *monotonicity*, and *countable subadditivity* of α^*, respectively.

Exercise 3.2.5 Show that $\alpha^*(C \cup D) = \alpha^*(C) + \alpha^*(D)$ for any subsets C and D of $\mathbf{R}$ such that $|x - y| \geq \eta$ for an $\eta > 0$ and all $x \in C$ and $y \in D$.

Exercise 3.2.6 Using [42, Chapter 4, Exercise 22, p. 101], prove that the equality in Exercise 3.2.5 holds for any sets $C, D \subset \mathbf{R}$ with $C^- \cap D^- = \emptyset$.

Remark 3.2.7 Any extended real-valued function α^* defined on the family of all subsets of $\mathbf{R}$ that has the properties listed in Proposition 3.2.4 is called a *measure* in $\mathbf{R}$. If it also has the property stated in Exercise 3.2.5, we call it a *metric measure* in $\mathbf{R}$. In the older terminology, a measure in $\mathbf{R}$ is usually referred to as an *outer measure* in $\mathbf{R}$.

Proposition 3.2.8 *If $E \subset \mathbf{R}$, then*

$$\alpha^*(E) = \inf \alpha^*(U)$$

where the infimum is taken over all open sets $U \subset \mathbf{R}$ containing E.

PROOF. If t denotes the right side of the desired equality, then $\alpha^*(E) \leq t$ by Proposition 3.2.4, 2. Assume that $\alpha^*(E) < t$ and find a positive function δ on E so that $\alpha^\delta(E) < t$. The open set $U = \bigcup_{x \in E} U(x, \delta(x))$ contains E, and we obtain a contradiction by showing that $\alpha^*(U) \leq \alpha^\delta(E)$. To this end, observe that there is a positive function θ on U such that for each $y \in U$ we can find an $x_y \in E$ with $U(y, \theta(y)) \subset U(x_y, \delta(x_y))$. If $P = \{(A_1, y_1), \ldots, (A_p, y_p)\}$ is a partition anchored in U that is θ-fine, then $\{(A_1, x_{y_1}), \ldots, (A_p, x_{y_p})\}$ is a partition anchored in E that is δ-fine. Thus $\sum_{i=1}^{p} \alpha(A_i) \leq \alpha^\delta(E)$, and as P is arbitrary, we have

$$\alpha^*(U) \leq \alpha^\theta(U) \leq \alpha^\delta(E). \quad \blacksquare$$

Let $\mathcal{C}$ be a family of cells. A *segmentation* of $\mathcal{C}$ is a finite family $\mathcal{D}$ of nonoverlapping cells such that each $D \in \mathcal{D}$ is contained in a $C \in \mathcal{C}$ and

$$C = \bigcup \{D \in \mathcal{D} : D \subset C\}$$

for every $C \in \mathcal{C}$. It is easy to see that a finite family $\mathcal{D}$ of nonoverlapping cells is a segmentation of $\mathcal{C}$ if and only if $\bigcup \mathcal{C} = \bigcup \mathcal{D}$ and $D \subset C$ whenever $D \in \mathcal{D}$ overlaps $C \in \mathcal{C}$.

Lemma 3.2.9 *Each finite family of cells has a segmentation.*

PROOF. Let $\{C_1, \ldots, C_k\}$ be a family of cells. Since each family consisting of fewer than two cells is a segmentation of itself, we may assume that $k \geq 2$. Proceeding inductively, suppose that $\{D_1, \ldots, D_n\}$ is a segmentation of $\{C_1, \ldots, C_{k-1}\}$. Each $(D_j - C_k)^-$ is the union of a finite collection $\mathcal{D}_j$ of nonoverlapping cells (we may assume that $\mathcal{D}_j$ contains at most two cells, but this is irrelevant), and there is a finite family $\mathcal{A}$ of nonoverlapping cells such that $(C_k - \bigcup_{j=1}^{n} D_j)^- = \bigcup \mathcal{A}$. If $\mathcal{B}$ is the collection of all cells among the intervals $C_k \cap D_1, \ldots, C_k \cap D_n$, it is easy to verify that

$$\mathcal{A} \cup \mathcal{B} \cup \left(\bigcup_{j=1}^{n} \mathcal{D}_j \right)$$

is a segmentation of $\{C_1, \ldots, C_k\}$. $\quad \blacksquare$

Proposition 3.2.10 *Let $\{A_1, \ldots, A_p\}$ and $\{B_1, \ldots, B_q\}$ be families of cells* such that $\bigcup_{i=1}^{p} A_i \subset \bigcup_{j=1}^{q} B_j$. If the cells $A_1, \ldots, A_p$ are nonoverlapping, then

$$\sum_{i=1}^{p} \alpha(A_i) \leq \sum_{j=1}^{q} \alpha(B_j).$$

PROOF. Let $A = \bigcup_{i=1}^{p} A_i$ and $B = \bigcup_{j=1}^{q} B_j$. If $\{D_1, \ldots, D_r\}$ is a segmentation of the family $\{A_1, \ldots, A_p, B_1, \ldots, B_q\}$, then

$$\sum_{i=1}^{p} \alpha(A_i) = \sum_{i=1}^{p} \sum_{D_k \subset A_i} \alpha(D_k) = \sum_{D_k \subset A} \alpha(D_k)$$

$$\leq \sum_{D_k \subset B} \alpha(D_k) \leq \sum_{j=1}^{q} \sum_{D_k \subset B_j} \alpha(D_k) = \sum_{j=1}^{q} \alpha(B_j). \quad \blacksquare$$

Lemma 3.2.11 *Let δ be a positive function on a set $E \subset \mathbf{R}$. If K is a compact subset of $\bigcup_{x \in E} U(x, \delta(x))$, then there is a partition $P = \{(A_1, x_1), \ldots, (A_p, x_p)\}$ anchored in E such that P is δ-fine and $K \subset (\bigcup_{i=1}^{p} A_i)^{\circ}$.*

PROOF. Recall that $U(x, r)$ is the interior of $U[x, r]$, and let $\delta_k = k\delta/(1+k)$ for $k = 1, 2 \ldots$. Since the family

$$\{U(z, \delta_k(z)) : z \in E, \ k = 1, 2, \ldots\}$$

is an open cover of the compact set K, there are $z_1, \ldots, z_n$ in E and positive integers $k_1, \ldots, k_n$ such that $K \subset \bigcup_{j=1}^{n} U(z_j, r_j)$ where $r_j = \delta_{k_j}(z_j)$. Let $\{A_1, \ldots, A_p\}$ be a segmentation of $\{U[z_1, r_1], \ldots, U[z_n, r_n]\}$. For $i = 1, \ldots, p$, select a $U[z_j, r_j]$ containing A_i and let $x_i = z_j$. Then $\{(A_1, x_1), \ldots, (A_p, x_p)\}$ is a partition anchored in E that is δ-fine, and

$$K \subset \bigcup_{j=1}^{n} U(z_j, r_j) \subset \left(\bigcup_{j=1}^{n} U[z_j, r_j] \right)^{\circ} = \left(\bigcup_{i=1}^{p} A_i \right)^{\circ}. \quad \blacksquare$$

Proposition 3.2.12 *If K is a compact subset of $\mathbf{R}$, then*

$$\alpha^*(K) = \inf \sum_{i=1}^{p} \alpha(A_i)$$

where the infimum is taken over all nonoverlapping collections $\{A_1, \ldots, A_p\}$ of cells for which $K \subset (\bigcup_{i=1}^{p} A_i)^{\circ}$.

PROOF. Denote by t the right side of the equality we want to establish, and assume that $\alpha^*(K) < t$. Then $\alpha^\delta(K) < t$ for a positive function δ on K. Applying Lemma 3.2.11 to the set $E = K$, we obtain a partition $P = \{(A_1, x_1), \ldots, (A_p, x_p)\}$ anchored in K that is δ-fine and such that $K \subset (\bigcup_{i=1}^p A_i)^\circ$. Thus

$$t \le \sum_{i=1}^p \alpha(A_i) \le \alpha^\delta(K)$$

and we have a contradiction.

Conversely, suppose that $t < \alpha^*(K)$ and find a nonoverlapping collection $\{A_1, \ldots, A_p\}$ of cells so that $K \subset (\bigcup_{i=1}^p A_i)^\circ$ and $\sum_{i=1}^p \alpha(A_i) < \alpha^*(K)$. There is a positive function δ on K such that $U(x, \delta(x)) \subset \bigcup_{i=1}^p A_i$ for each $x \in K$. If $\{(B_1, y_1), \ldots, (B_q, y_q)\}$ is a partition anchored in K that is δ-fine, then $\bigcup_{j=1}^q B_j \subset \bigcup_{i=1}^p A_i$, and we have $\sum_{j=1}^q \alpha(B_j) \le \sum_{i=1}^p \alpha(A_i)$ by Proposition 3.2.10. As $\{(B_1, y_1), \ldots, (B_q, y_q)\}$ is arbitrary, we obtain

$$\alpha^*(K) \le \alpha^\delta(K) \le \sum_{i=1}^p \alpha(A_i),$$

a contradiction. ∎

Exercise 3.2.13 If $A_1, \ldots, A_p$ are nonoverlapping cells, employ Propositions 3.2.10 and 3.2.12 to show that $\sum_{i=1}^p \alpha(A_i) \le \alpha^*(\bigcup_{i=1}^p A_i)$.

Proposition 3.2.14 *If U is an open subset of $\mathbf{R}$, then*

$$\alpha^*(U) = \sup \alpha^*(K)$$

where the supremum is taken over all compact sets $K \subset U$.

PROOF. If t denotes the right side of the equality we wish to prove, then $t \le \alpha^*(U)$ according to Proposition 3.2.4, 2. Find a positive function δ on U so that $U(x, \delta(x)) \subset U$ for each $x \in U$, and let $P = \{(A_1, x_1), \ldots, (A_p, x_p)\}$ be a partition anchored in U that is δ-fine. Since $K = \bigcup_{i=1}^p A_i$ is a compact subset of U, it follows from Exercise 3.2.13 that

$$\sum_{i=1}^p \alpha(A_i) \le \alpha^*(K) \le t.$$

The arbitrariness of P implies that $\alpha^*(U) \le \alpha^\delta(U) \le t$. ∎

Corollary 3.2.15 *If A is the union of a collection $\{A_1, \ldots, A_p\}$ of nonoverlapping cells, then*

$$\alpha^*(A^\circ) \leq \sum_{i=1}^{p} \alpha(A_i) \leq \alpha^*(A).$$

PROOF. If K is a compact subset of A°, then $\alpha^*(K) \leq \sum_{i=1}^{p} \alpha(A_i)$ according to Proposition 3.2.12. Hence $\alpha^*(A^\circ) \leq \sum_{i=1}^{p} \alpha(A_i)$ by Proposition 3.2.14, and the corollary follows from Exercise 3.2.13. ∎

Remark 3.2.16 If $p = 1$, then Corollary 3.2.15 is a direct consequence of Proposition 3.2.2 and Exercise 3.2.3. Note, however, that neither Proposition 3.2.2 nor Exercise 3.2.3 has been used in this section.

Exercise 3.2.17 Show that Corollary 3.2.15 remains valid when $A_1, \ldots, A_p$ are nonoverlapping intervals.

3.3 Measurable sets

We want to single out a family of subsets of $\mathbf{R}$ on which the α-measure α^* is *countably additive* in the sense of Corollary 3.3.3 below. Recall from Section 1.1 that $\alpha(B) = 0$ whenever B is a degenerate interval.

Definition 3.3.1 A set $E \subset \mathbf{R}$ is called α-*measurable* if, given $\varepsilon > 0$, there is a positive function δ on $\mathbf{R}$ such that

$$\sum_{i=1}^{p} \sum_{j=1}^{q} \alpha(A_i \cap B_j) < \varepsilon$$

for all δ-fine partitions $\{(A_1, x_1), \ldots, (A_p, x_p)\}$ and $\{(B_1, y_1), \ldots, (B_q, y_q)\}$ anchored in E and $\mathbf{R} - E$, respectively.

Intuitively speaking, a set $E \subset \mathbf{R}$ is α-measurable if E and its complement $\mathbf{R} - E$ are not "too entangled."

Proposition 3.3.2 *The following statements are true.*

1. *If $E \subset \mathbf{R}$ and $\alpha^*(E) = 0$, then E is α-measurable.*
2. *If E is an α-measurable set, then so is its complement $\mathbf{R} - E$.*
3. *If C and D are α-measurable sets, then so are the sets $C \cup D$, $C \cap D$, and $C - D$.*

4. *If C and D are disjoint subsets of* **R** *and one of them is α-measurable, then*

$$\alpha^*([C \cup D] \cap E) = \alpha^*(C \cap E) + \alpha^*(D \cap E)$$

for each set $E \subset \mathbf{R}$; in particular, $\alpha^*(C \cup D) = \alpha^*(C) + \alpha^*(D)$.

5. *If $C \subset D \subset \mathbf{R}$ and C is α-measurable with $\alpha^*(C) < +\infty$, then*

$$\alpha^*(D - C) = \alpha^*(D) - \alpha^*(C).$$

PROOF. Since

$$\sum_{i=1}^{p} \sum_{j=1}^{q} \alpha(A_i \cap B_j) \leq \sum_{i=1}^{p} \alpha(A_i)$$

for all partitions $\{(A_1, x_1), \ldots, (A_p, x_p)\}$ and $\{(B_1, y_1), \ldots, (B_q, y_q)\}$, the first claim follows from the definition of $\alpha^*(E)$. The second claim follows by symmetry.

Let C and D be α-measurable subsets of **R**, let $\varepsilon > 0$, and let δ_C and δ_D be positive functions on **R** associated with the pairs $(\varepsilon/2, C)$ and $(\varepsilon/2, D)$, respectively, according to Definition 3.3.1. Set $\delta = \min\{\delta_C, \delta_D\}$, and choose δ-fine partitions $\{(A_1, x_1), \ldots, (A_p, x_p)\}$ and $\{(B_1, y_1), \ldots, (B_q, y_q)\}$ anchored in $C \cup D$ and $\mathbf{R} - (C \cup D)$, respectively. Then

$$\sum_{i=1}^{p} \sum_{j=1}^{q} \alpha(A_i \cap B_j) \leq \sum_{x_i \in C} \sum_{j=1}^{q} \alpha(A_i \cap B_j) + \sum_{x_i \in D} \sum_{j=1}^{q} \alpha(A_i \cap B_j) < \varepsilon.$$

Thus $C \cup D$ is α-measurable and, in view of the second statement, so are the sets

$$C \cap D = \mathbf{R} - [(\mathbf{R} - C) \cup (\mathbf{R} - D)] \quad \text{and} \quad C - D = C \cap (\mathbf{R} - D).$$

This proves the third claim.

Let $D, E \subset \mathbf{R}$, and let C be an α-measurable set disjoint from D. Choose an $\varepsilon > 0$, and let Δ be a positive function on **R** associated with ε and C according to Definition 3.3.1. Select a positive function δ on $[C \cup D] \cap E$ and set $\theta(x) = \min\{\delta(x), \Delta(x)\}$ for each $x \in [C \cup D] \cap E$. Let $P_C = \{(A_1, x_1), \ldots, (A_p, x_p)\}$ and $P_D = \{(B_1, y_1), \ldots, (B_q, y_q)\}$ be partitions anchored in $C \cap E$ and $D \cap E$, respectively, and let $\{D_1, \ldots, D_r\}$ be a segmentation of the family

$$\{A_1, \ldots, A_p, B_1, \ldots, B_q\}.$$

After a suitable reordering, we may assume that $(A_i \cap D_k)^\circ = \emptyset$ for $i = 1, \ldots, p$ and $k = 1, \ldots, s$ where s is an integer with $0 \leq s \leq r$. Thus there is a partition

$$P = \{(A_1, x_1), \ldots, (A_p, x_p), (D_1, z_1), \ldots, (D_s, z_s)\}$$

such that $\{z_1, \ldots, z_s\} \subset \{y_1, \ldots, y_q\}$, $D_k \subset B_j$ whenever $z_k = y_j$, and

$$\left(\bigcup_{k=1}^{s} D_k\right) \cup \left(\bigcup_{i=1}^{p} \bigcup_{j=1}^{q} [A_i \cap B_j]\right) = \bigcup_{j=1}^{q} B_j.$$

If P_C and P_D are θ-fine, then so is P and we see that

$$\alpha^\delta([C \cup D] \cap E) \geq \sum_{i=1}^{p} \alpha(A_i) + \sum_{k=1}^{s} \alpha(D_k) = \sum_{i=1}^{p} \alpha(A_i) + \sum_{j=1}^{q} \alpha(B_j)$$

$$- \sum_{i=1}^{p} \sum_{j=1}^{q} \alpha(A_i \cap B_j) > \sum_{i=1}^{p} \alpha(A_i) + \sum_{j=1}^{q} \alpha(B_j) - \varepsilon.$$

As P_C and P_D are arbitrary, we obtain

$$\alpha^\delta([C \cup D] \cap E) \geq \alpha^\theta(C \cap E) + \alpha^\theta(D \cap E) - \varepsilon$$
$$\geq \alpha^*(C \cap E) + \alpha^*(D \cap E) - \varepsilon,$$

and since δ is arbitrary, this inequality yields

$$\alpha^*([C \cup D] \cap E) \geq \alpha^*(C \cap E) + \alpha^*(D \cap E) - \varepsilon.$$

Proposition 3.2.4, 3 and the arbitrariness of ε imply the fourth claim.

Finally, let $C \subset D \subset \mathbf{R}$ and assume that C is α-measurable. By the fourth claim, we obtain

$$\alpha^*(D) = \alpha^*(C) + \alpha^*(D - C).$$

If $\alpha^*(C) < +\infty$ then $\alpha^*(D - C) = \alpha^*(D) - \alpha^*(C)$, and the proposition is proved. ∎

Corollary 3.3.3 *If $\{E_n\}$ is a sequence of disjoint α-measurable sets, then*

$$\alpha^*\left(\bigcup_{n=1}^{\infty} [E \cap E_n]\right) = \sum_{n=1}^{\infty} \alpha^*(E \cap E_n)$$

for each set $E \subset \mathbf{R}$. In particular, $\alpha^*(\bigcup_{n=1}^{\infty} E_n) = \sum_{n=1}^{\infty} \alpha^*(E_n)$.

PROOF. By Propositions 3.3.2, 4 and 3.2.4, 2, we obtain that

$$\sum_{n=1}^{k} \alpha^*(E \cap E_n) = \alpha^*\left(\bigcup_{n=1}^{k} [E \cap E_n]\right) \leq \alpha^*\left(\bigcup_{n=1}^{\infty} [E \cap E_n]\right)$$

for $k = 1, 2, \ldots$, and hence $\sum_{n=1}^{\infty} \alpha^*(E \cap E_n) \leq \alpha^*(\bigcup_{n=1}^{\infty} [E \cap E_n])$. Applying Proposition 3.2.4, 3 and setting $E = \mathbf{R}$ completes the argument. ∎

Proposition 3.3.4 *The union of countably many α-measurable sets is α-measurable.*

PROOF. Let $E = \bigcup_{n=1}^{\infty} E_n$ where $E_1, E_2, \ldots$ are α-measurable sets. In view of Proposition 3.3.2, 3, it suffices to consider only the case when the sets E_n are disjoint.

Assume first that E is a subset of a cell. It follows from Corollary 3.3.3 and Propositions 3.2.4, 2 and 3.2.2 that the series $\sum_{n=1}^{\infty} \alpha^*(E_n)$ converges. Thus, given $\varepsilon > 0$, there is an integer $k \geq 1$ such that

$$\sum_{n=k+1}^{\infty} \alpha^*(E_n) < \frac{\varepsilon}{2}.$$

By Propositions 3.3.2, 3 and 3.2.4, 3, the set $C = \bigcup_{n=1}^{k} E_n$ is α-measurable and $\alpha^*(E - C) < \varepsilon/2$. Let δ be a positive function on $\mathbf{R}$ associated with $\varepsilon/2$ and C according to Definition 3.3.1. Making δ smaller, we may assume that $\alpha^\delta(E - C) < \varepsilon/2$. If $\{(A_1, x_1), \ldots, (A_p, x_p)\}$ and $\{(B_1, y_1), \ldots, (B_q, y_q)\}$ are δ-fine partitions anchored in E and $\mathbf{R} - E$, respectively, then

$$\sum_{i=1}^{p}\sum_{j=1}^{q}\alpha(A_i \cap B_j) = \sum_{x_i \in C}\sum_{j=1}^{q}\alpha(A_i \cap B_j) + \sum_{x_i \in E-C}\sum_{j=1}^{q}\alpha(A_i \cap B_j)$$

$$< \frac{\varepsilon}{2} + \sum_{x_i \in E-C}\alpha(A_i) \leq \frac{\varepsilon}{2} + \alpha^\delta(E - C) < \varepsilon,$$

establishing the α-measurability of E.

If the set E is arbitrary, the previous result shows that for all integers n the sets $C_n = E \cap [n, n+1]$ are α-measurable, and we have $E = \bigcup_{n=-\infty}^{\infty} C_n$. Given an integer n, let δ_n be a positive function on $\mathbf{R}$ associated with $\varepsilon/2^{|n|}$ and C_n according to Definition 3.3.1. With no loss of generality, we may assume that $\delta_n \leq 1$ and consequently, we may also assume that $\delta_n(x) = 1$ whenever x lies outside the cell $[n-2, n+3]$. It follows that $\delta = \inf \delta_n$ is a positive function on $\mathbf{R}$. Let $\{(A_1, x_1), \ldots, (A_p, x_p)\}$ and $\{(B_1, y_1), \ldots, (B_q, y_q)\}$ be δ-fine partitions anchored in E and $\mathbf{R} - E$, respectively. Then

$$\sum_{i=1}^{p}\sum_{j=1}^{q}\alpha(A_i \cap B_j) = \sum_{n=-\infty}^{\infty}\sum_{x_i \in C_n}\sum_{j=1}^{q}\alpha(A_i \cap B_j) < \sum_{n=-\infty}^{\infty}\varepsilon 2^{-|n|} = 3\varepsilon$$

and we see that E is again α-measurable. $\blacksquare$

Exercise 3.3.5 Show that the intersection of countably many α-measurable sets is α-measurable.

Remark 3.3.6 Let X be an arbitrary set. A family $\mathcal{M}$ of subsets of X is called a *σ-algebra* in X if it is closed with respect to complementations and countable unions. Since the empty union is countable, each σ-algebra in X contains the empty set; by complementation, it also contains X. A nonnegative extended real-valued function μ on a σ-algebra $\mathcal{M}$ in X is called a *measure* on $\mathcal{M}$ if it is *countably additive* in the following sense:

$$\mu\left(\bigcup \mathcal{E}\right) = \sum_{E \in \mathcal{E}} \mu(E)$$

for every countable disjoint family $\mathcal{E} \subset \mathcal{M}$. If μ is a measure on a σ-algebra $\mathcal{M}$ in X, then by taking $\mathcal{E} = \emptyset$, we see that $\mu(\emptyset) = 0$. It is important to distinguish between a measure on a σ-algebra in $\mathbf{R}$ and a measure in $\mathbf{R}$ (cf. Remark 3.2.7).

Using the terminology of Remark 3.3.6, we can conveniently summarize the most important properties of α-measurable subsets of $\mathbf{R}$.

Theorem 3.3.7 *The family $\mathcal{M}_\alpha$ of all α-measurable subsets of $\mathbf{R}$ is a σ-algebra in $\mathbf{R}$, and the α-measure α^* is a measure on $\mathcal{M}_\alpha$.* ∎

Our next goal is to show that α-measurable sets are plentiful. In view of Theorem 3.3.7, this will be established by showing that all open sets are α-measurable.

Lemma 3.3.8 *A set E is α-measurable if and only if for each $\varepsilon > 0$ there is an open set U and a closed set C such that*

$$C \subset E \subset U \quad \text{and} \quad \alpha^*(U - C) < \varepsilon.$$

PROOF. Let E be α-measurable and choose an $\varepsilon > 0$. Select a positive function δ on $\mathbf{R}$ associated with E and $\varepsilon/2$ according to Definition 3.3.1, and let

$$U = \bigcup_{x \in E} U(x, \delta(x)) \quad \text{and} \quad V = \bigcup_{x \in \mathbf{R}-E} U(x, \delta(x)).$$

Then U is open, $C = \mathbf{R} - V$ is closed, and $C \subset E \subset U$. If K is a compact subset of $U - C = U \cap V$, then applying Lemma 3.2.11 to the sets E and $\mathbf{R} - E$, we obtain δ-fine partitions $\{(A_1, x_1), \ldots, (A_p, x_p)\}$ and $\{(B_1, y_1), \ldots, (B_q, y_q)\}$ anchored in E and $\mathbf{R} - E$, respectively, and such that

$$K \subset \left(\bigcup_{i=1}^{p} A_i\right)^\circ \cap \left(\bigcup_{j=1}^{q} B_j\right)^\circ \subset \left(\bigcup_{i=1}^{p} \bigcup_{j=1}^{q} [A_i \cap B_j]\right)^\circ.$$

Now Proposition 3.2.12 implies that

$$\alpha^*(K) \leq \sum_{i=1}^{p} \sum_{j=1}^{q} \alpha(A_i \cap B_j) < \frac{\varepsilon}{2},$$

and as $U - C$ is an open set, the inequality $\alpha^*(U - C) < \varepsilon$ follows from Proposition 3.2.14.

Conversely, let U and C satisfy the conditions of the theorem for a given $\varepsilon > 0$. Choose a positive function δ on $\mathbf{R}$ so that $U(x, \delta(x)) \subset U$ for each $x \in E$ and $U(x, \delta(x)) \subset \mathbf{R} - C$ for each $x \in \mathbf{R} - E$. If $\{(A_1, x_1), \ldots, (A_p, x_p)\}$ and $\{(B_1, y_1), \ldots, (B_q, y_q)\}$ are δ-fine partitions anchored in E and $\mathbf{R} - E$, respectively, then $D = \bigcup_{i=1}^{p} \bigcup_{j=1}^{q} (A_i \cap B_j)$ is a subset of $U - C$. Thus, by Exercise 3.2.17 and Proposition 3.2.4, 2, we have

$$\sum_{i=1}^{p} \sum_{j=1}^{q} \alpha(A_i \cap B_j) \leq \alpha^*(D) \leq \alpha^*(U - C) < \varepsilon,$$

and the α-measurability of E is established. ∎

Exercise 3.3.9 If E is an α-measurable set, show that

$$\alpha^*(E) = \sup \alpha^*(K)$$

where the supremum is taken over all compact sets $K \subset E$.

Theorem 3.3.10 *Each open set is α-measurable.*

PROOF. Since every open subset of $\mathbf{R}$ is a countable union of bounded open sets, it suffices to show that an open set U with $\alpha^*(U) < +\infty$ is α-measurable. To this end, choose an $\varepsilon > 0$, and use Proposition 3.2.14 to find a compact set $K \subset U$ such that

$$\alpha^*(U) < \alpha^*(K) + \varepsilon.$$

Select a positive function δ on $U - K$ so that $U(x, \delta(x)) \subset U - K$ for every $x \in U - K$, and let $P = \{(A_1, x_1), \ldots, (A_p, x_p)\}$ be a partition anchored in $U - K$ that is δ-fine. Since $A = \bigcup_{i=1}^{p} A_i$ and K are disjoint compact subsets of U, it follows from Exercises 3.2.13 and 3.2.6 that

$$\sum_{i=1}^{p} \alpha(A_i) + \alpha^*(K) \leq \alpha^*(A) + \alpha^*(K) = \alpha^*(A \cup K)$$

$$\leq \alpha^*(U) < \alpha^*(K) + \varepsilon.$$

As P is arbitrary, we conclude that

$$\alpha^*(U - K) \leq \alpha^\delta(U - K) \leq \varepsilon,$$

and Lemma 3.3.8 implies the measurability of U. ∎

Exercise 3.3.11 As in Section 2.5, let γ be a *gap function* determined by the triple sequence $\{s_n, t_n, c_n\}$.

1. Use Exercise 3.2.3 to show that $\gamma^*(\mathbf{R}) = \sum_{n=1}^{\infty} c_n$.
2. Infer that every set $E \subset \mathbf{R}$ is γ-measurable with

$$\gamma^*(E) = \sum_{s_n \in E} \gamma^*(\{s_n\}).$$

Observe that these results do not depend on the fact that $\{s_n\}$ is a bounded sequence.

Note. The fact that all subsets of $\mathbf{R}$ are γ-measurable rests on special properties of the gap function γ. The reader should be aware that, in general, there could be sets which are not α-measurable (see [41, Chapter 3, Theorem 17, p. 64]).

A set $E \subset \mathbf{R}$ is called G_δ if it is the intersection of countably many open subsets of $\mathbf{R}$; it is called F_σ if it is the union of countably many closed subsets of $\mathbf{R}$. It follows from Theorems 3.3.10 and 3.3.7 that all G_δ sets and all F_σ sets are α-measurable. Moreover, Theorem 3.3.13 below shows that each α-measurable set differs from a G_δ set and an F_σ set by a set of α-measure zero.

Exercise 3.3.12 Prove the following facts.

1. A set E is G_δ if and only if its complement $\mathbf{R} - E$ is F_σ.
2. If U is a G_δ set, then $U = \bigcap_{n=1}^{\infty} U_n$ where U_n are open sets such that $U_{n+1} \subset U_n$ for $n = 1, 2, \ldots$.
3. Each closed set $C \subset \mathbf{R}$ is simultaneously F_σ and G_δ.
4. The intersection of countably many G_δ sets is a G_δ set.
5. Each F_σ set is a countable union of compact sets.

Theorem 3.3.13 *A set E is α-measurable if and only if there is a G_δ set U and an F_σ set C such that*

$$C \subset E \subset U \qquad and \qquad \alpha^*(U - C) = 0.$$

PROOF. If the condition of the theorem holds, then $E = C \cup (E - C)$ is a measurable set by Proposition 3.3.2, 1 and 3.

Conversely, suppose that E is α-measurable. By Lemma 3.3.8, there are open sets U_n and closed sets C_n such that

$$C_n \subset E \subset U_n \qquad \text{and} \qquad \alpha^*(U_n - C_n) < \frac{1}{n}$$

for $n = 1, 2, \ldots$. The sets $U = \bigcap_{n=1}^{\infty} U_n$ and $C = \bigcup_{n=1}^{\infty} C_n$ are, respectively, G_δ and F_σ. Moreover, $C \subset E \subset U$ and since

$$\alpha^*(U - C) \leq \alpha^*(U_n - C_n) < \frac{1}{n}$$

for $n = 1, 2, \ldots$, we have $\alpha^*(U - C) = 0$. $\blacksquare$

Lemma 3.3.14 *For each set $E \subset \mathbf{R}$ there is a G_δ set U such that $E \subset U$ and $\alpha^*(U) = \alpha^*(E)$.*

PROOF. If $\alpha^*(E) = +\infty$, it suffices to let $U = \mathbf{R}$. If $\alpha^*(E) < +\infty$, use Proposition 3.2.8 to find open sets U_n such that $E \subset U_n$ and $\alpha^*(U_n) < \alpha^*(E) + 1/n$ for $n = 1, 2, \ldots$. It follows from Proposition 3.2.4, 2 that $U = \bigcap_{n=1}^{\infty} U_n$ is the desired set. $\blacksquare$

Proposition 3.3.15 *If $E_1 \subset E_2 \subset \cdots \subset \mathbf{R}$, then*

$$\alpha^* \left(\bigcup_{n=1}^{\infty} E_n \right) = \lim \alpha^*(E_n).$$

PROOF. Let $E = \bigcup_{n=1}^{\infty} E_n$. By Lemma 3.3.14, there are α-measurable sets C_n (in fact, G_δ sets, but this is irrelevant) such that $E_n \subset C_n$ and $\alpha^*(C_n) = \alpha^*(E_n)$ for $n = 1, 2, \ldots$. Replacing C_n by $\bigcap_{k=n}^{\infty} C_k$, we may assume that $C_1 \subset C_2 \subset \cdots$ (see Exercise 3.3.5).

We let $C = \bigcup_{n=1}^{\infty} C_n$, and show first that $\lim \alpha^*(C_n) = \alpha^*(C)$. By Proposition 3.2.4, 2, the limit $\lim \alpha^*(C_n) \leq \alpha^*(C)$ exists, and the equality holds if it is infinite. Thus assume that $\lim \alpha^*(C_n) < +\infty$, and let $D_1 = C_1$ and $D_{k+1} = C_{k+1} - C_k$ for $k = 1, 2, \ldots$. Since the α-measurable sets D_k are disjoint and $\bigcup_{k=1}^{\infty} D_k = C$, Corollary 3.3.3 and Proposition 3.3.2, 5 yield

$$\alpha^*(C) = \sum_{k=1}^{\infty} \alpha^*(D_k)$$

$$= \alpha^*(C_1) + \lim_{n \to \infty} \sum_{k=1}^{n} [\alpha^*(C_{k+1}) - \alpha^*(C_k)] = \lim \alpha^*(C_n).$$

Now $E \subset C$, and so Proposition 3.2.4, 2 implies that

$$\alpha^*(E) \leq \alpha^*(C) = \lim \alpha^*(C_n) = \lim \alpha^*(E_n) \leq \alpha^*(E). \quad \blacksquare$$

Exercise 3.3.16 Let $\{E_n\}$ be a sequence of α-measurable sets such that $E_{n+1} \subset E_n$ for $n = 1, 2, \ldots$. Prove that

$$\alpha^* \left(\bigcap_{n=1}^{\infty} E_n \right) = \lim \alpha^*(E_n)$$

whenever $\alpha^*(E_1) < +\infty$, and give an example showing the necessity of the last assumption.

The following theorem is usually referred to as the *Carathéodory test* for measurability.

Theorem 3.3.17 *A set $E \subset \mathbf{R}$ is α-measurable if and only if*

$$\alpha^*(S) = \alpha^*(S \cap E) + \alpha^*(S - E)$$

for every set $S \subset \mathbf{R}$.

PROOF. As the converse follows from Proposition 3.3.2, 4, suppose that E satisfies the condition of the theorem, and let $E_n = E \cap U[0, n]$ for $n = 1, 2, \ldots$. Fix an integer $n \geq 1$ and use Lemma 3.3.14 to find a measurable set S such that $E_n \subset S$ and $\alpha^*(S) = \alpha^*(E_n)$. Replacing S by $S \cap U[0, n]$, we may assume that $S \subset U[0, n]$. Thus

$$+\infty > \alpha^*(S) = \alpha^*(S \cap E) + \alpha^*(S - E) = \alpha^*(E_n) + \alpha^*(S - E_n)$$

and we see that $\alpha^*(S - E_n) = 0$. By Proposition 3.3.2, 1 and 3, the set $E_n = S - (S - E_n)$ is α-measurable, and the measurability of $E = \bigcup_{n=1}^{\infty} E_n$ follows from Proposition 3.3.4. $\quad \blacksquare$

3.4 Calculating measures

For a set $E \subset \mathbf{R}$, we present a more customary way of evaluating the α-measure $\alpha^*(E)$; in particular, we bypass the calculation of $\alpha^\delta(E)$. Using Corollary 3.2.15, observe that if A is a cell with $\alpha^*(\partial A) = 0$, then

$$\alpha(A) = \alpha^*(A) = \alpha^*(A^\circ).$$

Proposition 3.4.1 *If $\mathcal{E}$ is a disjoint family of α-measurable sets, then the family $\mathcal{E}_+ = \{E \in \mathcal{E} : \alpha^*(E) > 0\}$ is countable.*

PROOF. Assume that $\mathcal{E}_+$ is uncountable, and let

$$\mathcal{E}_{p,q} = \left\{ E \in \mathcal{E} : \alpha^*(E \cap U[0,p]) > \frac{1}{q} \right\}$$

for $p, q = 1, 2 \ldots$. Since $\mathcal{E}_+ = \bigcup_{p,q=1}^{\infty} \mathcal{E}_{p,q}$, there are positive integers r and s such that the collection $\mathcal{E}_{r,s}$ is uncountable. Thus we can find disjoint α-measurable sets $A_n \subset U[0,r]$ so that $\alpha^*(A_n) > 1/s$ for $n = 1, 2, \ldots$. This is a contradiction, as

$$+\infty = \sum_{n=1}^{\infty} \alpha^*(A_n) = \alpha^* \left(\bigcup_{n=1}^{\infty} A_n \right) \leq \alpha^*(U[0,r]) . \quad \blacksquare$$

Corollary 3.4.2 $\alpha^*(\{x\}) = 0$ *for all but countably many* $x \in \mathbf{R}$. $\quad \blacksquare$

Note. Since any increasing function has only countably many discontinuities ([42, Theorem 4.30, p. 96]), Corollary 3.4.2 is also a direct consequence of Proposition 3.2.2.

Lemma 3.4.3 *If R and S are countable subsets of $\mathbf{R}$, then there is a real number z such that the sets R and $S + z = \{s + z : s \in S\}$ are disjoint.*

PROOF. Observe that

$$\{z \in \mathbf{R} : R \cap (S + z) \neq \emptyset\} = \{r - s : r \in R \text{ and } s \in S\} .$$

As the set on the right side is clearly countable, the lemma follows. $\quad \blacksquare$

Lemma 3.4.4 *Each nonempty open set $U \subset \mathbf{R}$ is the union of nonoverlapping cells $K_1, K_2, \ldots$ such that $\alpha^*(\partial K_n) = 0$ for all n. In particular,*

$$\alpha^*(U) = \sum_{n=1}^{\infty} \alpha(K_n) .$$

PROOF. A *dyadic cell* is the cell $[k2^{-n}, (k+1)2^{-n}]$ where k and n are integers with $n \geq 0$. The set of boundary points of all dyadic cells is the set of all dyadic rationals, which is countable and dense in $\mathbf{R}$. By Corollary 3.4.2 and Lemma 3.4.3, there is a real number z such that

$$\alpha^*(\partial[D + z]) = 0$$

for each dyadic cell D. If $\mathcal{D}$ is the family of all dyadic cells contained in the open set $V = U - z$, it is easy to show that $V = \bigcup \mathcal{D}$. Since any two dyadic cells are either nonoverlapping or one contains the other, the family

$\mathcal{D}$ contains a sequence $\{D_n\}$ of nonoverlapping cells whose union is still V. If $K_n = D_n + z$, then $\alpha^*(\partial K_n) = 0$ for all n and $U = \bigcup_{n=1}^{\infty} K_n$. As $K_1^{\circ}, K_2^{\circ}, \ldots$ are disjoint open sets, we have

$$\sum_{n=1}^{\infty} \alpha(K_n) = \sum_{n=1}^{\infty} \alpha^*(K_n^{\circ}) = \alpha^* \left(\bigcup_{n=1}^{\infty} K_n^{\circ} \right)$$

$$\leq \alpha^*(U) \leq \sum_{n=1}^{\infty} \alpha^*(K_n) = \sum_{n=1}^{\infty} \alpha(K_n),$$

and the lemma is proved. ∎

Proposition 3.4.5 *If $E \subset \mathbf{R}$, then*

$$\alpha^*(E) = \inf \sum_{i=1}^{\infty} \alpha(A_i)$$

where the infimum is taken over all sequences $\{A_i\}$ of cells whose interiors cover E.

PROOF. Leaving the trivial case to the reader, we assume that $E \neq \emptyset$. Denote by t the right side of the equality we want to establish. If $\{A_i\}$ is a sequence of cells such that $E \subset \bigcup_{i=1}^{\infty} A_i^{\circ}$, then

$$\alpha^*(E) \leq \sum_{i=1}^{\infty} \alpha^*(A_i^{\circ}) \leq \sum_{i=1}^{\infty} \alpha(A_i)$$

by Corollary 3.2.15 or Exercise 3.2.3. Hence $\alpha^*(E) \leq t$ and the equality holds whenever $\alpha^*(E) = +\infty$.

Thus suppose that $\alpha^*(E) < +\infty$ and choose an $\varepsilon > 0$. Employ Proposition 3.2.8 to find an open set $U \subset \mathbf{R}$ with $E \subset U$ and $\alpha^*(U) < \alpha^*(E) + \varepsilon$. By Lemma 3.4.4 there is a sequence $\{K_n\}$ of nonoverlapping cells such that

$$U = \bigcup_{n=1}^{\infty} K_n \quad \text{and} \quad \alpha^*(U) = \sum_{n=1}^{\infty} \alpha(K_n).$$

Furthermore, Proposition 3.2.2 and Corollary 3.4.2 imply the existence of cells L_n with

$$\alpha^*(\partial L_n) = 0, \quad K_n \subset L_n^{\circ}, \quad \text{and} \quad \alpha(L_n) < \alpha(K_n) + \varepsilon 2^{-n}$$

for $n = 1, 2, \ldots$. We conclude that $E \subset \bigcup_{n=1}^{\infty} L_n^{\circ}$ and

$$\sum_{n=1}^{\infty} \alpha(L_n) < \sum_{n=1}^{\infty} \alpha(K_n) + \varepsilon = \alpha^*(U) + \varepsilon < \alpha^*(E) + 2\varepsilon.$$

Therefore $t \leq \alpha^*(E) + 2\varepsilon$, and the proposition follows from the arbitrariness of ε. ∎

Remark 3.4.6 Proposition 3.4.5 shows that the α-measure α^* is equal to the usual outer measure induced by α (cf. [41, Chapter 3, Section 2]). Furthermore, in view of Theorem 3.3.17, the α-measurable sets coincide with the measurable sets obtained from α^* by the Carathéodory method (cf. [41, Chapter 3, Section 3]).

Exercise 3.4.7 Let $E \subset \mathbf{R}$, $c \in \mathbf{R}$, and $cE = \{cx : x \in E\}$.

1. Use Proposition 3.4.5 to show that

$$\lambda^*(E + c) = \lambda^*(E) \qquad \text{and} \qquad \lambda^*(cE) = |c|\lambda^*(E).$$

2. Use Lemma 3.3.8 or Theorem 3.3.13 to show that if E is λ-measurable, then so are the sets $E + c$ and cE.

3.5 Negligible sets

Definition 3.5.1 A set E of real numbers is called α-*negligible* whenever $\alpha^*(E) = 0$.

In view of Proposition 3.4.5, we have the following simple characterization of α-negligible sets.

Proposition 3.5.2 *A set $E \subset \mathbf{R}$ is α-negligible if and only if for each $\varepsilon > 0$ there is a sequence $\{A_i\}$ of cells such that*

$$E \subset \bigcup_{i=1}^{\infty} A_i^\circ \qquad \text{and} \qquad \sum_{i=1}^{\infty} \alpha(A_i) < \varepsilon.$$

The next proposition summarizes the main properties of α-negligible sets. Its simple proof is left to the reader.

Proposition 3.5.3 *The following statements are true.*

1. *Each α-negligible set is α-measurable.*
2. *Each subset of an α-negligible set is α-negligible.*
3. *A countable union of α-negligible sets is α-negligible.*
4. *A singleton $\{x\}$ is α-negligible if and only if α is continuous at x.*
5. *If α is continuous, then each countable set is α-negligible.* ∎

Definition 3.5.4 Let $\mathcal{C}$ be a claim depending on points of a set $E \subset \mathbf{R}$. We say that $\mathcal{C}$ holds *α-almost everywhere in E*, or for *α-almost all $x \in E$*, if the set of those $x \in E$ for which $\mathcal{C}(x)$ does not hold is α-negligible.

Definition 3.5.4 provides a significant simplification of the language. For instance, let f be a function defined on $\mathbf{R}$ and let $E \subset \mathbf{R}$. The statement *"there is an α-negligible set $N \subset E$ such that $f(x) > 0$ for each $x \in E - N$"* can be simplified to *"f is positive α-almost everywhere in E."*

Convention. Throughout Part I of this book, we agree that "α-almost everywhere" means "α-almost everywhere in $\mathbf{R}$."

3.6 Measurable functions

Definition 3.6.1 An extended real-valued function f defined on an α-measurable set E is called *α-measurable* whenever the set

$$\{x \in E : f(x) > c\}$$

is α-measurable for every $c \in \mathbf{R}$.

Exercise 3.6.2 Prove the following facts.

1. Any function on an α-negligible set is α-measurable.
2. A continuous function on an α-measurable set is α-measurable.
3. A monotone function on an α-measurable set is α-measurable.

Lemma 3.6.3 *Let f be an extended real-valued function defined on an α-measurable set E. Then f is α-measurable if and only if one (and hence each) of the sets*

$$\{x \in E : f(x) \geq c\}, \quad \{x \in E : f(x) < c\}, \quad or \quad \{x \in E : f(x) \leq c\}$$

is α-measurable for every $c \in \mathbf{R}$.

PROOF. If f is α-measurable and $c \in \mathbf{R}$, then the sets

$$\{x \in E : f(x) \geq c\} = \bigcap_{n=1}^{\infty} \left\{x \in E : f(x) > c - \frac{1}{n}\right\},$$

$$\{x \in E : f(x) \leq c\} = E - \{x \in E : f(x) > c\},$$

$$\{x \in E : f(x) < c\} = \bigcup_{n=1}^{\infty} \left\{x \in E : f(x) \leq c - \frac{1}{n}\right\},$$

are α-measurable by Exercise 3.3.5, and Propositions 3.3.2, 3 and 3.3.4. The lemma follows by symmetry. ∎

Exercise 3.6.4 Let f be an α-measurable extended real-valued function defined on an α-measurable set E. Show that the following extended real-valued functions are α-measurable.

1. $-f$, $|f|$, f^+, and f^-.
2. The restriction of f to any α-measurable subset of E.
3. The extension of f to $\mathbf{R}$ by a constant (finite or infinite).

Corollary 3.6.5 *Let f be an α-measurable extended real-valued function defined on an α-measurable set E. If A is an interval or a segment, then the set $\{x \in E : f(x) \in A\}$ is α-measurable.*

PROOF. If $A = [a, b)$ where $a \leq b$ are real numbers, then

$$\{x \in E : f(x) \in A\} = \{x \in E : f(x) \geq a\} \cap \{x \in E : f(x) < b\}$$

is α-measurable by Lemma 3.6.3 and Proposition 3.3.2, 3. The remaining cases are similar. ∎

Proposition 3.6.6 *If $\{f_n\}$ is a sequence of α-measurable extended real-valued functions defined on an α-measurable set E, then the extended real-valued functions*

$$\inf f_n, \quad \sup f_n, \quad \liminf f_n, \quad and \quad \limsup f_n$$

defined on E are also α-measurable.

PROOF. Since

$$\{x \in E : \sup f_n(x) > c\} = \bigcup_{n=1}^{\infty} \{x \in E : f_n(x) > c\}$$

for each $c \in \mathbf{R}$, we see that $\sup f_n$ is α-measurable. As

$$\inf f_n = -\sup(-f_n),$$

$$\limsup f_n = \inf_k \left(\sup_{n \geq k} f_n \right) \quad and \quad \liminf f_n = -\limsup(-f_n),$$

the proposition follows from Exercise 3.6.4, 1. ∎

Lemma 3.6.7 *Let V be an open subset of $\mathbf{R}^2 = \mathbf{R} \times \mathbf{R}$. There are sequences $\{I_n\}$ and $\{J_n\}$ of cells such that*

$$V = \bigcup_{n=1}^{\infty} (I_n \times J_n).$$

PROOF. Let $\mathcal{V}$ be the family of all products $I \times J \subset V$ where I and J are cells with rational endpoints. Since V is open, it is easy to see that $V = \bigcup \mathcal{V}$. Observing that the family $\mathcal{V}$ is countable completes the argument. ∎

Proposition 3.6.8 *Let f and g be finite α-measurable functions defined on an α-measurable set E, and let U be an open subset of $\mathbf{R}^2$ such that $(f(x), g(x)) \in U$ for each $x \in E$. If H is a continuous function defined on U, then the function $h : x \mapsto H(f(x), g(x))$ on E is α-measurable. In particular, the functions $f + g$ and fg are α-measurable, and if $g(x) \neq 0$ for every $x \in E$, then so is f/g.*

PROOF. Given $c \in \mathbf{R}$, the set $V = \{(r,s) \in U : H(r,s) > c\}$ is open by the continuity of H (Exercise 3.6.2, 2). Using Lemma 3.6.7, find cells I_n and J_n so that $V = \bigcup_{n=1}^{\infty} (I_n \times J_n)$. In view of Corollary 3.6.5, the set

$$\{x \in E : h(x) > c\} = \{x \in E : (f(x), g(x)) \in V\}$$

$$= \bigcup_{n=1}^{\infty} \{x \in E : (f(x), g(x)) \in I_n \times J_n\}$$

$$= \bigcup_{n=1}^{\infty} \left(\{x \in E : f(x) \in I_n\} \cap \{x \in E : g(x) \in J_n\} \right)$$

is α-measurable by Propositions 3.3.2, 3 and 3.3.4. Since the functions $H(r,s) = r + s$ and $H(r,s) = rs$ are continuous in $U = \mathbf{R}^2$ and the function $H(r,s) = r/s$ is continuous in $U = \{(r,s) \in \mathbf{R}^2 : s \neq 0\}$, the proof is completed. ∎

Proposition 3.6.9 *Let f and g be extended real-valued functions defined on an α-measurable set E, and let $f = g$ α-almost everywhere in E. Then f is α-measurable if and only if g is α-measurable.*

PROOF. If $N = \{x \in E : f(x) \neq g(x)\}$, then N is α-negligible and

$$\{x \in E - N : f(x) > c\} = \{x \in E - N : g(x) > c\}$$

for each $c \in \mathbf{R}$. Now it suffices to apply Proposition 3.5.3, 1 and 2. ∎

Definition 3.6.10 A finite α-measurable function on **R** is called α-*simple* if it assumes only finitely many values.

If E is a subset of the reals, we let

$$\chi_E(x) = \begin{cases} 1 \text{ if } x \in E, \\ 0 \text{ if } x \in \mathbf{R} - E, \end{cases}$$

and call the function χ_E the *characteristic function* of E.

Exercise 3.6.11 Show that a set $E \subset \mathbf{R}$ is α-measurable if and only if its characteristic function χ_E is α-measurable.

Proposition 3.6.12 *A function s on **R** is α-simple if and only if it can be written as a linear combination of characteristic functions of α-measurable sets.*

PROOF. Let s be α-simple and let $c_1, \ldots, c_n$ be the distinct values of s. If $E_i = \{x \in \mathbf{R} : s(x) = c_i\}$, then the sets $E_1, \ldots, E_n$ are α-measurable by Corollary 3.6.5, and it is easy to check that

$$s = \sum_{i=1}^{n} c_i \chi_{E_i} \,.$$

In view of Exercise 3.6.11 and Proposition 3.6.8, the converse is obvious. ∎

Remark 3.6.13 The reader should keep in mind that there are many ways in which an α-simple function can be written as a linear combination of characteristic functions of α-measurable sets. For example, we have

$$3\chi_{[0,1]} + 2\chi_{[1,2]} = 3\chi_{[0,2]} - \chi_{[1,2]} + 3\chi_{\{1\}} \,.$$

The next proposition indicates that the complexity of α-measurable functions is not substantially larger than that of α-measurable sets.

Proposition 3.6.14 *Let f be a nonnegative extended real-valued function on an α-measurable set E. Then f is α-measurable if and only if there is a sequence $\{s_n\}$ of nonnegative α-simple functions such that $s_n \nearrow f$.*

PROOF. As the converse follows from Proposition 3.6.6, assume that f is α-measurable. We use a construction similar to that employed in the proof of Theorem 2.3.11. For $n = 1, 2, \ldots$ and $k = 1, \ldots, n2^n$, let

$$E_n = \{x \in E : f(x) \geq n\},$$

$$E_{n,k} = \{x \in E : (k-1)2^{-n} \leq f(x) < k2^{-n}\}.$$

These sets are α-measurable by Corollary 3.6.5, so the functions

$$s_n = n\chi_{E_n} + \sum_{k=1}^{n2^n} (k-1)2^{-n}\chi_{E_{n,k}}$$

are α-simple for $n = 1, 2, \ldots$. It is easy to verify that $0 \leq s_1 \leq s_2 \leq \cdots$ and that for each $x \in E$, we have

$$0 \leq f(x) - s_n(x) \leq 2^{-n}$$

if $f(x) < n$, and $s_n(x) = n$ if $f(x) \geq n$. It follows that $s_n \nearrow f$. ∎

Exercise 3.6.15 Let f be a *bounded* α-measurable function on an α-measurable set E. Analyzing the proof of Proposition 3.6.14, show that there is a sequence of α-simple functions that converges to f *uniformly* (see [42, Definition 7.7, p. 147]). Show that the boundedness of f is essential.

3.7 The α_A-measure

If F is a function on a cell $A = [a, b]$, we define a function F_A on $\mathbf{R}$ by setting

$$F_A(x) = \begin{cases} F(a) & \text{if } x < a, \\ F(x) & \text{if } a \leq x \leq b, \\ F(b) & \text{if } b < x. \end{cases}$$

An easy calculation shows that $F_A(B) = F(A \cap B)$ for every interval B. In particular, α_A is an increasing function on $\mathbf{R}$ and

$$\alpha_A(B) = \alpha(A \cap B) \leq \alpha(B)$$

for every interval B. Consequently, $\alpha_A^*(E) \leq \alpha^*(E)$ for each set $E \subset \mathbf{R}$.

Remark 3.7.1 It is important to note that the increasing function α_A on $\mathbf{R}$ can be defined whenever α is an increasing function on A. The fact that α is defined and increasing on all of $\mathbf{R}$ is irrelevant.

Proposition 3.7.2 *If c is an interior point of a cell $[a, b]$, then*

$$\alpha^*_{[a,b]}(E) = \alpha^*_{[a,c]}(E) + \alpha^*_{[c,b]}(E)$$

for each set $E \subset \mathbf{R}$.

PROOF. Let $E \subset \mathbf{R}$ and let δ_a and δ_b be positive functions on E. Set $\delta = \min\{\delta_a, \delta_b\}$ and choose a partition $P = \{(C_1, z_1), \ldots, (C_n, z_n)\}$ anchored in E that is δ-fine. We have

$$\sum_{i=1}^{n} \alpha_{[a,b]}(C_i) = \sum_{i=1}^{n} \alpha_{[a,c]}(C_i) + \sum_{i=1}^{n} \alpha_{[c,b]}(C_i) \leq \alpha^{\delta_a}_{[a,c]}(E) + \alpha^{\delta_b}_{[c,b]}(E),$$

and the inequality

$$\alpha^*_{[a,b]}(E) \leq \alpha^{\delta}_{[a,b]}(E) \leq \alpha^{\delta_a}_{[a,c]}(E) + \alpha^{\delta_b}_{[c,b]}(E)$$

follows from the arbitrariness of P. Since δ_a and δ_b are arbitrary, we conclude that

$$\alpha^*_{[a,b]}(E) \leq \alpha^*_{[a,c]}(E) + \alpha^*_{[c,b]}(E).$$

Proceeding towards a contradiction, suppose that the previous inequality is strict and find a positive function Δ on E so that

$$\alpha^{\Delta}_{[a,b]}(E) < \alpha^*_{[a,c]}(E) + \alpha^*_{[c,b]}(E) \leq \alpha^{\Delta}_{[a,c]}(E) + \alpha^{\Delta}_{[c,b]}(E).$$

There are partitions $\{(A_1, x_1), \ldots, (A_p, x_p)\}$ and $\{(B_1, y_1), \ldots, (B_q, y_q)\}$ anchored in E that are Δ-fine and such that

$$\alpha^{\Delta}_{[a,b]}(E) < \sum_{i=1}^{p} \alpha_{[a,c]}(A_i) + \sum_{j=1}^{q} \alpha_{[c,b]}(B_j)$$

$$= \sum_{i=1}^{p} \alpha_{[a,b]}([a, c] \cap A_i) + \sum_{j=1}^{q} \alpha_{[a,b]}([c, b] \cap B_j).$$

If Q consists of all pairs $([a, c] \cap A_i, x_i)$ and $([c, b] \cap B_j, y_j)$ where $[a, c] \cap A_i$ and $[c, b] \cap B_j$ are cells, then Q is a partition anchored in E that is Δ-fine. This contradicts the last inequality. ∎

Proposition 3.7.3 *Let A be a cell and let E be any subset of $\mathbf{R}$. Then*

$$\alpha^*_A(E) \leq \alpha^*(A \cap E)$$

and the equality occurs whenever $\alpha^*(E \cap \partial A) = 0$. In particular,

$$\alpha^*_A(E) = \alpha^*_A(A \cap E).$$

PROOF. Assuming that $\alpha^*(A \cap E) < \alpha_A^*(E)$, there is a positive function δ on $A \cap E$ such that $\alpha^\delta(A \cap E) < \alpha_A^*(E)$. Extend δ to a positive function Δ on E so that $U(x, \Delta(x)) \subset \mathbf{R} - A$ for each $x \in E - A$. Since

$$\alpha^\delta(A \cap E) < \alpha_A^*(E) \leq \alpha_A^\Delta(E),$$

we can find a partition $\{(A_1, x_1), \ldots, (A_p, x_p)\}$ anchored in E that is Δ-fine and such that

$$\alpha^\delta(A \cap E) < \sum_{i=1}^p \alpha_A(A_i) = \sum_{i=1}^p \alpha(A \cap A_i).$$

This is a contradiction, as by the choice of Δ, the collection

$$\{(A \cap A_i, x_i) : A^\circ \cap A_i^\circ \neq \emptyset\}$$

is a partition anchored in $A \cap E$ that is δ-fine.

If $\alpha^*(E \cap \partial A) = 0$, choose an $\varepsilon > 0$ and find a positive function θ on $E \cap \partial A$ so that $\alpha^\theta(E \cap \partial A) < \varepsilon$. There is a positive function δ on E such that $\alpha_A^\delta(E) < \alpha_A^*(E) + \varepsilon$. Without loss of generality, we may assume that $\delta(x) \leq \theta(x)$ for each $x \in E \cap \partial A$, and that $U(x, \delta(x)) \cap \partial A = \emptyset$ for all $x \in E - \partial A$. If $Q = \{(B_1, y_1), \ldots, (B_q, y_q)\}$ is a partition anchored in $A \cap E$ that is δ-fine, then

$$\sum_{j=1}^q \alpha(B_j) = \sum_{y_j \in A^\circ} \alpha_A(B_j) + \sum_{y_j \in \partial A} \alpha(B_j)$$

$$\leq \alpha_A^\delta(E) + \alpha^\theta(E \cap \partial A) < \alpha_A^*(E) + 2\varepsilon$$

and consequently,

$$\alpha^*(A \cap E) \leq \alpha^\delta(A \cap E) \leq \alpha_A^*(E) + 2\varepsilon.$$

The proposition follows from the arbitrariness of ε. $\blacksquare$

Proposition 3.7.4 *If $E \subset \mathbf{R}$, then*

$$\alpha^*(E) = \lim_{n \to \infty} \alpha_{U[0,n]}^*(E).$$

PROOF. For $n = 1, 2, \ldots$, let $A_n = U[0, n]$ and observe that

$$\alpha^*(A_n \cap E) = \alpha_{A_{n+1}}^*(A_n \cap E) \leq \alpha_{A_{n+1}}^*(E) \leq \alpha^*(E)$$

by Proposition 3.7.3. Now it suffices to apply Proposition 3.3.15 to the sequence $\{A_n \cap E\}$. $\blacksquare$

The precise relationship between the measures α^* and α_A^* is given by the following proposition.

Proposition 3.7.5 *If $A = [a, b]$ is a cell, then*

$$\alpha_A^*(E) = \alpha^*(A \cap E) - \Big(\chi_E(a)[\alpha(a) - \alpha(a-)] + \chi_E(b)[\alpha(b+) - \alpha(b)]\Big)$$

for each set $E \subset \mathbf{R}$.

PROOF. In view of Proposition 3.7.3, we may assume that E is a subset of A. If $C = E \cap A^\circ$, then $\alpha_A^*(C) = \alpha^*(C)$ by Proposition 3.7.3. If $D = E \cap \partial A$, then it follows from Exercise 3.2.5 and Proposition 3.2.2 that

$$\alpha_A^*(D) = \alpha^*(D) - \Big(\chi_E(a)[\alpha(a) - \alpha(a-)] + \chi_E(b)[\alpha(b+) - \alpha(b)]\Big).$$

Since the sets A° and ∂A are α-measurable as well as α_A-measurable, an application of Proposition 3.3.2, 4 completes the argument. ∎

Note. The correction term

$$\chi_E(a)[\alpha(a) - \alpha(a-)] + \chi_E(b)[\alpha(b+) - \alpha(b)]$$

in Proposition 3.7.5 reconciles the different concepts of additivity associated with the α-length and α-measure: the former being additive with respect to *nonoverlapping cells* (Section 1.1), the latter with respect to *disjoint sets* (Proposition 3.3.2, 4). If $c \in \mathbf{R}$, intuitively one may think of $\alpha(c) - \alpha(c-)$ and $\alpha(c+) - \alpha(c)$ as the "portions" of $\alpha^*(\{c\}) = \alpha(c+) - \alpha(c-)$ allocated to the cells $[c - 1, c]$ and $[c, c + 1]$, respectively. Such an allocation, however, depends on the increasing function α that generates the measure α^*; it cannot be determined by the measure α^* alone.

Exercise 3.7.6 Let $\{A_n\}$ be a sequence of nonoverlapping cells such that

$$\bigcup_{n=1}^{\infty} A_n = \mathbf{R} \quad \text{and} \quad \inf_n \lambda(A_n) > 0.$$

1. Prove that a set $E \subset \mathbf{R}$ is α-measurable if and only if it is α_A-measurable for each cell A.
2. Prove that
$$\alpha^*(E) = \sum_{n=1}^{\infty} \alpha_{A_n}^*(E)$$
   for each $E \subset \mathbf{R}$.
3. Show by example that without the assumption $\inf_n \lambda(A_n) > 0$ the equality in the previous paragraph may be false.

Chapter 4
Integrable functions

Using the concept of measure introduced in Chapter 3, we study the structure of integrable functions. Throughout, we tacitly assume that on every cell A in question there is an increasing function α (the convention of Section 1.1), and that α_A is the associated increasing function on $\mathbf{R}$ defined in Section 3.7 (cf. Remark 3.7.1).

4.1 Integral and measure

Theorem 4.1.1 *Let A be a cell. A set $E \subset \mathbf{R}$ is α_A-measurable if and only if χ_E belongs to $\mathcal{R}(A, \alpha)$, in which case*

$$\alpha_A^*(E) = \int_A \chi_E \, d\alpha \,.$$

PROOF. Recall that $\alpha_A(B) = \alpha(A \cap B)$ for each cell $B \subset \mathbf{R}$, and choose an $\varepsilon > 0$ that will be used throughout the proof.

Assume first that E is α_A-measurable, and find a positive function δ on $\mathbf{R}$ associated with E and $\varepsilon/2$ according to Definition 3.3.1. If $P = \{(A_1, x_1), \ldots, (A_n, x_n)\}$ and $Q = \{(A_1, y_1), \ldots, (A_n, y_n)\}$ are δ-fine partitions in A, then

$$\{(A_i, x_i) : x_i \in E\} \qquad \text{and} \qquad \{(A_i, y_i) : y_i \in E\}$$

are δ-fine partitions anchored in E, while

$$\{(A_j, x_j) : x_j \in \mathbf{R} - E\} \qquad \text{and} \qquad \{(A_j, y_j) : y_j \in \mathbf{R} - E\}$$

are δ-fine partitions anchored in $\mathbf{R} - E$. Thus

$$\sum_{i=1}^{n} |\chi_E(x_i) - \chi_E(y_i)| \alpha(A_i) = \sum \{\alpha(A_i) : x_i \in E,\ y_i \in \mathbf{R} - E\}$$

$$+ \sum \{\alpha(A_i) : y_i \in E,\ x_i \in \mathbf{R} - E\}$$

$$= \sum_{x_i \in E} \sum_{y_j \in \mathbf{R}-E} \alpha(A_i \cap A_j) + \sum_{y_i \in E} \sum_{x_j \in \mathbf{R}-E} \alpha(A_i \cap A_j) < \varepsilon,$$

and $\chi_E \in \mathcal{R}(A, \alpha)$ by Lemma 2.2.2.

Conversely, assume that $\chi_E \in \mathcal{R}(A, \alpha)$, and use Lemma 2.2.2 to find a positive function δ on A so that

$$\sum_{i=1}^{n} |\chi_E(x_i) - \chi_E(y_i)| \alpha(A_i) < \varepsilon$$

for all partitions $\{(A_1, x_1), \ldots, (A_n, x_n)\}$ and $\{(A_1, y_1), \ldots, (A_n, y_n)\}$ in A that are δ-fine. Let θ be a positive function on $\mathbf{R}$ such that $\theta(x) \leq \delta(x)$ for each $x \in A$ and $U(x, \theta(x)) \subset \mathbf{R} - A$ for every $x \in \mathbf{R} - A$. Choose θ-fine partitions $\{(B_1, u_1), \ldots, (B_p, u_p)\}$ and $\{(C_1, v_1), \ldots, (C_q, v_q)\}$ anchored in E and $\mathbf{R} - E$, respectively. For $i = 1, \ldots, p$ and $j = 1, \ldots, q$, set

$$A_{i,j} = A \cap B_i \cap C_j, \qquad x_{i,j} = u_i, \qquad y_{i,j} = v_j,$$

and denote by N the collection of all ordered pairs (i, j) for which $A_{i,j}$ is a cell (i.e., $(A_{i,j})^{\circ} \neq \emptyset$). By the choice of θ, the families

$$\{(A_{i,j}, x_{i,j}) : (i, j) \in N\} \qquad \text{and} \qquad \{(A_{i,j}, y_{i,j}) : (i, j) \in N\}$$

are δ-fine partitions in A. Therefore

$$\sum_{i=1}^{p} \sum_{j=1}^{q} \alpha_A(B_i \cap C_j) = \sum_{(i,j) \in N} \alpha(A_{i,j})$$

$$= \sum_{(i,j) \in N} |\chi_E(x_{i,j}) - \chi_E(y_{i,j})| \alpha(A_{i,j}) < \varepsilon$$

and the measurability of E is established.

Finally, assume again that $\chi_E \in \mathcal{R}(A, \alpha)$ and choose a positive function δ on E so that $\alpha_A^{\delta}(E) < \alpha_A^*(E) + \varepsilon$. There is a positive function θ on A such that $|\sigma(\chi_E, P) - \int_A \chi_E| < \varepsilon$ for each θ-fine partition P of A. With no loss of generality, we may assume that $\theta(x) \leq \delta(x)$ for every $x \in A \cap E$. Now if $P = \{(A_1, x_1), \ldots, (A_p, x_p)\}$ is a θ-fine partition of A, then

$$\int_A \chi_E - \varepsilon < \sigma(\chi_E, P) = \sum_{x_i \in E} \alpha(A_i) \leq \alpha_A^\delta(E) < \alpha_A^*(E) + \varepsilon \,.$$

On the other hand, there is a positive function Δ on E such that $\Delta(x) \leq \theta(x)$ for each $x \in A \cap E$ and $U(x, \Delta(x)) \subset \mathbf{R} - A$ for each $x \in E - A$. Let $Q = \{(B_1, y_1), \ldots, (B_q, y_q)\}$ be a partition anchored in E that is Δ-fine. Using the choice of Δ and Corollary 1.2.5, it is easy to construct a θ-fine partition P of A so that

$$\sum_{j=1}^q \alpha_A(B_j) \leq \sigma(\chi_E, P) < \int_A \chi_E + \varepsilon \,.$$

As Q is arbitrary, we obtain

$$\alpha_A^*(E) \leq \alpha_A^\Delta(E) \leq \int_A \chi_E + \varepsilon \,.$$

Consequently $|\alpha_A^*(E) - \int_A \chi_E| < 2\varepsilon$, and the theorem follows from the arbitrariness of ε. ∎

Corollary 4.1.2   *Let A be a cell and let s be an α_A-simple function. Then s belongs to $\mathcal{R}(A, \alpha)$, and if $s = \sum_{i=1}^n c_i \chi_{E_i}$ where $c_1, \ldots, c_n$ are real numbers and $E_1, \ldots, E_n$ are α_A-measurable sets, then*

$$\int_A s \, d\alpha = \sum_{i=1}^n c_i \alpha_A^*(E_i) \,. \quad ∎$$

While the corollary is an immediate consequence of Theorem 4.1.1 and Proposition 2.1.3, its claim is nontrivial: as we stressed in Remark 3.6.13, the linear combination $\sum_{i=1}^n c_i \chi_{E_i}$ is not determined uniquely by s, yet the number $\sum_{i=1}^n c_i \alpha_A^*(E_i)$ is unique. It takes an effort to prove this fact directly.

Corollary 4.1.3   *Let f be an α_A-measurable function on a cell A. If there is a $g \in \mathcal{R}(A, \alpha)$ such that $|f| \leq g$, then $f \in \mathcal{R}(A, \alpha)$. In particular, each bounded α_A-measurable function on A belongs to $\mathcal{R}(A, \alpha)$.*

PROOF.   Using the dominated convergence theorem, the corollary follows from Corollary 4.1.2 and Proposition 3.6.14. ∎

Corollary 4.1.4   *Let A be a cell. A set $E \subset \mathbf{R}$ is α_A-negligible if and only if χ_E belongs to $\mathcal{R}(A, \alpha)$ and $\int_A \chi_E \, d\alpha = 0$.* ∎

In the remainder of this section we establish several *almost everywhere* type statements related to Definition 3.5.4. Note that for a cell A, the expressions "α_A-almost everywhere" and "α_A-almost everywhere in A" are interchangeable (see the convention in Section 3.5 and Proposition 3.7.3).

Theorem 4.1.5 *Let f be a function defined on a cell A. If $f = 0$ α_A-almost everywhere, then $f \in \mathcal{R}(A, \alpha)$ and $\int_A f \, d\alpha = 0$.*

PROOF. Let $E = \{x \in A : f(x) \neq 0\}$, and for $n = 1, 2, \ldots$, let $f_n = \min\{|f|, n\chi_E\}$. Then $f_n \nearrow |f|$, and as $\sigma(f_n, P) \leq n\sigma(\chi_E, P)$ for every partition P of A, it follows from Corollary 4.1.4 that $f_n \in \mathcal{R}(A, \alpha)$ and $\int_A f_n = 0$. By the monotone convergence theorem, $|f| \in \mathcal{R}(A, \alpha)$ and $\int_A |f| = 0$. The proof is completed by observing that $|\sigma(f, P)| \leq \sigma(|f|, P)$ for each partition P of A. ∎

Corollary 4.1.6 *Let f and g be functions on a cell A, and let $f = g$ α_A-almost everywhere. Then f belongs to $\mathcal{R}(A, \alpha)$ if and only if g does, in which case $\int_A f \, d\alpha = \int_A g \, d\alpha$.*

PROOF. By Theorem 4.1.5, the function $h = f - g$ belongs to $\mathcal{R}(A, \alpha)$ and $\int_A h = 0$. The corollary follows from Proposition 2.1.3. ∎

Theorem 4.1.7 *Let f be a nonnegative integrable function on a cell A. Then $\int_A f \, d\alpha = 0$ if and only if $f = 0$ α_A-almost everywhere.*

PROOF. As the converse has been established in Theorem 4.1.5, assume that $\int_A f = 0$ and let $E = \{x \in A : f(x) > 0\}$. If $f_n = \min\{nf, \chi_E\}$ for $n = 1, 2, \ldots$, then $f_n \nearrow \chi_E$. Since $\sigma(f_n, P) \leq n\sigma(f, P)$ for every partition P of A, we see that $f_n \in \mathcal{R}(A, \alpha)$ and $\int_A f_n = 0$. An application of the monotone convergence theorem and Corollary 4.1.4 completes the proof. ∎

Theorem 4.1.8 *Let f be an integrable function on a cell A. Then $f = 0$ α_A-almost everywhere if and only if $\int_B f \, d\alpha = 0$ for every cell $B \subset A$.*

PROOF. Let $\int_B f = 0$ for every cell $B \subset A$, and let $\varepsilon > 0$. By Henstock's lemma there is a positive function δ on A such that $\sigma(|f|, P) < \varepsilon$ for each δ-fine partition P of A. Consequently, $|f| \in \mathcal{R}(A, \alpha)$ and $\int_A |f| = 0$. Now $f = 0$ α_A-almost everywhere by Theorem 4.1.8. The converse follows from Theorem 4.1.5. ∎

The following definition is a simple but very useful extension of Definition 2.1.1.

Definition 4.1.9 Let α be an increasing function on a cell A. We say that an extended real-valued function f defined α_A-almost everywhere in A is *integrable* over A with respect to α if there is a function $g \in \mathcal{R}(A, \alpha)$ such that $f = g$ α_A-almost everywhere. The real number $\int_A g \, d\alpha$ is called the *integral* of f over A with respect to α, denoted by $\int_A f \, d\alpha$.

According to Corollary 4.1.6, the integrability of f and $\int_A f \, d\alpha$ depend only on f and not on the choice of g. We denote by $\overline{\mathcal{R}}(A, \alpha)$ the family of all extended real-valued functions defined α_A-almost everywhere in A that are integrable over A with respect to α. Clearly,

$$\mathcal{R}(A, \alpha) \subset \overline{\mathcal{R}}(A, \alpha)$$

and the integral on $\overline{\mathcal{R}}(A, \alpha)$ is an extension of that on $\mathcal{R}(A, \alpha)$. By definition, each $f \in \overline{\mathcal{R}}(A, \alpha)$ is *finite* α_A-almost everywhere.

Extending properties of $\mathcal{R}(A, \alpha)$ to $\overline{\mathcal{R}}(A, \alpha)$ is generally easy, and we leave it to the reader. As an example, we extend the monotone convergence theorem to $\overline{\mathcal{R}}(A, \alpha)$.

Proposition 4.1.10 *Let f be an extended real-valued function defined α_A-almost everywhere in a cell A, and let $\{f_n\}$ be a sequence in $\overline{\mathcal{R}}(A, \alpha)$ such that $f_n \leq f_{n+1}$ α_A-almost everywhere for $n = 1, 2, \ldots$. If $\lim f_n = f$ α_A-almost everywhere and $\lim \int_A f_n \, d\alpha < +\infty$, then $f \in \overline{\mathcal{R}}(A, \alpha)$ and*

$$\int_A f \, d\alpha = \lim \int_A f_n \, d\alpha \, .$$

PROOF. Using Proposition 3.5.3, 3, it is easy to exhibit an α_A-negligible set $N \subset A$ such that $\{f_n\}$ is an increasing sequence of real-valued functions defined on $A - N$ and $\lim f_n(x) = f(x)$ for each $x \in A - N$. In view of Corollary 4.1.6, we may assume that $f_n(x) = f(x) = 0$ for all $x \in N$ and $n = 1, 2, \ldots$, and consequently that $\{f_n\}$ is a sequence in $\mathcal{R}(A, \alpha)$ with $f_n \nearrow f$.

Next we consider the set $A_+ = \{x \in A : f(x) = +\infty\}$. By the dominated convergence theorem, the function $g_k = \min\{f^+, k\}$ belongs to $\mathcal{R}(A, \alpha)$ and

$$\int_A g_k = \lim_{n \to \infty} \int_A \min\{f_n^+, k\} \leq \lim \int_A f_n^+$$

$$= \lim \int_A f_n + \lim \int_A f_n^- \leq \lim \int_A f_n + \int_A f_1^- < +\infty$$

for $k = 1, 2, \ldots$. Since $\lim(g_k/k) = \chi_{A_+}$, the dominated convergence theorem and Corollary 4.1.4 imply that A_+ is α_A-negligible.

Applying Corollary 4.1.6 again, we may assume that $f_n(x) = f(x) = 0$ for all $x \in A_+$ and $n = 1, 2, \ldots$. The proposition follows from the monotone convergence theorem. ∎

Exercise 4.1.11 Let A be a cell, and let $\{f_n\}$ be a sequence in $\overline{\mathcal{R}}(A, \alpha)$ such that $\sum_{n=1}^{\infty} \int_A |f_n| < +\infty$. Show that $\sum_{n=1}^{\infty} f_n$ converges absolutely α_A-almost everywhere.

4.2 Semicontinuous functions

The continuity of an extended real-valued function f on a cell A is defined in the obvious way: we say that f is *continuous* at $x \in A$ if $\lim f(x_n) = f(x)$ for each sequence $\{x_n\}$ in A that converges to x. When f is continuous at each $x \in A$, we say that it is *continuous*.

Definition 4.2.1 An extended real-valued function f on a cell A is called, respectively, *lower* or *upper semicontinuous* if for each $x \in A$ and each $c < f(x)$ or $c > f(x)$ there is a $\delta > 0$ such that $c < f(y)$ or $c > f(y)$ for every $y \in A$ with $|x - y| < \delta$.

A standard argument shows that an extended real-valued function on a cell A is continuous if and only if it is both lower and upper semicontinuous. Since an extended real-valued function f on a cell A is lower semicontinuous if and only if $-f$ is upper semicontinuous, it suffices to investigate the lower semicontinuous extended real-valued functions. A canonical example of a lower semicontinuous function on any cell is the characteristic function χ_U of an open set $U \subset \mathbf{R}$.

Exercise 4.2.2 Let f and g be lower semicontinuous extended real-valued functions on a cell A. Prove the following statements.

1. For each $c \in \overline{\mathbf{R}}$, the set $\{x \in A : f(x) > c\}$ is a relatively open subset of A.
2. If $\{x_n\}$ is a sequence in A converging to an $x \in A$, then
$$\liminf f(x_n) \geq f(x).$$
3. There is a $z \in A$ such that $\inf_{x \in A} f(x) = f(z)$.
4. If $f > -\infty$, then $f > c$ for a $c \in \mathbf{R}$.
5. The functions $\min\{f, g\}$ and $\max\{f, g\}$ are lower semicontinuous.
6. If $f > -\infty$ and $g > -\infty$, then $f + g$ is a lower semicontinuous function.

Observe that either of the conditions 1 and 2 is equivalent to the lower semicontinuity of f.

Proposition 4.2.3 *Let $f > -\infty$ be an extended real-valued function on a cell A. Then f is lower semicontinuous if and only if there is a sequence $\{f_n\}$ of real-valued continuous functions on A such that $f_n \nearrow f$.*

PROOF. Let $\{f_n\}$ be a sequence of real-valued continuous functions on A with $f_n \nearrow f$. Choose $x \in A$ and $c < f(x)$. Then $f_n(x) > c$ for some integer $n \geq 1$, and by the continuity of f_n, there is a $\delta > 0$ such that $f(y) \geq f_n(y) > c$ for each $y \in A$ with $|x - y| < \delta$. This proves the lower semicontinuity of f.

To prove the converse, assume first that f is lower semicontinuous and bounded. For each $x \in A$ and $n = 1, 2, \ldots$, let

$$f_n(x) = \inf_{z \in A}[\,f(z) + n|x - z|\,].$$

Clearly, $-\infty < f_n \leq f_{n+1} \leq f < +\infty$; in particular, $\lim f_n \leq f$. Since $z \mapsto f(z) + n|x - z|$ is a real-valued semicontinuous function on A, there is an $x_n \in A$ such that $f_n(x) = f(x_n) + n|x - x_n|$ (Exercise 4.2.2, 3). Thus we obtain

$$f_n(y) - f_n(x) \leq f(x_n) + n|y - x_n| - [\,f(x_n) + n|x - x_n|\,]$$
$$\leq n|x - y|\,.$$

By symmetry $|f_n(y) - f_n(x)| \leq n|y - x|$, and the continuity of f_n follows. As f is bounded and

$$f(x_n) + n|x - x_n| = f_n(x) \leq f(x)\,,$$

we see that $\lim x_n = x$. Now Exercise 4.2.2, 2 implies that

$$f(x) \geq \lim f_n(x) = \lim[\,f(x_n) + n|x - x_n|\,] \geq \liminf f(x_n) \geq f(x)\,,$$

and hence $f_n \nearrow f$.

If $f > -\infty$ is an arbitrary lower semicontinuous function, then by Exercise 4.2.2, 5 and 4, the functions $f_k = \min\{f, k\}$ are lower semicontinuous and bounded for $k = 1, 2, \ldots$. Using our earlier result, we can find sequences $\{f_{k,n}\}_n$ of real-valued continuous functions such that $f_{k,n} \nearrow f_k$ as $n \to \infty$. For $i = 1, 2, \ldots$, the functions

$$g_i = \max\{f_{k,n} : k, n = 1, \ldots, i\}$$

are real-valued and continuous. Moreover, $g_i \leq g_{i+1} \leq f$ and so $g_i \nearrow g$ for some extended real-valued function $g \leq f$. On the other hand, $g_i \geq f_{k,i}$ for all $i \geq k$, and consequently

$$g = \lim_{i \to \infty} g_i \geq \lim_{i \to \infty} f_{k,i} = f_k$$

for $k = 1, 2, \ldots$. It follows that $g \geq f$ and the proposition is proved. ∎

Exercise 4.2.4 The sequence $\{f_n\}$ we constructed in the proof of Proposition 4.2.3 consists of Lipschitz functions on A. Using this observation in conjunction with Exercise 2.2.7 and the monotone convergence theorem, solve Exercise 2.3.9 without involving the Weierstrass approximation theorem.

Corollary 4.2.5 *Each bounded lower semicontinuous function on a cell A belongs to $\mathcal{R}(A, \alpha)$.*

PROOF. Applying the monotone convergence theorem, the corollary follows from Theorem 2.2.8 and Proposition 4.2.3. ∎

4.3 The Perron test

Let α be an increasing function on a cell A, let F be any function defined on A, and let $x \in A$. If $\delta > 0$, we denote by A_δ the family of all cells $B \subset A \cap U(x, \delta)$ for which the ratio $F(B)/\alpha(B)$ is defined, and set

$$\mathcal{D}_\alpha F(x) = \sup_{\delta > 0} \left[\inf_{B \in A_\delta} \frac{F(B)}{\alpha(B)} \right] \quad \text{and} \quad \mathcal{D}^\alpha F(x) = \inf_{\delta > 0} \left[\sup_{B \in A_\delta} \frac{F(B)}{\alpha(B)} \right].$$

The extended real-valued functions

$$\mathcal{D}_\alpha F : x \mapsto \mathcal{D}_\alpha F(x) \quad \text{and} \quad \mathcal{D}^\alpha F : x \mapsto \mathcal{D}^\alpha F(x)$$

are called, respectively, the *lower* and *upper Lebesgue α-derivates* of F in A.

Exercise 4.3.1 If F is a function defined on a cell A and $x \in A$, show that the following hold.

1. $\mathcal{D}_\alpha(-F)(x) = -\mathcal{D}^\alpha F(x)$.
2. $\mathcal{D}^\alpha F(x) < \mathcal{D}_\alpha F(x)$ if and only if there is a $\delta > 0$ such that both functions α and F are constant in $A \cap U(x, \delta)$; in this case

$$\mathcal{D}_\alpha F(x) = -\mathcal{D}^\alpha F(x) = +\infty.$$

Exercise 4.3.2 Let α and F be *continuously* differentiable functions on a cell A, and let $\alpha'(x) > 0$ for each $x \in A$. Using the generalized mean value theorem ([42, Theorem 5.5, p. 107]), show that

$$\mathcal{D}_\alpha F(x) = \mathcal{D}^\alpha F(x) = \frac{F'(x)}{\alpha'(x)}$$

for every $x \in A$; in particular, $\mathcal{D}_\lambda F(x) = \mathcal{D}^\lambda F(x) = F'(x)$.

Exercise 4.3.3 Let F be the differentiable function of Example 2.4.12. Show that $\mathcal{D}_\lambda F(0) = -\infty$ and $\mathcal{D}^\lambda F(0) = +\infty$. Thus, in Exercise 4.3.2, continuous differentiability cannot be replaced by mere differentiability.

Proposition 4.3.4 *If F is a function on a cell A, then the Lebesgue lower derivate $\mathcal{D}_\alpha F$ of F is lower semicontinuous.*

PROOF. Choose an $x \in A$ and a $c < \mathcal{D}_\alpha F(x)$. By definition, there is a $\delta > 0$ such that $\inf[F(B)/\alpha(B)] > c$ where the infimum is taken over all cells $B \subset A \cap U(x, \delta)$ for which the ratio $F(B)/\alpha(B)$ is defined. If $y \in A \cap U(x, \delta)$, find an $\eta > 0$ so that $U(y, \eta) \subset U(x, \delta)$, and observe that

$$\mathcal{D}_\alpha F(y) \geq \inf \frac{F(B)}{\alpha(B)} > c$$

where the infimum is taken over all cells $B \subset A \cap U(y, \eta)$ for which the ratio $F(B)/\alpha(B)$ is defined. ∎

Definition 4.3.5 Let α be an increasing function on a cell A, and let f be an extended real-valued function on A. We say that a finite function F on A is an α-*majorant* or α-*minorant* of f in A whenever

$$-\infty \neq \mathcal{D}_\alpha F \geq f \qquad \text{or} \qquad +\infty \neq \mathcal{D}^\alpha F \leq f ,$$

respectively.

Let f be an extended real-valued function on a cell A. In A, a function F is an α-majorant of f if and only if $-F$ is an α-minorant of $-f$. If F is an α-majorant of f in A, then F is also an α-majorant of f in any subcell of A.

If f is an extended real-valued function on A, we set

$$U(f, A; \alpha) = \inf M(A) \qquad \text{and} \qquad L(f, A; \alpha) = \sup m(A)$$

where the infimum and supremum are taken over all α-majorants M and all α-minorants m of f in A, respectively. When no confusion is possible, we usually drop the character α from the symbols $U(f, A; \alpha)$ and $L(f, A; \alpha)$. In view of the previous paragraph, $U(-f, A) = -L(f, A)$.

Lemma 4.3.6 *Let F be a function on a cell A. If $\mathcal{D}_\alpha F \geq 0$, then F is increasing.*

PROOF. Choose an $\varepsilon > 0$, and find a positive function δ on A so that given $x \in A$, we have $F(B) > -\varepsilon\alpha(B)$ for each cell $B \subset A \cap U(x, \delta(x))$. If C is a subcell of A, use Proposition 1.2.4 to find a δ-fine partition $\{(C_1, z_1), \ldots, (C_n, z_n)\}$ of C and observe that

$$F(C) = \sum_{i=1}^{n} F(C_i) > -\varepsilon \sum_{i=1}^{n} \alpha(C_i) \geq -\varepsilon\alpha(A).$$

As ε is arbitrary, $F(C) \geq 0$ and the lemma follows. ∎

Corollary 4.3.7 *If f is an extended real-valued function defined on a cell A, then*

$$L(f, A; \alpha) \leq U(f, A; \alpha).$$

PROOF. If M and m are, respectively, an α-majorant and α-minorant of f in A, then

$$\mathcal{D}_\alpha(M - m) \geq \mathcal{D}_\alpha M - \mathcal{D}^\alpha m \geq 0;$$

to show that $\mathcal{D}_\alpha M(x) - \mathcal{D}^\alpha m(x) \geq 0$ for all $x \in A$, one must argue separately according to whether $f(x)$ is finite or infinite. By Lemma 4.3.6, the function $M - m$ is increasing. In particular, $m(A) \leq M(A)$ and the proposition follows. ∎

Lemma 4.3.8 *Let E be an α_A-negligible subset of a cell A. Given $\varepsilon > 0$, there is a nonnegative increasing function β on A such that $\beta(A) < \varepsilon$ and $\mathcal{D}_\alpha\beta(x) = +\infty$ for each $x \in E$.*

PROOF. By Proposition 3.5.2, there are sequences $\{A_{i,k}\}_{i=1}^{\infty}$ of cells such that

$$E \subset \bigcup_{i=1}^{\infty} (A_{i,k})^\circ \quad \text{and} \quad \sum_{i=1}^{\infty} \alpha_A(A_{i,k}) < \varepsilon 2^{-k}$$

for $k = 1, 2, \ldots$. If B is a subinterval of A, let

$$\beta(B) = \sum_{i,k=1}^{\infty} \alpha(B \cap A_{i,k})$$

and observe that

$$\beta(A) = \sum_{i,k=1}^{\infty} \alpha_A(A_{i,k}) < \sum_{k=1}^{\infty} \varepsilon 2^{-k} = \varepsilon.$$

Select an $x \in E$ and for $k = 1, 2, \ldots$, find a positive integer i_k with $x \in (A_{i_k,k})^{\circ}$. Given an integer $n \geq 1$, there is a $\delta > 0$ such that $U(x, \delta) \subset \bigcap_{k=1}^{n} A_{i_k,k}$. Thus, for each cell $B \subset A \cap U(x, \delta)$, we have

$$\beta(B) \geq \sum_{k=1}^{\infty} \alpha(B \cap A_{i_k,k}) = n\alpha(B) + \sum_{k=n+1}^{\infty} \alpha(B \cap A_{i_k,k}) \geq n\alpha(B).$$

It follows that $\mathcal{D}_\alpha \beta(x) \geq n$, and as n is arbitrary, $\mathcal{D}_\alpha \beta(x) = +\infty$.

The nonnegative function $\beta : B \mapsto \beta(B)$ defined on all subintervals of A is additive in the sense of Section 1.1. Thus, if $A = [a, b]$, it is easy to see that $x \mapsto \beta([a, x])$ is the desired function on A. ∎

Proposition 4.3.9 *Let f and g be extended real-valued functions defined on a cell A. If $f = g$ α_A-almost everywhere, then*

$$U(f, A; \alpha) = U(g, A; \alpha) \qquad and \qquad L(f, A; \alpha) = L(g, A; \alpha).$$

PROOF. Choose an $\varepsilon > 0$, and let β be the function of Lemma 4.3.8 associated with ε and the α_A-negligible set $E = \{x \in A : f(x) \neq g(x)\}$. Let M be an α-majorant of f in A. As $\mathcal{D}_\alpha(M + \beta) \geq \mathcal{D}_\alpha M + \mathcal{D}_\alpha \beta$, the function $M + \beta$ is an α-majorant of g in A, and

$$U(g, A) \leq M(A) + \beta(A) \leq M(A) + \varepsilon.$$

Since M and ε are arbitrary, $U(g, A) \leq U(f, A)$ and by symmetry, $U(g, A) = U(f, A)$. Applying this result to $-f$ and $-g$ yields

$$L(g, A) = -U(-g, A) = -U(-f, A) = L(f, A). \quad ∎$$

Lemma 4.3.10 *A real-valued function f on a cell A belongs to $\mathcal{R}(A, \alpha)$ if and only if*

$$L(f, A; \alpha) = U(f, A; \alpha) \neq \pm\infty,$$

in which case this common value equals $\int_A f \, d\alpha$.

PROOF. Let $I = L(f, A) = U(f, A)$ be a finite number. Choose an $\varepsilon > 0$, and find an α-majorant M and an α-minorant m of f in A so that

$$M(A) < I + \varepsilon \qquad and \qquad m(A) > I - \varepsilon.$$

There is a positive function δ on A such that

$$M(B) \geq [f(x) - \varepsilon]\alpha(B) \qquad \text{and} \qquad m(B) \leq [f(x) + \varepsilon]\alpha(B)$$

for each $x \in A$ and each cell $B \subset A \cap U(x, \delta(x))$. If $P = \{(A_1, x_1), \ldots, (A_p, x_p)\}$ is a δ-fine partition of A, then

$$m(A_i) - \varepsilon\alpha(A_i) \leq f(x_i)\alpha(A_i) \leq M(A_i) + \varepsilon\alpha(A_i)$$

for $i = 1, \ldots, p$. Summing up these inequalities yields

$$I - \varepsilon[1 + \alpha(A)] < m(A) - \varepsilon\alpha(A) \leq \sigma(f, P)$$
$$\leq M(A) + \varepsilon\alpha(A) < I + \varepsilon[1 + \alpha(A)],$$

and we see that $f \in \mathcal{R}(A, \alpha)$ and $\int_A f = I$.

Conversely, assume that $f \in \mathcal{R}(A, \alpha)$. Choose an $\varepsilon > 0$ and use Henstock's lemma to find a positive function δ on A so that

$$\sum_{i=1}^{p} \left| f(x_i)\alpha(A_i) - \int_{A_i} f \right| < \varepsilon$$

for each δ-fine partition $\{(A_1, x_1), \ldots, (A_p, x_p)\}$ in A. If B is a subcell of A, let Π_B be the family of all δ-fine partitions $Q = \{(B_1, y_1), \ldots, (B_q, y_q)\}$ in A for which $\bigcup_{j=1}^{q} B_j = B$, and let

$$S(B) = \sup_{Q \in \Pi_B} \sigma(f, Q) \qquad \text{and} \qquad s(B) = \inf_{Q \in \Pi_B} \sigma(f, Q).$$

Since

$$\int_B f - \varepsilon \leq s(B) \leq S(B) \leq \int_B f + \varepsilon$$

by the choice of δ, we can define finite functions M and m on A as follows:

$$M(x) = \begin{cases} S([a, x]) & \text{if } x \in (a, b], \\ 0 & \text{if } x = a, \end{cases}$$

$$m(x) = \begin{cases} s([a, x]) & \text{if } x \in (a, b], \\ 0 & \text{if } x = a. \end{cases}$$

In particular,

$$M(A) - m(A) = S(A) - s(A) \leq 2\varepsilon.$$

Let C and D be nonoverlapping subcells of A whose union is a cell. If $Q_C \in \Pi_C$ and $Q_D \in \Pi_D$, then $Q_C \cup Q_D$ belongs to $\Pi_{C \cup D}$ and

$$\sigma(f, Q_C) + \sigma(f, Q_D) = \sigma(f, Q_C \cup Q_D) \le S(C \cup D).$$

As Q_C and Q_D are arbitrary, $S(C) + S(D) \le S(C \cup D)$ and we conclude that $M(B) \ge S(B)$ for each cell $B \subset A$. Indeed, if $B = [a, x]$ then $M(B) = S(B)$; if $B = [x, y]$ and $a < x$, then

$$M(B) = M(y) - M(x) = S([a, y]) - S([a, x]) \ge S(B),$$

since $[a, y]$ is the union of nonoverlapping cells $[a, x]$ and B. Now choose an $x \in A$ and a cell $B \subset A \cap U(x, \delta(x))$. As $Q = \{(B, x)\}$ belongs to Π_B, we have

$$M(B) \ge S(B) \ge \sigma(f, Q) = f(x)\alpha(B)$$

and hence $\mathcal{D}_\alpha M(x) \ge f(x) > -\infty$. Thus M is an α-majorant of f in A, and analogously we can show that m is an α-minorant of f in A. From this and Corollary 4.3.7, we obtain that $U(f, A)$ and $L(f, A)$ are real numbers that satisfy the inequality

$$0 \le U(f, A) - L(f, A) \le M(A) - m(A) \le 2\varepsilon.$$

The lemma follows from the arbitrariness of ε. ∎

Corollary 4.3.11 *Let M be a real-valued function on a cell A such that $\mathcal{D}_\alpha M > -\infty$. Then $\mathcal{D}_\alpha M \in \overline{\mathcal{R}}(A, \alpha)$ and $\int_A \mathcal{D}_\alpha M \, d\alpha \le M(A)$; in particular, $\mathcal{D}_\alpha M < +\infty$ α_A-almost everywhere.*

PROOF. By Propositions 4.3.4 and 4.2.3, there is a sequence $\{f_n\}$ of finite continuous functions on A such that $f_n \nearrow \mathcal{D}_\alpha M$. As M is an α-majorant in A of each f_n, Lemma 4.3.10 implies that $\lim \int_A f_n \le M(A) < +\infty$, and the corollary follows from the monotone convergence theorem (Proposition 4.1.10). ∎

The next theorem will be referred to as the *Perron test* for integrability.

Theorem 4.3.12 *An extended real-valued function f on a cell A belongs to $\overline{\mathcal{R}}(A, \alpha)$ if and only if*

$$L(f, A; \alpha) = U(f, A; \alpha) \ne \pm\infty,$$

in which case this common value equals $\int_A f \, d\alpha$.

PROOF. It follows from Definition 4.1.9 and Corollary 4.3.11 that in either case the function f is finite α_A-almost everywhere. Thus the theorem is a consequence of Proposition 4.3.9. ∎

Exercise 4.3.13 Let F be a function on a cell A such that $\mathcal{D}_\alpha F(x) > -\infty$ for all $x \in A$ and $\mathcal{D}_\alpha F(x) \geq 0$ for α_A-almost all $x \in A$. Show that F is increasing.

4.4 Approximations

Lemma 4.4.1 Let $A = [a, b]$ be a cell and let $h \in \overline{\mathcal{R}}(A, \alpha)$. Set $H(x) = \int_a^x h \, d\alpha$ for each $x \in [a, b]$. If h is lower semicontinuous and $h > -\infty$, then H is an α-majorant of h in A.

PROOF. Let $x \in A$ and $c < h(x)$. There is a $\delta > 0$ such that $c < h(y)$ for each $y \in A$ with $|x - y| < \delta$. If B is a subcell of $A \cap U(x, \delta)$, then

$$H(B) = \int_B h \geq c\alpha(B),$$

from which we conclude that $\mathcal{D}_\alpha H(x) \geq c$. The arbitrariness of c implies that $\mathcal{D}_\alpha H(x) \geq h(x) > -\infty$. ∎

Proposition 4.4.2 Let f be an extended real-valued function defined on a cell A. Then

$$U(f, A; \alpha) = \inf \int_A h \, d\alpha \qquad and \qquad L(f, A; \alpha) = \sup \int_A g \, d\alpha$$

where the infimum is taken over all lower semicontinuous $h \in \overline{\mathcal{R}}(A, \alpha)$ for which $f \leq h \neq -\infty$, and the supremum is taken over all upper semicontinuous $g \in \overline{\mathcal{R}}(A, \alpha)$ for which $+\infty \neq g \leq f$.

PROOF. Denote by $\mathcal{L}$ the family of all lower semicontinuous functions h in $\overline{\mathcal{R}}(A, \alpha)$ for which $f \leq h \neq -\infty$, and let $c = \inf_{h \in \mathcal{L}} \int_A h$. If $A = [a, b]$ and $h \in \mathcal{L}$, set $H(x) = \int_a^x h$ for each $x \in [a, b]$. By Lemma 4.4.1, in the cell A the function H is an α-majorant of h and hence of f. Consequently

$$\int_A h = H(A) \geq U(f, A),$$

and we see from the arbitrariness of h that $c \geq U(f, A)$. On the other hand, if M is an α-majorant of f in A, it follows from Proposition 4.3.4 and Corollary 4.3.11 that $\mathcal{D}_\alpha M \in \mathcal{L}$ and

$$c \leq \int_A \mathcal{D}_\alpha M \leq M(A).$$

As M is arbitrary, $c \leq U(f, A)$. Therefore $c = U(f, A)$, and applying this to the function $-f$ completes the argument. ∎

The following approximation result is the important *Vitali–Carathéodory theorem*. It says that an extended real-valued function is integrable if and only if it can be sandwiched in between upper and lower semicontinuous extended real-valued functions whose integrals differ by a small amount.

Theorem 4.4.3 *Let f be an extended real-valued function on a cell A. Then $f \in \overline{\mathcal{R}}(A, \alpha)$ if and only if for each $\varepsilon > 0$ there are $g < +\infty$ and $h > -\infty$ in $\overline{\mathcal{R}}(A, \alpha)$ such that g is upper semicontinuous, h is lower semicontinuous, $g \leq f \leq h$, and $\int_A (h - g)\, d\alpha < \varepsilon$.*

PROOF. Let $f \in \overline{\mathcal{R}}(A, \alpha)$ and $\varepsilon > 0$. By Perron's test and Proposition 4.4.2, in $\overline{\mathcal{R}}(A, \alpha)$ there is a lower semicontinuous $h > -\infty$ and an upper semicontinuous $g < +\infty$ such that $g \leq f \leq h$ and

$$\int_A f - \frac{\varepsilon}{2} < \int_A g \leq \int_A h < \int_A f + \frac{\varepsilon}{2}.$$

It follows that $\int_A (h - g) < \varepsilon$.

Conversely, choose an $\varepsilon > 0$, and find $g < +\infty$ and $h > -\infty$ in $\overline{\mathcal{R}}(A, \alpha)$ so that g is upper semicontinuous, h is lower semicontinuous, $g \leq f \leq h$, and $\int_A (h - g) < \varepsilon$. According to Proposition 4.4.2 and Corollary 4.3.7, we obtain

$$-\infty < \int_A g \leq L(f, A) \leq U(f, A) \leq \int_A h < +\infty$$

and consequently, $0 \leq U(f, A) - L(f, A) < \varepsilon$. As ε is arbitrary, $U(f, A) = L(f, A)$ is a real number, and f belongs to $\overline{\mathcal{R}}(A, \alpha)$ by the Perron test. ∎

Note. The Vitali–Carathéodory theorem cannot be proved within the framework of *finite integrable functions*. Even when f is finite, the extended real-valued functions g and h of Theorem 4.4.3 *may not* be finite. For an example and further discussion we refer to [12, Example 1] and [35, Corollary 5]. On the other hand, Exercise 4.2.2, 5 shows that if $f \in \mathcal{R}(A, \alpha)$ is *bounded*, then g and h may be assumed to have the same bounds as f.

Proposition 4.4.4 *Let A be a cell and $f \in \overline{\mathcal{R}}(A, \alpha)$. For each $\varepsilon > 0$ there is a real-valued continuous function φ on A such that $\int_A |f - \varphi|\, d\alpha < \varepsilon$.*

PROOF. Choose a $\psi \in \mathcal{R}(A, \alpha)$ with $\psi = f$ α_A-almost everywhere. Given $\varepsilon > 0$, use the Vitali–Carathéodory theorem to find $g < +\infty$ and $h > -\infty$

in $\overline{\mathcal{R}}(A, \alpha)$ so that g is upper semicontinuous, h is lower semicontinuous, $g \leq \psi \leq h$, and $\int_A (h - g) < \varepsilon/2$. By Proposition 4.2.3 and the monotone convergence theorem (Proposition 4.1.10), there is a finite continuous function φ on A such that $\varphi \leq h$ and $\int_A (h - \varphi) < \varepsilon/2$. It follows that

$$\int_A |f - \varphi| = \int_A |\psi - \varphi| \leq \int_A |\psi - h| + \int_A |h - \varphi|$$

$$< \int_A (h - g) + \frac{\varepsilon}{2} < \varepsilon. \quad \blacksquare$$

Proposition 4.4.5 *Let A be a cell and $f \in \overline{\mathcal{R}}(A, \alpha)$. There is a sequence $\{f_n\}$ of finite continuous functions on A such that $f_n \to f$ α_A-almost everywhere and $\lim \int_A f_n \, d\alpha = \int_A f \, d\alpha$.*

PROOF. Use Proposition 4.4.4 to find a sequence $\{f_n\}$ of finite continuous functions on A so that $\int_A |f - f_n| < 2^{-n}$ for $n = 1, 2, \ldots$. Then $\lim \int_A f_n = \int_A f$ because

$$\lim \left| \int_A f - \int_A f_n \right| \leq \lim \int_A |f - f_n| = 0.$$

By Exercise 4.1.11, the series $\sum_{n=1}^{\infty} (f - f_n)$ converges α_A-almost everywhere, and the proposition follows from [42, Theorem 3.23, p. 60]. $\quad \blacksquare$

Corollary 4.4.6 *If A is a cell, then each function $f \in \overline{\mathcal{R}}(A, \alpha)$ is α_A-measurable.*

PROOF. In view of Exercise 3.6.2, 2, the corollary is an easy consequence of Propositions 3.6.6 and 4.4.5. $\quad \blacksquare$

We call the following theorem the *Lebesgue test* for the integrability of nonnegative functions.

Theorem 4.4.7 *Let f be a nonnegative extended real-valued function on a cell A, and let S be the family of all nonnegative α_A-simple functions $s \leq f$. Then $f \in \overline{\mathcal{R}}(A, \alpha)$ if and only if f is α_A-measurable and $\sup_{s \in S} \int_A s \, d\alpha$ is a finite number, in which case*

$$\int_A f \, d\alpha = \sup_{s \in S} \int_A s \, d\alpha.$$

PROOF. If $f \in \overline{\mathcal{R}}(A, \alpha)$, then $\sup_{s \in S} \int_A s \, d\alpha \leq \int_A f < +\infty$ by Proposition 2.1.3 and Corollary 4.1.6. That the equality occurs follows from Corollary 4.4.6 and Propositions 3.6.14 and 4.1.10. Another application of Propositions 3.6.14 and 4.1.10 yields the converse. $\quad \blacksquare$

Corollary 4.4.8 *A bounded function on a cell A belongs to $\mathcal{R}(A, \alpha)$ if and only if it is α_A-measurable.* ∎

Note. It follows from the Lebesgue test and Corollaries 4.1.2 and 2.2.5 that over a cell A the McShane integral with respect to α coincides with the usual *Lebesgue integral* with respect to the measure α_A^* (cf. [41, Chapter 11, Section 3]).

Chapter 5
Descriptive definition

Let α be an increasing function on a cell $A = [a, b]$ and let $f \in \overline{\mathcal{R}}(A, \alpha)$. An *indefinite integral* of f is any function F on A such that $F(B) = \int_B f \, d\alpha$ for each cell $B \subset A$. It follows from the discussion in Section 1.1 that any two indefinite integrals of f differ only by an additive constant. In view of Propositions 2.1.9 and 2.1.10, a specific indefinite integral F of f is given by the formula

$$F(x) = \int_a^x f \, d\alpha$$

for each $x \in A$. In the special case when $\alpha = \lambda$, we characterize all functions on A that are indefinite integrals of members of $\overline{\mathcal{R}}(A, \lambda)$.

Since the λ-measure λ^*, called the *Lebesgue measure* in **R**, is particularly important, throughout Part I of this book we agree that the words "measure," "measurable," and "negligible" always refer to this measure. The same applies to the expressions "almost all" and "almost everywhere." We also agree to write $\lambda(E)$ instead of $\lambda^*(E)$ for every set $E \subset \mathbf{R}$. As $\lambda^*(A) = \lambda(A)$ for each interval A (Proposition 3.2.2), this simplification will cause no confusion.

5.1 AC functions

A *division* of a cell A is a finite collection $\mathcal{D}$ of nonoverlapping cells whose union is A.

Definition 5.1.1 Let F be a function on a cell A. The *variation* of F on a cell $B \subset A$ is the extended real number

$$V(F, B) = \sup \sum_{D \in \mathcal{D}} |F(D)|$$

where the supremum is taken over all divisions $\mathcal{D}$ of B. If $V(F, A) < +\infty$, we say that F is of *bounded variation*.

Convention. It is useful to let $V(F, B) = 0$ whenever B is a degenerate interval.

Proposition 5.1.2   *Let F be a function on a cell $[a, b]$. If $c \in [a, b]$, then*

$$V(F, [a, b]) = V(F, [a, c]) + V(F, [c, b]).$$

PROOF.   Avoiding a triviality, assume that $c \in (a, b)$. If $\mathcal{D}_a$ and $\mathcal{D}_b$ are divisions of $[a, c]$ and $[c, b]$, respectively, then $\mathcal{D} = \mathcal{D}_a \cup \mathcal{D}_b$ is a division of $[a, b]$ and

$$\sum_{A \in \mathcal{D}_a} |F(A)| + \sum_{B \in \mathcal{D}_b} |F(B)| = \sum_{D \in \mathcal{D}} |F(D)| \leq V(F, [a, b]).$$

From the arbitrariness of $\mathcal{D}_a$ and $\mathcal{D}_b$, we obtain

$$V(F, [a, c]) + V(F, [c, b]) \leq V(F, [a, b]).$$

On the other hand, if $\mathcal{D}$ is a division of $[a, b]$, let $\mathcal{E}$ be a segmentation of $\mathcal{D} \cup \{[a, c], [c, b]\}$ (see Lemma 3.2.9). Then $\mathcal{E}_a = \{E \in \mathcal{E} : E \subset [a, c]\}$ and $\mathcal{E}_b = \{E \in \mathcal{E} : E \subset [c, b]\}$ are divisions of $[a, c]$ and $[c, b]$, respectively, and $\mathcal{E} = \mathcal{E}_a \cup \mathcal{E}_b$. Thus, letting $\mathcal{E}_D = \{E \in \mathcal{E} : E \subset D\}$ for each $D \in \mathcal{D}$, we obtain

$$\sum_{D \in \mathcal{D}} |F(D)| = \sum_{D \in \mathcal{D}} \left| \sum_{E \in \mathcal{E}_D} F(E) \right| \leq \sum_{D \in \mathcal{D}} \sum_{E \in \mathcal{E}_D} |F(E)| = \sum_{E \in \mathcal{E}} |F(E)|$$

$$= \sum_{A \in \mathcal{E}_a} |F(A)| + \sum_{B \in \mathcal{E}_b} |F(B)| \leq V(F, [a, c]) + V(F, [c, b]).$$

The arbitrariness of $\mathcal{D}$ yields

$$V(F, [a, b]) \leq V(F, [a, c]) + V(F, [c, b]),$$

and the proposition is proved. ∎

Corollary 5.1.3   *Let F be a function on a cell $A = [a, b]$, and let*

$$VF(x) = V(F, [a, x])$$

for each $x \in A$. If F is of bounded variation, then VF is a finite function on A such that $|F(B)| \leq VF(B)$ for each interval $B \subset A$. In particular, the functions VF and $VF - F$ are increasing.

PROOF. Let $V(F, A) < +\infty$. Since

$$VF(x) \leq VF(x) + V(F, [x, b]) = V(F, A)$$

for each $x \in A$, we see that VF is a finite function. Let $B = [x, y]$ be a subcell of A. Observing that $\{B\}$ is a division of B and using Proposition 5.1.2, we obtain

$$|F(B)| \leq V(F, B) = VF(y) - VF(x) = VF(B).$$

In particular, $VF(B) \geq 0$ and $(VF - F)(B) = VF(B) - F(B) \geq 0$, which implies that the functions VF and $VF - F$ are increasing. ∎

Exercise 5.1.4 Let F be a function on a cell A. Prove that $V|F| \leq VF$, and show by example that this inequality can be strict.

Corollary 5.1.5 *A function F on a cell A is of bounded variation if and only if it is the difference of two increasing functions on A.*

PROOF. If $F = G - H$ where G and H are increasing functions on A, it is easy to see that

$$V(F, A) \leq V(G, A) + V(H, A) = G(A) + H(A) < +\infty.$$

The converse follows from Corollary 5.1.3, since $F = VF - (VF - F)$. ∎

Definition 5.1.6 A function F on a cell A is called AC (an abbreviation for *absolutely continuous*) if, given $\varepsilon > 0$, there is an $\eta > 0$ such that

$$\sum_{k=1}^{n} |F(B_k)| < \varepsilon$$

whenever $\{B_1, \ldots, B_n\}$ is a collection of nonoverlapping subcells of A with $\sum_{k=1}^{n} \lambda(B_k) < \eta$.

It follows immediately from the definition that each absolutely continuous function is continuous; we shall see later (Example 5.3.11 and Corollary 5.3.14) that the converse is not true. A simple example of an AC function is any Lipschitz function (see Exercise 2.2.7); many more examples are provided by Corollary 5.1.10 below.

Proposition 5.1.7 *Each AC function on a cell A is of bounded variation.*

PROOF. Let F be an AC function on A, and let $\eta > 0$ be associated with $\varepsilon = 1$ according to Definition 5.1.6. Fix a division $\mathcal{A} = \{A_1, \ldots, A_n\}$ of

A so that $\lambda(A_i) < \eta$ for $i = 1, \ldots, n$. If $\mathcal{D}$ is any division of A, find a segmentation $\mathcal{C}$ of $\mathcal{A} \cup \mathcal{D}$. Setting $\mathcal{C}_i = \{C \in \mathcal{C} : C \subset A_i\}$ for $i = 1, \ldots, n$ and $\mathcal{C}_D = \{C \in \mathcal{C} : C \subset D\}$ for each $D \in \mathcal{D}$, we obtain

$$\sum_{D \in \mathcal{D}} |F(D)| = \sum_{D \in \mathcal{D}} \left| \sum_{C \in \mathcal{C}_D} F(C) \right| \leq \sum_{C \in \mathcal{C}} |F(C)| = \sum_{i=1}^{n} \sum_{C \in \mathcal{C}_i} |F(C)| < n.$$

As $\mathcal{D}$ is arbitrary, $V(F, A) \leq n$. $\blacksquare$

Proposition 5.1.8 *An increasing function α on a cell A is AC if and only if each negligible set E is α_A-negligible.*

PROOF. Suppose α is AC and choose a negligible set E and $\varepsilon > 0$. Select an $\eta > 0$ associated with ε according to Definition 5.1.6, and use Proposition 3.2.8 to find an open set U with $E \subset U$ and $\lambda(U) < \eta$. By Lemma 3.4.4, there is a countable collection $\mathcal{K}$ of nonoverlapping cells such that $\bigcup \mathcal{K} = U$ and $\sum_{K \in \mathcal{K}} \alpha_A(K) = \alpha_A^*(U)$. Since $\sum_{K \in \mathcal{K}} \lambda(K) = \lambda(U) < \eta$, we have

$$\sum_{K \in \mathcal{C}} \alpha_A(K) = \sum_{K \in \mathcal{C}} \alpha(A \cap K) < \varepsilon$$

for each finite family $\mathcal{C} \subset \mathcal{K}$. Thus $\alpha_A^*(E) \leq \alpha_A^*(U) \leq \varepsilon$, and E is α_A-negligible by the arbitrariness of ε.

Conversely, suppose that α is not AC. Then there is an $\varepsilon > 0$ such that for each $n = 1, 2, \ldots$, we can find a set $E_n \subset A$ that is the union of nonoverlapping cells $A_1, \ldots, A_k$ with

$$\lambda(E_n) = \sum_{i=1}^{k} \lambda(A_i) < 2^{-n} \qquad \text{and} \qquad \varepsilon \leq \sum_{i=1}^{k} \alpha(A_i) \leq \alpha_A^*(E_n)$$

(the last inequality is provided by Corollary 3.2.15). The set

$$E = \bigcap_{n=1}^{\infty} \bigcup_{j=n}^{\infty} E_j$$

is negligible because

$$\lambda(E) \leq \lambda \left(\bigcup_{j=n}^{\infty} E_j \right) \leq \sum_{j=n}^{\infty} \lambda(E_j) < \sum_{j=n}^{\infty} 2^{-j} = 2^{-n+1}$$

for $n = 1, 2, \ldots$, and yet $\alpha_A^*(E) \geq \varepsilon$ by Exercise 3.3.16. $\blacksquare$

Proposition 5.1.9 *If F is a function on a cell A, then the following statements are equivalent.*

1. *The function F is AC.*
2. *The function VF is AC.*
3. *Given a negligible set $E \subset A$ and $\varepsilon > 0$, there is a positive function δ on E such that $\sum_{i=1}^{p} |F(A_i)| < \varepsilon$ for each partition $\{(A_1, x_1), \ldots, (A_p, x_p)\}$ in A anchored in E that is δ-fine.*

PROOF. $(1 \Rightarrow 3)$ Choose a negligible set $E \subset A$ and $\varepsilon > 0$. If η is associated with ε according to Definition 5.1.6, use Proposition 3.2.8 to find an open set U such that $E \subset U$ and $\lambda(U) < \eta$. There is a positive function δ on E with $U(x, \delta(x)) \subset U$ for every $x \in E$. If $\{(A_1, x_1), \ldots, (A_p, x_p)\}$ is a partition in A anchored in E that is δ-fine, then $\bigcup_{i=1}^{p} A_i \subset U$. Thus $\sum_{i=1}^{p} |F(A_i)| < \varepsilon$ by the choice of η.

$(3 \Rightarrow 2)$ Let E be a negligible subset of A. In view of Propositions 3.7.3 and 5.1.8, it suffices to show that E is $(VF)_A$-negligible. To this end, choose an $\varepsilon > 0$ and find a positive function δ on E associated with ε and E according to the third statement of the proposition. Select a partition $\{(A_1, x_1), \ldots, (A_p, x_p)\}$ anchored in E that is δ-fine. If $\mathcal{D}_i$ is a division of $A \cap A_i$ for $i = 1, \ldots, p$, then $\{(D, x_i) : D \in \mathcal{D}_i, i = 1, \ldots, p\}$ is a partition in A anchored in E that is δ-fine. Thus

$$\sum_{i=1}^{p} \sum_{D \in \mathcal{D}_i} |F(D)| < \varepsilon,$$

and it follows from the arbitrariness of the divisions $\mathcal{D}_i$ that

$$\sum_{i=1}^{p} (VF)_A(A_i) = \sum_{i=1}^{p} V(F, A \cap A_i) \leq \varepsilon.$$

Consequently $(VF)_A^*(E) \leq (VF)_A^\delta(E) \leq \varepsilon$ and the $(VF)_A$-negligibility of E is established.

$(2 \Rightarrow 1)$ This implication is a direct consequence of Corollary 5.1.3. $\blacksquare$

Corollary 5.1.10 *Let A be a cell and let $f \in \overline{\mathcal{R}}(A, \lambda)$. If F is an indefinite integral of f, then F is AC.*

PROOF. Choose a negligible set $E \subset A$ and $\varepsilon > 0$. In view of Corollary 4.1.6, we may assume that f is a finite function on A with $f(x) = 0$ for every $x \in E$.

By Henstock's lemma, there is a positive function δ on A such that

$$\sum_{i=1}^{p} |f(x_i)\lambda(A_i) - F(A_i)| < \varepsilon$$

for each δ-fine partition $P = \{(A_1, x_1), \ldots, (A_p, x_p)\}$ in A. If P is also anchored in E, then $\sum_{i=1}^{p} |F(A_i)| < \varepsilon$ and F is AC by Proposition 5.1.9, 3. ∎

Exercise 5.1.11 Show that a function F on a cell A is AC whenever either of the following conditions is satisfied.

1. Given $\varepsilon > 0$, there is an $\eta > 0$ such that $|\sum_{k=1}^{n} F(B_k)| < \varepsilon$ whenever $\{B_1, \ldots, B_n\}$ is a collection of nonoverlapping subcells of A with $\sum_{k=1}^{n} \lambda(B_k) < \eta$.
2. Given a negligible set $E \subset A$ and $\varepsilon > 0$, there is a positive function δ on E such that $|\sum_{i=1}^{p} F(A_i)| < \varepsilon$ for each partition $\{(A_1, x_1), \ldots, (A_p, x_p)\}$ in A anchored in E that is δ-fine.

5.2 Covering theorems

We note that every family of nonoverlapping, in particular disjoint, cells is countable because the interior of each cell contains a rational number. If $A = [a, b]$ is a cell, then

$$c_A = \frac{a+b}{2} \qquad \text{and} \qquad r_A = \frac{b-a}{2}$$

are called the *center* and *radius* of A, respectively. Clearly, $A = U[c_A, r_A]$, and we set $A^{\bullet} = U[c_A, 5r_A]$.

Lemma 5.2.1 *Let $\mathcal{E}$ be a family of subcells of a cell A. There is a disjoint family $\mathcal{F} \subset \mathcal{E}$ such that for each $B \in \mathcal{E}$ we can find a $C \in \mathcal{F}$ with $B \cap C \neq \emptyset$ and $B \subset C^{\bullet}$. In particular,*

$$\bigcup \mathcal{E} \subset \bigcup \{C^{\bullet} : C \in \mathcal{F}\}.$$

PROOF. If $r = r_A$ is the radius of A, set

$$\mathcal{E}_n = \{B \in \mathcal{E} : r2^{-n} < r_B \leq r2^{-n+1}\}$$

for $n = 1, 2, \ldots$, and observe that each $\mathcal{E}_n$ contains at most $2^n - 1$ disjoint cells. Thus it is easy to prove by induction that each family $\mathcal{C} \subset \mathcal{E}_n$ contains a *maximal* disjoint subfamily $\mathcal{D}$, i.e., a disjoint family $\mathcal{D} \subset \mathcal{C}$ such that $\bigcup \mathcal{D}$ meets each element of $\mathcal{C}$.

Let $\mathcal{F}_1$ be a maximal disjoint subfamily of $\mathcal{E}_1$, and assuming the families $\mathcal{F}_1 \subset \mathcal{E}_1, \ldots, \mathcal{F}_{n-1} \subset \mathcal{E}_{n-1}$ have been defined so that $\bigcup_{i=1}^{n-1} \mathcal{F}_i$ is disjoint, let $\mathcal{F}_n$ be a maximal disjoint subfamily of

$$\left\{ B \in \mathcal{E}_n : B \cap \left[\bigcup_{i=1}^{n-1} \left(\bigcup \mathcal{F}_i \right) \right] = \emptyset \right\}.$$

We claim that $\mathcal{F} = \bigcup_{n=1}^{\infty} \mathcal{F}_n$ is the desired family. Indeed, $\mathcal{F}$ is disjoint by construction. If $B \in \mathcal{E}$, then $B \in \mathcal{E}_n$ for a unique integer $n \geq 1$, and it follows from the maximality of $\mathcal{F}_n$, that B meets a cell $C \in \bigcup_{i=1}^{n} \mathcal{F}_i$. Since $r_B \leq r2^{-n+1} < 2r_C$, we see that $B \subset C^{\bullet}$. ∎

Definition 5.2.2 A family $\mathcal{E}$ of cells is called a *Vitali cover* of a set $E \subset \mathbf{R}$ if for each $x \in E$ and each $\eta > 0$ there is a $B \in \mathcal{E}$ with $x \in B$ and $B \subset U(x, \eta)$.

We refer to the following proposition as *Vitali's covering theorem*.

Proposition 5.2.3 *If a family $\mathcal{E}$ of subcells of a cell A is a Vitali cover of a set $E \subset A$, then there is a disjoint family $\mathcal{F} \subset \mathcal{E}$ such that $E - \bigcup \mathcal{F}$ is negligible.*

PROOF. Let $\{C_1, C_2, \ldots\}$ be an enumeration of a family $\mathcal{F} \subset \mathcal{E}$ that satisfies the conditions of Lemma 5.2.1.

Claim. For $n = 1, 2, \ldots$, we have

$$E - \bigcup_{i=1}^{n} C_i \subset \bigcup_{i>n} C_i^{\bullet}.$$

Proof. Since $\bigcup_{i=1}^{n} C_i$ is a closed set and $\mathcal{E}$ is a Vitali cover of E, for each $x \in E - \bigcup_{i=1}^{n} C_i$ there is a $B \in \mathcal{E}$ such that $x \in B$ and $B \cap (\bigcup_{i=1}^{n} C_i) = \emptyset$. By the choice of $\mathcal{F}$, we can find a C_k with $B \cap C_k \neq \emptyset$ and $B \subset C_k^{\bullet}$. It follows that $k > n$, so $x \in \bigcup_{i>n} C_i^{\bullet}$.

Now if $\mathcal{F}$ is finite, the claim implies that $E \subset \bigcup \mathcal{F}$, and the proposition is proved. If $\mathcal{F}$ is infinite, then

$$\lambda \left(E - \bigcup \mathcal{F} \right) \leq \lambda \left(E - \bigcup_{i=1}^{n} C_i \right) \leq \lambda \left(\bigcup_{i=n+1}^{\infty} C_i^{\bullet} \right) \leq 5 \sum_{i=n+1}^{\infty} \lambda(C_i)$$

for $n = 1, 2, \ldots$. Since $\sum_{i=1}^{\infty} \lambda(C_i) \leq \lambda(A) < +\infty$, the proposition follows. ∎

Lemma 5.2.1 applies well to the Lebesgue measure λ because $\lambda(A^\bullet) = 5\lambda(A)$ for each cell A (see Exercise 3.4.7, 1). For a general measure α^*, which does not have this property, we need a more complicated covering result.

Lemma 5.2.4 *Let r be a bounded positive function defined on a bounded set $E \subset \mathbf{R}$. There are subsets $E_1, \ldots, E_{14}$ of E such that for $k = 1, \ldots, 14$, the family $\{U[x, r(x)] : x \in E_k\}$ is disjoint and*

$$E \subset \bigcup_{k=1}^{14} \bigcup_{x \in E_k} U(x, r(x)).$$

PROOF. Choose an $x_1 \in E$ so that $r(x_1) \geq \frac{3}{4}\sup\{r(x) : x \in E\}$. Proceeding inductively, assume that $x_1, \ldots, x_n$ have been chosen in E so that x_i lies outside $\bigcup_{j=1}^{i-1} U(x_j, r(x_j))$ and

$$r(x_i) \geq \frac{3}{4}\sup\left\{r(x) : x \in E - \bigcup_{j=1}^{i-1} U(x_i, r(x_i))\right\}$$

for $i = 1, \ldots, n$. If $E \subset \bigcup_{i=1}^{n} U(x_i, r(x_i))$, the induction stops. Otherwise, select an $x_{n+1} \in E - \bigcup_{i=1}^{n} U(x_i, r(x_i))$ so that

$$r(x_{n+1}) \geq \frac{3}{4}\sup\left\{r(x) : x \in E - \bigcup_{i=1}^{n} U(x_i, r(x_i))\right\}.$$

The induction produces a finite or infinite sequence $\{x_1, x_2, \ldots\}$ of distinct points of E, and we let $r_n = r(x_n)$.

Claim 1. If $i \neq j$, then the cells $U[x_i, r_i/3]$ and $U[x_j, r_j/3]$ are disjoint.

Proof. If $i < j$, then $x_j \notin U(x_i, r_i)$ and $r_i \geq 3r_j/4$. Hence

$$\frac{r_i}{3} + \frac{r_j}{3} \leq \frac{1}{3}\left(r_i + \frac{4}{3}r_i\right) < r_i \leq |x_i - x_j|$$

and we see that $U[x_i, r_i/3] \cap U[x_j, r_j/3] = \emptyset$. The claim follows by symmetry.

Claim 2. $E \subset \bigcup_n U(x_n, r_n)$.

Proof. As this is merely a condition of the construction when the sequence $x_1, x_2, \ldots$ is finite, assume that infinitely many points x_n have been produced. If there is an $x \in E - \bigcup_{n=1}^{\infty} U(x_n, r_n)$, then $r_n \geq 3r(x)/4 > 0$ for $n = 1, 2 \ldots$. Since E and r are bounded, each $U[x_n, r_n]$ is contained in a cell A and a contradiction follows from Claim 1.

Claim 3. Each $U[x_n, r_n]$ intersects at most eleven $U[x_i, r_i]$ with $i < n$ and $r_i \leq 3r_n/2$.

Proof. Let T be the set of all positive integers $i < n$ for which $r_i \leq 3r_n/2$ and $U[x_i, r_i] \cap U[x_n, r_n] \neq \emptyset$, and let t be the number of elements in T. If $i \in T$ then $|x_i - x_n| \leq r_i + r_n \leq 5r_n/2$, and hence

$$|x - x_n| \leq |x - x_i| + |x_i - x_n| \leq \frac{r_i}{3} + \frac{5}{2}r_n \leq \frac{r_n}{2} + \frac{5}{2}r_n = 3r_n$$

whenever $|x - x_i| \leq r_i/3$. It follows that $U[x_i, r_i/3] \subset U[x_n, 3r_n]$ for each $i \in T$. Since $r_i \geq 3r_n/4$ for every $i < n$, Claim 1 implies that

$$t\frac{r_n}{2} \leq \sum_{i \in T} \lambda(U[x_i, r_i/3]) < \lambda(U[x_n, 3r_n]) = 6r_n .$$

Consequently $t < 12$ as required.

Claim 4. Each $U[x_n, r_n]$ intersects no more than two $U[x_i, r_i]$ with $i < n$ and $r_i > 3r_n/2$.

Proof. Let $i < j < n$ be such that both $U[x_i, r_i]$ and $U[x_j, r_j]$ intersect $U[x_n, r_n]$ and $\min\{r_i, r_j\} > 3r_n/2$. Then $r_i \geq 3r_j/4$, and setting

$$a_i = |x_i - x_n|, \qquad a_j = |x_j - x_n|, \qquad a = |x_i - x_j|,$$

we have $a_i \leq r_i + r_n$ and $a_j \leq r_j + r_n$. Since x_n lies outside $U(x_i, r_i) \cup U(x_j, r_j)$ and x_j lies outside $U(x_i, r_i)$, we also have $r_i \leq a_i$, $r_j \leq a_j$, and $r_i \leq a$. Combining these inequalities yields

$$a_i - a_j \leq r_i + r_n - r_j < r_i + \frac{2}{3}r_j - r_j < r_i \leq a,$$

$$a_j - a_i \leq r_j + r_n - r_i < \frac{4}{3}r_i + \frac{2}{3}r_i - r_i = r_i \leq a.$$

Therefore $|a_i - a_j| < a$, and x_n lies in between x_i and x_j. This establishes the claim.

In view of Claims 3 and 4, at most thirteen sets $U[x_i, r_i]$ with $i < n$ meet $U[x_n, r_n]$. Define families $E_1, \ldots, E_{14}$ inductively by viewing them as boxes and deciding for each x_n into which box it is placed. Place x_i into E_i for $i = 1, \ldots, 14$, and assume that for an integer $n \geq 14$, the points $x_1, \ldots, x_n$ have been placed into $E_1, \ldots, E_{14}$ so that for $k = 1, \ldots, 14$, each family $\mathcal{E}_k = \{U[x_i, r_i] : x_i \in E_k, i \leq n\}$ consists of disjoint sets. Since $U[x_{n+1}, r_{n+1}]$ meets at most thirteen sets $U[x_i, r_i]$ with $i \leq n$, there is a positive integer $p \leq 14$ such that $U[x_{n+1}, r_{n+1}]$ meets no member of $\mathcal{E}_p$. Placing x_{n+1} into E_p completes the proof. ∎

Definition 5.2.5 A family $\mathcal{E}$ of cells is called a *centered Vitali cover* of a set $E \subset \mathbf{R}$ if for each $x \in E$ and each $\eta > 0$ there is a $U[x, r] \in \mathcal{E}$ with $r < \eta$.

Proposition 5.2.6 *Let α be an increasing function in $\mathbf{R}$. If a family $\mathcal{E}$ of cells is a centered Vitali cover of a bounded set $E \subset \mathbf{R}$, then there is a disjoint family $\mathcal{F} \subset \mathcal{E}$ such that $E - \bigcup \mathcal{F}$ is α-negligible.*

PROOF. As E is bounded, $\alpha^*(E) < +\infty$. If $\alpha^*(E) = 0$ there is nothing to prove, so we assume that $\alpha^*(E) > 0$. There is a positive bounded function r_1 on E such that $U[x, r_1(x)] \in \mathcal{E}$ for each $x \in E$. Let $E_1, \ldots, E_{14}$ be subsets of E associated with r_1 according to Lemma 5.2.4. Then

$$\alpha^*(E) \le \sum_{k=1}^{14} \alpha^* \left(E \cap \bigcup_{x \in E_k} U[x, r_1(x)] \right),$$

and we can find a positive integer $k \le 14$ with

$$\frac{1}{14} \alpha^*(E) \le \alpha^* \left(E \cap \bigcup_{x \in E_k} U[x, r_1(x)] \right).$$

Since E_k is a countable set, it follows from Proposition 3.3.15 that there are points $x_1, \ldots, x_n$ in E_k such that

$$\frac{1}{15} \alpha^*(E) < \alpha^* \left(E \cap \bigcup_{i=1}^{n} U[x_i, r_1(x_i)] \right).$$

The collection $\mathcal{F}_1 = \{U[x_i, r_1(x_i)] : i = 1, \ldots, n\}$ is a disjoint subfamily of $\mathcal{E}$ and the set $F_1 = \bigcup \mathcal{F}_1$ is closed. By Proposition 3.3.2, 4, we have

$$\alpha^*(E) = \alpha^*(E \cap F_1) + \alpha^*(E - F_1) > \frac{1}{15} \alpha^*(E) + \alpha^*(E - F_1)$$

and hence $\alpha^*(E - F_1) < c\alpha^*(E)$ where $c = 14/15$.

If $\alpha^*(E - F_1) = 0$ we are done; if $\alpha^*(E - F_1) > 0$ we continue as follows. Since the set F_1 is closed, there is a positive bounded function r_2 on $E - F_1$ such that $U[x, r_2(x)] \in \mathcal{E}$ and $U[x, r_2(x)] \cap F_1 = \emptyset$ for each $x \in E - F_1$. Arguing as before, we can find a finite disjoint family $\mathcal{F}_2 \subset \mathcal{E}$ such that the closed set $F_2 = \bigcup \mathcal{F}_2$ is disjoint from F_1 and

$$\alpha^*(E - [F_1 \cup F_2]) = \alpha^*([E - F_1] - F_2) < c\alpha^*(E - F_1) < c^2 \alpha^*(E).$$

Proceeding inductively, we construct subfamilies $\mathcal{F}_1, \mathcal{F}_2, \ldots$ whose union $\mathcal{F}$ is a disjoint family such that

$$\alpha^*\left(E - \bigcup \mathcal{F}\right) \le \alpha^*\left(E - \bigcup_{n=1}^{p}\left[\bigcup \mathcal{F}_n\right]\right) < c^p \alpha^*(E)$$

for $p = 1, 2, \ldots$. As $0 < c < 1$, the proposition follows. ∎

Note. Lemmas 5.2.1 and 5.2.4 are of independent importance. Many results are easier to obtain directly from these lemmas than from the corresponding Propositions 5.2.3 and 5.2.6. Good examples are Theorems 5.3.5 and 5.4.9 as well as Lemma 5.4.3.

5.3 Differentiation

Let α be an increasing function on a cell A, let F be any function defined on A, and let $x \in A$. If $\delta > 0$, we denote by A_δ^* the family of all cells $B \subset A \cap U(x, \delta)$ for which $x \in B$ and the ratio $F(B)/\alpha(B)$ is defined. The numbers

$$D_\alpha F(x) = \sup_{\delta > 0}\left[\inf_{B \in A_\delta^*} \frac{F(B)}{\alpha(B)}\right] \quad \text{and} \quad D^\alpha F(x) = \inf_{\delta > 0}\left[\sup_{B \in A_\delta^*} \frac{F(B)}{\alpha(B)}\right]$$

are called the *lower* and *upper* α-*derivates* of F at x. If

$$D_\alpha F(x) = D^\alpha F(x) \ne \pm\infty,$$

we denote this common value by $F_\alpha'(x)$ and say that F is α-*derivable* at x.

Exercise 5.3.1 Let F be a function on a cell A, and let $x \in A$. Recall the definition of the Lebesgue lower and upper derivates (Section 4.3) and prove that
$$\mathcal{D}_\alpha F(x) \le D_\alpha F(x) \quad \text{and} \quad D^\alpha F(x) \le \mathcal{D}^\alpha F(x).$$
Show by example that these inequalities can be sharp.

Exercise 5.3.2 Let F be a function on a cell A such that $D_\alpha F \ge 0$. Following the proof of Lemma 4.3.6, show that F is increasing.

The derivates have a closer connection with the usual derivative than the Lebesgue derivates of Section 4.3 (cf. Proposition 5.3.3 below with Exercise 4.3.3).

Proposition 5.3.3   *Let F be a function defined on a cell A. Then F is λ-derivable at $x \in A$ if and only if F is differentiable at x, in which case $F'_\lambda(x) = F'(x)$.*

PROOF. It is easy to see that F is λ-derivable at x if and only if

$$\lim \frac{F(z) - F(y)}{z - y} = c \neq \pm\infty$$

where the limit is taken as $y, z \in A$ approach x so that $y \leq x \leq z$; in this case $F'_\lambda(x) = c$. Thus, if F is λ-derivable at x, then it is differentiable at x and $F'(x) = F'_\lambda(x)$.

Conversely, suppose that F is differentiable at x with $F'(x) = c$, and choose $y, z \in A$ so that $y < x < z$. Then

$$\left| \frac{F(z) - F(y)}{z - y} - c \right| \leq \left| \frac{F(z) - F(x)}{z - x} - c \right| \cdot \frac{z - x}{z - y}$$

$$+ \left| \frac{F(y) - F(x)}{y - x} - c \right| \cdot \frac{x - y}{z - y}$$

$$\leq \left| \frac{F(z) - F(x)}{z - x} - c \right| + \left| \frac{F(y) - F(x)}{y - x} - c \right|$$

and the λ-derivability of F at x follows. ∎

Exercise 5.3.4   Let F be a function defined on a cell A, and let $x \in A$ be such that $0 < \alpha'(x) < +\infty$. Show that F is α-derivable at x if and only if it is differentiable at x, in which case $F'_\alpha(x) = F'(x)/\alpha'(x)$.

We use the covering results of Section 4.5 to prove some classical differentiability theorems. Recall that the concept of an indefinite integral has been introduced at the beginning of this chapter.

Theorem 5.3.5   *Let A be a cell and $f \in \overline{\mathcal{R}}(A, \lambda)$. If F is an indefinite integral of f, then for almost all $x \in A$ the function F is differentiable at x and $F'(x) = f(x)$.*

PROOF. We may assume that $f \in \mathcal{R}(A, \lambda)$. Let E be the set of all $x \in A$ for which either F is not differentiable at x or $F'(x) \neq f(x)$. Given $x \in E$,

we can find a $\gamma(x) > 0$ so that for each $\beta > 0$ there is a cell $B \subset A \cap U(x, \beta)$ with $x \in B$ and
$$|f(x)\lambda(B) - F(B)| \geq \gamma(x)\lambda(B).$$

Indeed, in view of Proposition 5.3.3, the negation of the previous statement implies that F is differentiable at x and $F'(x) = f(x)$.

Fix an integer $n \geq 1$, let $E_n = \{x \in E : \gamma(x) \geq 1/n\}$, and choose an $\varepsilon > 0$. Using Henstock's lemma, find a positive function δ on A so that

$$\sum_{i=1}^{p} |f(x_i)\lambda(A_i) - F(A_i)| < \frac{\varepsilon}{n}$$

for each δ-fine partition $\{(A_1, x_1), \ldots, (A_p, x_p)\}$ in A. Let $\mathcal{E}$ be the collection of all cells $B \subset A$ such that $B \subset U(x_B, \delta(x_B))$ for a point $x_B \in B \cap E_n$ and

$$|f(x_B)\lambda(B) - F(B)| \geq \frac{1}{n}\lambda(B).$$

By Lemma 5.2.1, there is a disjoint subfamily $\mathcal{F}$ of $\mathcal{E}$ for which

$$E_n \subset \bigcup \{C^\bullet : C \in \mathcal{F}\}.$$

For each finite family $\mathcal{T} \subset \mathcal{F}$, the collection $\{(C, x_C) : C \in \mathcal{T}\}$ is a δ-fine Perron partition in A. Hence

$$\sum_{C \in \mathcal{T}} \lambda(C) \leq n \sum_{C \in \mathcal{T}} |f(x_C)\lambda(C) - F(C)| < \varepsilon,$$

and consequently

$$\lambda(E_n) \leq \sum_{C \in \mathcal{F}} \lambda(C^\bullet) = 5 \sum_{C \in \mathcal{F}} \lambda(C) \leq 5\varepsilon.$$

It follows from the arbitrariness of ε that E_n is negligible, and according to Proposition 3.5.3, 3, so is $E = \bigcup_{n=1}^{\infty} E_n$. $\blacksquare$

Remark 5.3.6 Let A be a cell. Theorem 5.3.5 shows that each $f \in \overline{\mathcal{R}}(A, \lambda)$ is the derivative almost everywhere in A of its indefinite integrals. Thus, among all functions F on A that are differentiable almost everywhere in A, we would like to characterize those satisfying the following condition:

- $F' \in \overline{\mathcal{R}}(A, \lambda)$ and F is an indefinite integral of F'.

We call a characterization of this kind a *partial descriptive definition* of the McShane integral. It turns out, however, that for the McShane integral we can do better (cf. Theorem 5.3.15). Among all functions F on A (whether differentiable or not), we shall characterize those satisfying the following conditions:

- F is differentiable almost everywhere in A;
- $F' \in \overline{\mathcal{R}}(A, \lambda)$ and F is an indefinite integral of F'.

Such a characterization is called a *full descriptive definition* of the McShane integral. Partial and full descriptive definitions are very similar; the difference is that the former applies to a restricted family of functions.

Lemma 5.3.7   *Let α be an increasing function defined on a cell A. If $D^\lambda\alpha(x) > c \geq 0$ for every x in a set $E \subset A$, then $\alpha(B) > c\lambda(E)$ for each cell $B \subset A$ containing E.*

PROOF.   Let $B \subset A$ be a cell containing E. The family $\mathcal{E}$ of all cells $C \subset B$ for which $\alpha(C) > c\lambda(C)$ is a Vitali cover of $E \cap B°$. By Vitali's covering theorem, there is a disjoint family $\mathcal{F} \subset \mathcal{E}$ such that $E \cap B° - \bigcup \mathcal{F}$ is a negligible set. Thus

$$c\lambda(E) = c\lambda(E \cap B°) \leq c\lambda\left(\bigcup \mathcal{F}\right)$$

$$= c\sum_{C \in \mathcal{F}} \lambda(C) < \sum_{C \in \mathcal{F}} \alpha(C) \leq \alpha(B),$$

where the last inequality holds, since $\sum_{C \in \mathcal{T}} \alpha(C) \leq \alpha(B)$ for each finite family $\mathcal{T} \subset \mathcal{F}$. ∎

Theorem 5.3.8   *If α is an increasing function on a cell A, then α is differentiable almost everywhere in A.*

PROOF.   Since α increases, $0 \leq D_\lambda \alpha \leq D^\lambda \alpha$. We show first that

$$N = \{x \in A : D_\lambda\alpha(x) < D^\lambda\alpha(x)\}$$

is a negligible set. If $\mathcal{Q}$ is the collection of all pairs (r, s) where $r < s$ are positive rationals, then $\mathcal{Q}$ is countable and

$$N = \bigcup_{(r,s) \in \mathcal{Q}} \{x \in A : D_\lambda\alpha(x) < r < s < D^\lambda\alpha(x)\}.$$

Thus it suffices to choose an $(r, s) \in \mathcal{Q}$ and prove that the set

$$E = \{x \in A : D_\lambda\alpha(x) < r < s < D^\lambda\alpha(x)\}$$

is negligible. To this end, choose an $\varepsilon > 0$ and use Proposition 3.2.8 to find an open set $U \subset \mathbf{R}$ such that $E \subset U$ and $\lambda(U) < \lambda(E) + \varepsilon$. The family $\mathcal{E}$ of all cells $B \subset A \cap U$ for which $\alpha(B) < r\lambda(B)$ is a Vitali cover of E. By Vitali's covering theorem, there is a disjoint family $\mathcal{F} \subset \mathcal{E}$ such that $E - \bigcup\mathcal{F}$ is a negligible set. Since $D^\lambda\alpha(x) > s$ for each $x \in E$, Lemma 5.3.7 implies that $\alpha(B) > s\lambda(E \cap B)$ whenever $B \in \mathcal{F}$. Therefore

$$s\lambda(E) = s\lambda\left[E \cap \left(\bigcup\mathcal{F}\right)\right] \le s \sum_{B \in \mathcal{F}} \lambda(E \cap B) < \sum_{B \in \mathcal{F}} \alpha(B)$$

$$< r \sum_{B \in \mathcal{F}} \lambda(B) = r\lambda\left(\bigcup\mathcal{F}\right) \le r\lambda(U) < r\lambda(E) + r\varepsilon,$$

and consequently $\lambda(E) < \varepsilon r/(s-r)$. Hence E is negligible by the arbitrariness of ε. We conclude that $D_\lambda\alpha = D^\lambda\alpha$ almost everywhere in A, and complete the proof by observing that the set

$$H = \{x \in A : D^\lambda\alpha(x) = +\infty\}$$

is negligible. Indeed, this is true, since $n\lambda(H) < \alpha(A)$ for $n = 1, 2, \ldots$ according to Lemma 5.3.7. $\blacksquare$

Corollary 5.3.9 *A function F of bounded variation on a cell A is differentiable almost everywhere in A. In particular, each AC function on A is differentiable almost everywhere in A.* $\blacksquare$

The following theorem sharpens the result of Theorem 2.4.3.

Theorem 5.3.10 *If α is an increasing function on a cell A, then α' belongs to $\overline{\mathcal{R}}(A, \lambda)$ and*

$$\int_A \alpha' \, d\lambda \le \alpha(A).$$

PROOF. For $x \in \mathbf{R}$ and $n = 1, 2, \ldots$, let

$$\alpha_n(x) = \frac{\alpha_A\left(x + \frac{1}{n}\right) - \alpha_A(x)}{\frac{1}{n}}$$

and observe that $\alpha_n \ge 0$ and $\alpha_n \to \alpha'$ almost everywhere in A. Furthermore, $\alpha_n \in \mathcal{R}(A, \lambda)$ by Theorem 2.3.11 and Proposition 2.1.3. If $A = [a, b]$, then by applying Theorem 2.4.2, we obtain

$$\liminf \int_a^b \alpha_n \, d\lambda = \liminf\left(n \int_b^{b+\frac{1}{n}} \alpha_A \, d\lambda - n \int_a^{a+\frac{1}{n}} \alpha_A \, d\lambda\right) \le \alpha(b) - \alpha(a)$$

(cf. the proof of Theorem 2.4.3). The theorem follows from the generalized Fatou lemma, which is easy to derive from Proposition 4.1.10. ∎

Example 5.3.11   We show that the inequality in Theorem 5.3.10 may be strict even when the increasing function α is *continuous*. To this end, let C be a set obtained from the cell $[0,1]$ by successively removing the open middle segments of length 3^{-r} for $r = 1, 2, \ldots$. More specifically, at the first step remove the open middle segment $\Delta_{1,1} = (3^{-1}, 2 \cdot 3^{-1})$, at the second step remove the open middle segments

$$\Delta_{2,1} = (3^{-2}, 2 \cdot 3^{-2}) \quad \text{and} \quad \Delta_{2,2} = (7 \cdot 3^{-2}, 8 \cdot 3^{-2}),$$

and at the kth step remove the open middle segments $\Delta_{r,1}, \ldots, \Delta_{r,2^{r-1}}$, each of length 3^{-r}. The closed set

$$C = [0,1] - \bigcup_{r=1}^{\infty} \bigcup_{s=1}^{2^{r-1}} \Delta_{r,s}$$

is negligible, since

$$\lambda(C) = \lambda([0,1]) - \sum_{r=1}^{\infty} \sum_{s=1}^{2^{r-1}} \lambda(\Delta_{r,s}) = 1 - \frac{1}{2} \sum_{r=1}^{\infty} \left(\frac{2}{3}\right)^r = 0.$$

Denote by $c_{r,s}$ the midpoint of the open segment $\Delta_{r,s}$, and define a function α on $[0,1] - C$ by setting $\alpha(x) = c_{r,s}$ for each $x \in \Delta_{r,s}$. It is not difficult to show that α has a unique extension to a continuous increasing function on the cell $[0,1]$, still denoted by α. Clearly, the function α is differentiable at each $x \in [0,1] - C$ with $\alpha'(x) = 0$. Yet

$$\int_0^1 \alpha' \, d\lambda = 0 < 1 = \alpha(1) - \alpha(0).$$

Remark 5.3.12   The set C and function α of Example 5.3.11 are called, respectively, the *Cantor ternary set* and *Cantor ternary function* (cf. [42, Section 2.44, p. 41] and [41, Chapter 2, Problems 36 and 44, pp. 44 and 47]).

Lemma 5.3.13   *If F is an AC function on a cell A and $F' \geq 0$ almost everywhere in A, then F is increasing.*

PROOF.   The proof is similar to that of Lemma 4.3.6. There is a negligible set $E \subset A$ such that $F'(x) \geq 0$ for each $x \in A - E$. Choose an $\varepsilon > 0$, and use Proposition 5.3.3 to find a positive function δ on $A - E$ so that given $x \in A - E$, we have $F(B) > -\varepsilon\lambda(B)$ for each cell $B \subset A \cap U(x, \delta(x))$ with $x \in B$. By Proposition 5.1.9, 3, there is a positive function η on E such that $\sum_{i=1}^{p} |F(A_i)| < \varepsilon$ for each partition $\{(A_1, x_1), \ldots, (A_p, x_p)\}$ in A anchored in E that is η-fine. Define a positive function θ on A by setting

$$\theta(x) = \begin{cases} \delta(x) & \text{if } x \in A - E, \\ \eta(x) & \text{if } x \in E. \end{cases}$$

If C is a subcell of A, find a θ-fine P-partition $\{(C_1, z_1), \ldots, (C_n, z_n)\}$ of C, whose existence is guaranteed by Proposition 1.2.4. We have

$$F(C) = \sum_{z_i \in E} F(C_i) + \sum_{z_i \in A-E} F(C_i)$$

$$\geq -\varepsilon - \sum_{z_i \in A-E} \varepsilon\lambda(C_i) \geq -\varepsilon[1 + \lambda(A)],$$

and the lemma follows from the arbitrariness of ε. ∎

Corollary 5.3.14 *The Cantor ternary function α is not AC.* ∎

The next theorem, which is a full descriptive definition of the McShane integral (see Remark 5.3.6), shows that the pathology exhibited in Example 5.3.11 does not occur for AC functions.

Theorem 5.3.15 *A function F on a cell A is AC if and only if the following conditions are satisfied:*

1. *F is differentiable almost everywhere in A;*
2. *$F' \in \overline{\mathcal{R}}(A, \lambda)$ and F is an indefinite integral of F'.*

PROOF. As the converse follows from Corollary 5.1.10, assume that F is AC. By Corollary 5.3.9, the function F is differentiable almost everywhere in A. According to Proposition 5.1.7, Corollary 5.1.5, and Theorem 5.3.10, the derivative F' belongs to $\overline{\mathcal{R}}(A, \lambda)$. Now let G be an indefinite integral of F'. In view of Corollary 5.1.10 and Theorem 5.3.5, the function $F - G$ is AC and $(F - G)' = 0$ almost everywhere in A. It follows from Lemma 5.3.13 that $F - G$ is a constant function, and consequently F is an indefinite integral of F'. ∎

Remark 5.3.16   Using Theorems 2.4.1 and 5.3.15, it is easy to see that Corollaries 2.4.9 and 2.4.10 remain valid when "$\mathcal{R}$" and "differentiable" are replaced by "$\overline{\mathcal{R}}$" and "AC," respectively.

5.4 Singular functions

We shall not attempt the technical task of generalizing Theorem 5.3.15 to the case of an arbitrary length α. We only prove the Lebesgue decomposition theorem and give two sample results.

Definition 5.4.1   An increasing function α on a cell A is called *singular* whenever $\alpha' = 0$ almost everywhere in A.

Examples of singular functions are the gap function (Section 2.5) and the Cantor ternary function (Remark 5.3.12). By Lemma 5.3.13, the only functions that are simultaneously singular and AC are the constant functions.

Lemma 5.4.2   *Let α be an increasing function on a cell A and let E be an α_A-negligible subset of A. Then $\alpha' = 0$ almost everywhere in E.*

PROOF.   Choose an $\varepsilon > 0$ and for an integer $n \geq 1$, let $E_n = \{x \in E : \alpha'(x) > 1/n\}$. By Proposition 3.5.2, there is a sequence $\{A_i\}$ of cells such that

$$E_n \subset \bigcup_{i=1}^{\infty} A_i^{\circ} \quad \text{and} \quad \sum_{i=1}^{\infty} \alpha_A(A_i) < \frac{\varepsilon}{n}.$$

Applying Lemma 5.3.7, we obtain

$$\lambda(E_n) \leq \sum_{i=1}^{\infty} \lambda(E_n \cap A_i) < n \sum_{i=1}^{\infty} \alpha_A(A_i) < \varepsilon,$$

and E_n is negligible by the arbitrariness of ε. Since $\{x \in E : \alpha'(x) > 0\}$ is the union of the sets $E_1, E_2, \ldots$, the lemma is proved. ∎

Lemma 5.4.3   *If α is an increasing function on a cell A, then the set of all $x \in A$ with $\alpha'(x) = 0$ is α_A-negligible.*

PROOF.   Let $E = \{x \in A : \alpha'(x) = 0\}$. Given $\varepsilon > 0$, there is a bounded positive function r on E such that

$$\alpha_A\Big(U[x, r(x)] \Big) = \alpha\Big(A \cap U[x, r(x)] \Big) < \frac{\varepsilon}{14\lambda(A)} \lambda\Big(A \cap U[x, r(x)] \Big)$$

for each $x \in E$. Let $E_1, \ldots, E_{14}$ be subsets of E associated with r according to Lemma 5.2.4. Then

$$\sum_{k=1}^{14} \sum_{x \in E_k} \alpha_A \Big(U[x, r(x)] \Big) < \frac{\varepsilon}{14\lambda(A)} \sum_{k=1}^{14} \sum_{x \in E_k} \lambda \Big(A \cap U[x, r(x)] \Big)$$

$$\leq \frac{\varepsilon}{14\lambda(A)} \sum_{k=1}^{14} \lambda \Big(A \cap \bigcup_{x \in E_k} U[x, r(x)] \Big) \leq \varepsilon$$

and the proposition follows from Proposition 3.5.2. ∎

Exercise 5.4.4 Prove Lemma 5.4.3 by employing Lemma 5.2.1 instead of Lemma 5.2.4.

Combining Lemmas 5.4.2 and 5.4.3, we obtain the following characterization of singular functions.

Proposition 5.4.5 *An increasing function α on a cell A is singular if and only if there is an α_A-negligible set $E \subset A$ such that $\lambda(A - E) = 0$.* ∎

The next result is called the *Lebesgue decomposition theorem*. It allows us to separate the AC and singular "parts" of an increasing function.

Theorem 5.4.6 *Let α be an increasing function on a cell A. There is a pair of increasing functions β and γ on A such that β is AC, γ is singular, and $\alpha = \beta + \gamma$. Up to an additive constant, the functions β and γ are unique.*

PROOF. If $A = [a, b]$, let $\beta(x) = \int_a^x \alpha' \, d\lambda$ for each $x \in A$. Since $\alpha' \geq 0$ almost everywhere in A, the function β increases. It follows from Theorem 5.3.10 that $\gamma = \alpha - \beta$ is also an increasing function. Moreover, Theorem 5.3.15 implies that β is AC and γ is singular. Let μ and ν be another pair of increasing functions on A such that μ is AC, ν is singular, and $\alpha = \mu + \nu$. Then $F = \beta - \mu = \nu - \gamma$ is an AC function and $F' = 0$ almost everywhere in A. Thus F is constant by Lemma 5.3.13, which establishes the uniqueness. ∎

Theorem 5.4.7 *Let α be an AC increasing function on a cell A. A function F on A is AC if and only if the following conditions are satisfied:*

1. *F is α-derivable α_A-almost everywhere;*
2. *$F'_\alpha \in \overline{\mathcal{R}}(A, \alpha)$ and F is an indefinite integral of F'_α.*

PROOF.   Let F be AC. By Corollary 5.3.9 and Proposition 5.1.8, both functions α and F are differentiable α_A-almost everywhere. According to Lemma 5.4.3 and Exercise 5.3.4, for α_A-almost all x, the function F is α-derivable at x and $F'(x) = F'_\alpha(x)\alpha'(x)$. Theorem 5.3.15 implies that $F' \in \overline{\mathcal{R}}(A, \lambda)$ and that F is an indefinite integral of F'. In view of Remark 5.3.16, the second condition follows from Corollary 2.4.9.

Conversely, assume that conditions 1 and 2 of the theorem are satisfied. For each $x \in A$, let

$$f(x) = \begin{cases} F'_\alpha(x) & \text{if } F \text{ is } \alpha\text{-derivable at } x, \\ 0 & \text{otherwise,} \end{cases}$$

and observe that $f \in \mathcal{R}(A, \alpha)$. From Corollary 2.4.9 and Remark 5.3.16 we infer that $f\alpha' \in \overline{\mathcal{R}}(A, \lambda)$ and that F is an indefinite integral of $f\alpha'$. Thus the function F is AC by Corollary 5.1.10. ∎

Let α be an increasing function on a cell A. We say that a function F on A is *symmetrically α-derivable* at $x \in A$ if $\alpha_A(U[x, r]) > 0$ for all sufficiently small $r > 0$ and a finite limit

$$F_\alpha^{sym}(x) = \lim_{r \to 0+} \frac{F_A(U[x, r])}{\alpha_A(U[x, r])}$$

exists; the number $F_\alpha^{sym}(x)$ is called the *symmetric α-derivate* of F at x. Clearly, if F is α-derivable at $x \in A$, it is symmetrically α-derivable at x and $F_\alpha^{sym}(x) = F'_\alpha(x)$. On the other hand, if $0 \in A^\circ$ and $F(x) = |x|$ for all $x \in A$, then F is not differentiable at 0, and yet $F_\lambda^{sym}(0) = 0$ exists.

Exercise 5.4.8 If α is an increasing function on a cell A, show that the set $\{x \in A : \alpha_\lambda^{sym}(x) = 0\}$ is α_A-negligible.

Theorem 5.4.9 *Let α be an increasing function on a cell A, and let $f \in \mathcal{R}(A, \alpha)$. If F is an indefinite integral of f, then for α_A-almost all $x \in A$ the function F is symmetrically α-derivable at x and $F_\alpha^{sym}(x) = f(x)$.*

PROOF.   Assuming that $f \in \mathcal{R}(A, \alpha)$, we employ the same idea used in the proof of Theorem 5.3.5. Let E be the set of all $x \in A$ for which either F is not symmetrically α-derivable at x, or $F_\alpha^{sym}(x) \neq f(x)$. Given $x \in E$, we can find a $\gamma(x) > 0$ so that for each $\beta > 0$ there is a positive $r < \beta$ with

$$|f(x)\alpha_A(U[x, r]) - F_A(U[x, r])| \geq \gamma(x)\alpha_A(U[x, r]).$$

Fix an integer $n \geq 1$, let $E_n = \{x \in E : \gamma(x) \geq 1/n\}$, and choose an $\varepsilon > 0$. By Henstock's lemma, there is a positive function δ on A such that

$$\sum_{i=1}^{p} |f(x_i)\alpha(A_i) - F(A_i)| < \frac{\varepsilon}{n}$$

for each δ-fine partition $\{(A_1, x_1), \ldots, (A_p, x_p)\}$ in A. On E_n we can find a positive function $\rho < \delta$ so that if $U_x = U[x, \rho(x)]$, then

$$|f(x)\alpha(A \cap U_x) - F(A \cap U_x)| \geq \frac{1}{n}\alpha_A(U_x)$$

for each $x \in E_n$. Let $E_{n,1}, \ldots, E_{n,14}$ be subsets of E_n associated with ρ according to Lemma 5.2.4. For each finite set $T \subset E_{n,k}$, the collection

$$\{(A \cap U_x, x) : x \in T\}$$

is a δ-fine Perron partition in A. Hence

$$\sum_{x \in T} \alpha_A(U_x) \leq n \sum_{x \in T} |f(x)\alpha(A \cap U_x) - F(A \cap U_x)| < \varepsilon$$

and we conclude that

$$\sum_{k=1}^{14} \sum_{x \in E_{n,k}} \alpha_A(U_x) \leq 14\varepsilon .$$

It follows from Proposition 3.5.2 that E_n is α_A-negligible, and by Proposition 3.5.3, so is $E = \bigcup_{n=1}^{\infty} E_n$. ∎

Chapter 6
The Henstock–Kurzweil integral

The McShane integral was defined by means of arbitrary partitions. Using only Perron partitions in Definition 2.1.1, we obtain the *generalized Riemann integral of Henstock and Kurzweil* or simply the *Henstock–Kurzweil integral* (cf. the note preceding Proposition 2.1.3). We shall see that the effect of this inconspicuous change is dramatic.

6.1 The P-integral

We remind the reader that a Perron partition, abbreviated as P-partition, is a partition $\{(A_1, x_1), \ldots, (A_p, x_p)\}$ where $x_i \in A_i$ for $i = 1, \ldots, p$ (cf. Definition 1.2.1). Recall that the Stieltjes sum $\sigma(f, P; \alpha)$ has been defined in Section 1.3.

Definition 6.1.1 Let α be an increasing function on a cell A. A function f on A is called *P-integrable* over A with respect to α if there is a real number I having the following property: given $\varepsilon > 0$, we can find a positive function δ on A such that

$$|\sigma(f, P; \alpha) - I| < \varepsilon$$

for each δ-fine Perron partition P of A.

Proposition 1.2.4 implies that the number I from Definition 6.1.1 is determined uniquely by the P-integrable function f (cf. Remark 2.1.2). We call it the *P-integral* of f over A with respect to α, denoted by $\int_A f \, d\alpha$ or $\int_a^b f \, d\alpha$ if $A = [a, b]$. Using the same notation for the integral and P-integral is legitimate: from Definitions 2.1.1 and 6.1.1 it is clear that each integrable function is P-integrable and that the integral and P-integral are equal. For the same reason we shall use the terms "integral" and "P-integral" interchangeably, usually reserving the term "P-integral" for emphasis only. The family of all functions that are P-integrable over A with respect to α is denoted by $\mathcal{R}_*(A, \alpha)$. The obvious inclusion $\mathcal{R}(A, \alpha) \subset \mathcal{R}_*(A, \alpha)$ is generally proper (see Theorem 6.1.2 below and Example 2.4.12).

Since the Henstock–Kurzweil integral is an extension of the McShane integral, it inherits automatically some of the McShane integral's properties. For example, Theorems 2.2.8, 2.3.10, and 2.3.11 are trivially true for the Henstock–Kurzweil integral, and we leave it to the reader to identify the remaining trivial cases. Many other properties of the McShane integral can be established for the Henstock–Kurzweil integral by virtually identical proofs. This is true for all statements in Section 2.1 and for the statements in Section 2.3 preceding Corollary 2.3.7; naturally, in Proposition 2.1.8 and Lemma 2.3.1 only P-partitions are allowed.

Note. Repeating the proof of Theorem 2.3.4 is not the only way to obtain the monotone convergence theorem for the P-integral. In Proposition 6.3.8 below we show that this result is, in fact, a consequence of Theorem 2.3.4 and the α_A-measurability of P-integrable functions.

Theorem 6.1.2 *If F is a differentiable function on a cell $[a, b]$, then F' belongs to $\mathcal{R}_*([a, b], \lambda)$ and*

$$\int_a^b F' \, d\lambda = F(b) - F(a).$$

PROOF. In view of Lemma 2.4.7, this principal generalization of the fundamental theorem of calculus (Theorem 2.4.8) follows immediately from Definition 6.1.1. ∎

Remark 6.1.3 Theorem 6.1.2 sets the difference between the McShane and Henstock–Kurzweil integrals. In conjunction with Example 2.4.12, it shows that the absolute value of a P-integrable function need not be P-integrable, a sharp contrast to Proposition 2.2.3. We refer to this fact by saying that the P-integral is *conditionally convergent* (cf. Proposition 6.1.9).

Theorem 2.4.1 is true for the P-integral but requires a more elaborate proof.

Theorem 6.1.4 *Let $A = [a, b]$ be a cell, and let $g \in \mathcal{R}_*(A, \alpha)$ be a nonnegative function. Setting $\beta(x) = \int_a^x g \, d\alpha$ for each $x \in [a, b]$ defines an increasing* *function β on A. A function f on A belongs to $\mathcal{R}_*(A, \beta)$ if and only if fg belongs to $\mathcal{R}_*(A, \alpha)$, in which case*

$$\int_A f \, d\beta = \int_A fg \, d\alpha.$$

PROOF.   Choose an $\varepsilon > 0$, and use the Henstock lemma for the P-integral to find positive functions δ_n, $n = 1, 2, \ldots$, so that

$$\sum_{i=1}^p \left| g(x_i)\alpha(A_i) - \int_{A_i} g \, d\alpha \right| < \frac{\varepsilon}{n2^n}$$

for each δ_n-fine P-partition $\{(A_1, x_1), \ldots, (A_p, x_p)\}$ in A. If

$$E_n = \{x \in A : n - 1 \le |f(x)| < n\},$$

then A is the disjoint union of the sets $E_1, E_2, \ldots$. Thus we can define a positive function δ on A by setting $\delta(x) = \delta_n(x)$ whenever $x \in E_n$. If $P = \{(A_1, x_1), \ldots, (A_p, x_p)\}$ is a δ-fine P-partition of A, then the collection $\{(A_i, x_i) : x_i \in E_n\}$ is a δ_n-fine P-partition in A for $n = 1, 2, \ldots$. Hence

$$|\sigma(fg, P; \alpha) - \sigma(f, P; \beta)| \le \sum_{i=1}^p |f(x_i)| \cdot \left| g(x_i)\alpha(A_i) - \int_{A_i} g \, d\alpha \right|$$

$$\le \sum_{n=1}^\infty n \sum_{x_i \in E_n} \left| g(x_i)\alpha(A_i) - \int_{A_i} g \, d\alpha \right|$$

$$< \sum_{n=1}^\infty n \frac{\varepsilon}{n2^n} = \varepsilon$$

and the theorem follows from Definition 6.1.1. ∎

Since Theorem 2.4.2 holds for the P-integral (with the same proof), Theorems 6.1.2 and 6.1.4 imply that so do Corollaries 2.4.9 and 2.4.10. For the reader's convenience, we list in a separate proposition all statements from the first four sections of Chapter 2 that are true for the P-integral. When no confusion can arise, we refer to them simply by referring to the corresponding statements about the McShane integral.

Proposition 6.1.5 *Replacing partitions by P-partitions when appropriate, the following statements are true for the P-integral:*

- *Lemma 2.3.1;*

- *Propositions 2.1.3, 2.1.5, 2.1.8, 2.1.9, and 2.1.10;*
- *Theorems 2.2.8, 2.3.4, 2.3.10, 2.3.11, 2.4.1, and 2.4.2;*
- *Corollaries 2.1.4, 2.1.6, 2.3.2, 2.4.9, and 2.4.10.* ∎

The next theorem shows that no *improper* Henstock–Kurzweil integrals exist. An analogous statement for the McShane integral is false (see Example 6.1.7 below).

Theorem 6.1.6 *Let f be a function defined on a cell $[a, b]$, and let f belong to $\mathcal{R}_*([a, c], \alpha)$ for each $c \in (a, b)$. If a finite limit $\lim_{c \to b-} \int_a^c f \, d\alpha = I$ exists, then $f \in \mathcal{R}_*([a, b], \alpha)$ and*

$$\int_a^b f \, d\alpha = I + f(b)[\alpha(b) - \alpha(b-)].$$

In particular, $\int_a^b f \, d\alpha = I$ whenever α is left continuous at b.

PROOF. Choose an $\varepsilon > 0$, and find a $\gamma \in (a, b)$ so that

$$|f(b)| \cdot [\alpha(b-) - \alpha(t)] < \varepsilon \quad \text{and} \quad \left| \int_a^t f - I \right| < \varepsilon$$

whenever $\gamma < t < b$. Select a strictly increasing sequence $\{c_n\}_{n=0}^\infty$ in $[a, b)$ with $c_0 = a$ and $\lim c_n = b$. By Henstock's lemma, there is a positive function δ_n on $[c_{n-1}, c_n]$ such that

$$\sum_{i=1}^p \left| f(x_i)\alpha(A_i) - \int_{A_i} f \right| < \varepsilon 2^{-n}$$

for each δ_n-fine P-partition $\{(A_1, x_1), \ldots, (A_p, x_p)\}$ in $[c_{n-1}, c_n]$. With no loss of generality, we may assume that $\delta_1(c_0) \leq c_1 - c_0$ and that

$$\delta_{n+1}(c_n) = \delta_n(c_n) \leq \min\{c_n - c_{n-1}, c_{n+1} - c_n\},$$

$$\delta_n(x) \leq \min\{x - c_{n-1}, c_n - x\}$$

for every $x \in (c_{n-1}, c_n)$ and $n = 1, 2, \ldots$. On $[a, b]$ define a positive function δ by setting

$$\delta(x) = \begin{cases} \delta_n(x) & \text{if } x \in [c_{n-1}, c_n] \text{ and } n = 1, 2, \ldots, \\ b - \gamma & \text{if } x = b, \end{cases}$$

and choose a δ-fine P-partition $P = \{(A_1, x_1), \ldots, (A_p, x_p)\}$ of $[a, b]$. After a suitable reordering, we may assume that $A_i = [t_{i-1}, t_i]$ where $a = t_0 < \cdots < t_p = b$. Replacing $([t_{i-1}, t_i], x_i)$ by the pair $\{([t_{i-1}, x_i], x_i), ([x_i, t_i], x_i)\}$ whenever $t_{i-1} < x_i < t_i$, we obtain a δ-fine P-partition Q of $[a, b]$ for which $\sigma(f, Q) = \sigma(f, P)$. Thus, with no loss of generality, we may also assume that

$$\{x_1, \ldots, x_p\} \subset \{t_0, \ldots, t_p\}.$$

If $P_n = \{(A_i, x_i) : A_i \subset [c_{n-1}, c_n]\}$ and k is the first positive integer with $c_k \geq t_{p-1}$, then the choice of δ implies the following facts.

1. P_n is a δ_n-fine P-partition of $[c_{n-1}, c_n]$ for $n = 1, \ldots, k-1$.
2. P_k is a δ_k-fine P-partition of $[c_{k-1}, t_{p-1}]$; in particular, P_k is a δ_k-fine P-partition in $[c_{k-1}, c_k]$.
3. $P = (\bigcup_{n=1}^{k} P_n) \cup \{([t_{p-1}, b], b)\}$.

Using conditions 1–3 and observing that

$$\int_a^{t_{p-1}} f = \sum_{n=1}^{k-1} \int_{c_{n-1}}^{c_n} f + \int_{c_{k-1}}^{t_{p-1}} f,$$

we obtain

$$\left| \sigma(f, P) - I - f(b)[\alpha(b) - \alpha(b-)] \right|$$

$$\leq \left| f(b)[\alpha(b) - \alpha(t_{p-1})] - f(b)[\alpha(b) - \alpha(b-)] \right|$$

$$+ \sum_{n=1}^{k-1} \left| \sigma(f, P_n) - \int_{c_{n-1}}^{c_n} f \right| + \left| \sigma(f, P_k) - \int_{c_{k-1}}^{t_{p-1}} f \right| + \left| \int_a^{t_{p-1}} f - I \right|$$

$$< |f(b)| \cdot [\alpha(b-) - \alpha(t_{p-1})] + \sum_{n=1}^{k} \varepsilon 2^{-n} + \varepsilon < 3\varepsilon,$$

which proves the theorem. ∎

Example 6.1.7 Given a real number s, let $f(x) = x^{-s} \cos x^{-1}$ for each $x \in (0, 1]$, and set $f(0) = 0$. If $0 < a < 1$, then $f \in \mathcal{R}_*([a, 1], \lambda)$ by Theorem 2.2.8. Moreover, Corollary 2.4.10 implies that

$$\int_a^1 x^{-s} \cos x^{-1} \, dx = \int_1^{1/a} t^{s-2} \cos t \, dt$$

where dx and dt stand for $d\lambda(x)$ and $d\lambda(t)$, respectively. It follows from Theorem 6.1.6 and Example 2.6.7 that $f \in \mathcal{R}_*([0,1], \lambda)$ if and only if $s < 2$.

On the other hand, applying the result of Example 2.4.11 and using the technique introduced in Example 2.4.12, it is not difficult to show that $f \in \mathcal{R}([0,1], \lambda)$ if and only if $s < 1$.

The following exercise is important, as it shows that the multiplicative property of Theorem 2.3.12 is false for the P-integral.

Exercise 6.1.8 Show that the P-integral $\int_0^1 x^{-3/2} \cos x^{-1}\, dx$ exists, while the P-integral $\int_0^1 x^{-3/2} \cos^2 x^{-1}\, dx$ does not. *Hint.* Use Example 6.1.7 and the identity $\cos 2t = 1 - 2\cos^2 t$.

The difference between the McShane and Henstock–Kurzweil integrals is particularly transparent when we integrate with respect to a *gap function* (see Section 2.5). Many properties of the P-integral can be appreciated by comparing the next proposition with Proposition 2.5.2.

Proposition 6.1.9 *Let $A = [a, b]$ be a cell, and let γ be a gap function determined by the triple sequence $\{s_n, t_n, c_n\}$. If $a < s_1 < s_2 < \cdots < b$, then a function f on A belongs to $\mathcal{R}_*(A, \gamma)$ if and only if the series $\sum_{n=1}^{\infty} c_n f(s_n)$ converges, in which case*

$$\int_A f\, d\gamma = \sum_{n=1}^{\infty} c_n f(s_n).$$

PROOF. If $s = \lim s_n < b$, then γ is constant on the cell $[s, b]$. Thus, with no loss of generality, we may assume that $s = b$. Given $t \in [a, b)$, it follows from Proposition 2.5.2 that $f \in \mathcal{R}_*([a, t], \gamma)$ and

$$\int_a^t f\, d\gamma = \sum_{s_n < t} c_n f(s_n) + f(t)[\gamma(t) - \gamma(t-)].$$

Indeed, this is clear if $t \neq s_n$ for $n = 1, 2, \ldots$; otherwise, select a $t' < t$ so that no s_n lies in the segment $[t', t)$, and observe that

$$\int_t^{t'} f\, d\gamma = f(t)[\gamma(t) - \gamma(t-)].$$

As γ is left-continuous at b, it is easy to show that a finite limit

$$\lim_{t \to b-} \int_a^t f\, d\gamma$$

exists if and only if the series $\sum_{n=1}^{\infty} c_n f(s_n)$ converges, in which case

$$\lim_{t \to b-} \int_a^t f \, d\gamma = \sum_{n=1}^{\infty} c_n f(s_n) \, .$$

An application of Theorem 6.1.6 completes the proof. ∎

6.2 Integration by parts

Since the P-integral has no immediate multiplicative property (see Exercise 6.1.8), it is important that the results of Section 2.6 still hold, although their proofs have to be slightly modified.

Throughout this section, $A = [a, b]$ is a fixed cell, α is an increasing function on A, and f is a function in $\mathcal{R}_*(A, \alpha)$. We define a function F on A by setting $F(x) = \int_a^x f \, d\alpha$ for each $x \in [a, b]$. Observe that F is continuous at $x \in A$ whenever α is continuous at x or $f(x) = 0$ (Corollary 2.3.2 and Remark 2.3.3).

Theorem 6.2.1   *If F is continuous and G is an increasing function on A, then fG belongs to $\mathcal{R}_*(A, \alpha)$, F belongs to $\mathcal{R}(A, G)$, and*

$$\int_a^b fG \, d\alpha = F(b)G(b) - \int_a^b F \, dG \, .$$

PROOF.   Choose an $\varepsilon > 0$, and observe that $F \in \mathcal{R}(A, G)$ according to Theorem 2.2.8. Using the uniform continuity of F in A ([42, Theorem 4.19, p. 91]), find a $\Delta > 0$ so that $|F(x) - F(y)| < \varepsilon$ for all x and y in A with $|x - y| < 2\Delta$. By Henstock's lemma, there is a positive function δ on A such that

$$\sum_{i=1}^{p} \left| f(x_i)[\alpha(t_i) - \alpha(t_{i-1})] - [F(t_i) - F(t_{i-1})] \right| < \varepsilon$$

for each δ-fine P-partition $\{([t_0, t_1], x_1), \dots, ([t_{p-1}, t_p], x_p)\}$ of A. Making δ smaller, we may assume that $\delta \leq \Delta$, and that

$$\left| \sigma(F, P; G) - \int_a^b F \, dG \right| < \varepsilon$$

for each δ-fine partition P of A. Let $P = \{([t_0, t_1], x_1), \dots, ([t_{p-1}, t_p], x_p)\}$

be a δ-fine P-partition of A. Expanding $F(b)G(b)$ as in the proof of Theorem 2.6.1, we obtain

$$|\sigma(fG, P; \alpha) + \sigma(F, P; G) - F(b)G(b)|$$

$$= \left| \sum_{i=1}^{p} \Big(f(x_i)G(x_i)[\alpha(t_i) - \alpha(t_{i-1})] + F(x_i)[G(t_i) - G(t_{i-1})] \right.$$

$$\left. - G(t_{i-1})[F(t_i) - F(t_{i-1})] - F(t_i)[G(t_i) - G(t_{i-1})] \Big) \right|$$

$$= \left| \sum_{i=1}^{p} \Big(f(x_i)G(x_i)[\alpha(t_i) - \alpha(t_{i-1})] - G(x_i)[F(t_i) - F(t_{i-1})] \right.$$

$$+ [G(x_i) - G(t_{i-1})] \cdot [F(t_i) - F(t_{i-1})]$$

$$\left. + [F(x_i) - F(t_i)] \cdot [G(t_i) - G(t_{i-1})] \Big) \right|$$

$$\leq \sum_{i=1}^{p} |G(x_i)| \cdot \Big| f(x_i)[\alpha(t_i) - \alpha(t_{i-1})] - [F(t_i) - F(t_{i-1})] \Big|$$

$$+ \sum_{i=1}^{p} |G(x_i) - G(t_{i-1})| \cdot |F(t_i) - F(t_{i-1})|$$

$$+ \sum_{i=1}^{p} |F(x_i) - F(t_i)| \cdot [G(t_i) - G(t_{i-1})] < \varepsilon G(b)$$

$$+ \varepsilon \sum_{i=1}^{p} |G(x_i) - G(t_{i-1})| + \varepsilon \sum_{i=1}^{p} [G(t_i) - G(t_{i-1})] \leq \varepsilon[G(b) + 2G(A)].$$

Consequently

$$\left| \sigma(fG, P; \alpha) + \int_a^b F\, dG - F(b)G(b) \right| \leq \left| \sigma(F, P; G) - \int_a^b F\, dG \right|$$

$$+ |\sigma(fG, P; \alpha) + \sigma(F, P; G) - F(b)G(b)| < \varepsilon[1 + G(b) + 2G(A)],$$

and the theorem is proved. $\blacksquare$

Remark 6.2.2 The first and second mean value theorems of integral calculus (Theorems 2.6.5 and 2.6.6) remain true for the P-integral. The first holds trivially, and in view of Theorem 6.2.1, the proof of the second is identical to that of Theorem 2.6.6.

Now assume that $g \in \mathcal{R}_*(A, \alpha)$, and define a function G on A by setting $G(x) = \int_a^x g\, d\alpha$ for each $x \in [a, b]$.

Theorem 6.2.3 *If F and G are continuous in A, then $fG + Fg$ belongs to* $\mathcal{R}_*(A, \alpha)$ *and*

$$\int_a^b (fG + Fg)\, d\alpha = F(b)G(b)\,.$$

PROOF.   Let c be a positive constant such that $|F| \leq c$, $|G| \leq c$, and $\alpha(A) \leq c$. Given $\varepsilon > 0$, choose a positive function δ on A so that

$$|f(x)| \cdot |G(x) - G(t)| \leq \frac{\varepsilon}{4c} \qquad \text{and} \qquad |g(x)| \cdot |F(x) - F(t)| \leq \frac{\varepsilon}{4c}$$

for each x and t in A with $|t - x| < \delta(x)$, and that

$$\sum_{i=1}^p \left| f(x_i)[\alpha(t_i) - \alpha(t_{i-1})] - [F(t_i) - F(t_{i-1})] \right| < \frac{\varepsilon}{4c},$$

$$\sum_{i=1}^p \left| g(x_i)[\alpha(t_i) - \alpha(t_{i-1})] - [G(t_i) - G(t_{i-1})] \right| < \frac{\varepsilon}{4c}$$

for each δ-fine P-partition $P = \{([t_0, t_1], x_1), \ldots, ([t_{p-1}, t_p], x_p)\}$ of A. Expanding $F(b)G(b)$ as in the proof of Theorem 2.6.1, for such a partition P we obtain

$$|\sigma(fG + Fg, P; \alpha) - F(b)G(b)|$$

$$\leq \sum_{i=1}^p \left| f(x_i)G(x_i)[\alpha(t_i) - \alpha(t_{i-1})] - G(t_{i-1})[F(t_i) - F(t_{i-1})] \right|$$

$$+ \sum_{i=1}^p \left| F(x_i)g(x_i)[\alpha(t_i) - \alpha(t_{i-1})] - F(t_i)[G(t_i) - G(t_{i-1})] \right|$$

$$\leq \sum_{i=1}^p |G(t_{i-1})| \cdot \left| f(x_i)[\alpha(t_i) - \alpha(t_{i-1})] - [F(t_i) - F(t_{i-1})] \right|$$

$$+ \sum_{i=1}^p |f(x_i)| \cdot |G(x_i) - G(t_{i-1})| \cdot [\alpha(t_i) - \alpha(t_{i-1})]$$

$$+ \sum_{i=1}^p |F(t_i)| \cdot \left| g(x_i)[\alpha(t_i) - \alpha(t_{i-1})] - [G(t_i) - G(t_{i-1})] \right|$$

$$+ \sum_{i=1}^p |g(x_i)| \cdot |F(x_i) - F(t_i)| \cdot [\alpha(t_i) - \alpha(t_{i-1})]$$

$$< 2\left[c\frac{\varepsilon}{4c} + \frac{\varepsilon}{4c}\alpha(A) \right] \leq \varepsilon. \quad \blacksquare$$

Corollary 6.2.4 *If F and G are continuous in A and $Fg \in \mathcal{R}_*(A, \alpha)$, then $fG \in \mathcal{R}_*(A, \alpha)$ and*

$$\int_a^b fG \, d\alpha = F(b)G(b) - \int_a^b Fg \, d\alpha. \quad \blacksquare$$

Proposition 6.2.5 *Let u and v be differentiable functions on A. If $u'v$ belongs to $\mathcal{R}_*(A, \lambda)$, then so does uv' and*

$$\int_a^b u'v \, d\lambda = u(b)v(b) - u(a)v(a) - \int_a^b uv' \, d\lambda. \quad \blacksquare$$

The following example shows that Proposition 6.2.5, whose proof is similar to that of Proposition 2.6.3, is false if u and v are merely differentiable functions. In particular, the assumptions of Theorem 6.2.3 alone do not imply that either of the functions Fg and fG belongs to $\mathcal{R}_*(A, \alpha)$.

Example 6.2.6 For $x \in (0, 1]$ let

$$u(x) = x^2 \sin x^{-4} \quad \text{and} \quad v(x) = x^2 \cos x^{-4},$$

and set $u(0) = v(0) = 0$. Then u and v are differentiable functions on $[0, 1]$, and an easy calculation reveals that

$$u(x)v'(x) = x^3 \sin 2x^{-4} - 2x^{-1} \cos 2x^{-4} + 2x^{-1}$$

for each $x \in (0, 1]$. Using Examples 2.4.11 and 6.1.7 together with Corollary 2.4.10, it is an easy exercise to show that $uv' \notin \mathcal{R}_*([0, 1], \lambda)$.

6.3 Connections with measures

We prove first a proposition similar to Corollary 4.1.6.

Proposition 6.3.1 *Let f and g be functions on a cell A, and let $f = g$ α_A-almost everywhere. Then f belongs to $\mathcal{R}_*(A, \alpha)$ if and only if g does, in which case $\int_A f \, d\alpha = \int_A g \, d\alpha$.*

PROOF. Since $\mathcal{R}(A, \alpha) \subset \mathcal{R}_*(A, \alpha)$, it follows from Theorem 4.1.5 that the function $h = f - g$ belongs to $\mathcal{R}_*(A, \alpha)$ and $\int_A h = 0$. An application of Proposition 2.1.3 completes the proof. $\blacksquare$

Let A be a cell, and let α be an increasing function on A. In view of Proposition 6.3.1, we can extend the P-integral to the family $\overline{\mathcal{R}}_*(A, \alpha)$ defined in the obvious way modeled by Definition 4.1.9. If $f \in \overline{\mathcal{R}}_*(A, \alpha)$, we say that a function F on A is an *indefinite integral* (or *indefinite P-integral* when emphasizing is appropriate) of f whenever

$$F(B) = \int_B f \, d\alpha$$

for each cell $B \subset A$. The reader can readily verify that Theorems 5.3.5 and 5.4.9, together with their proofs, hold also for the P-integral.

Lemma 6.3.2   *Let α be an increasing function on $\mathbf{R}$. If a function F on $\mathbf{R}$ has only countably many discontinuities, then F is α-measurable.*

PROOF.   Let $c \in \mathbf{R}$ and $E = \{x \in \mathbf{R} : F(x) > c\}$. If $x \in E$ and F is continuous at x, then $x \in E^\circ$. Thus E differs from its interior by a countable set, and the lemma follows from Theorems 3.3.10 and 3.3.7. ∎

Theorem 6.3.3   *If A is a cell, then each $f \in \overline{\mathcal{R}}_*(A, \alpha)$ is α_A-measurable.*

PROOF.   Choose an indefinite integral F of f and set

$$G_n(x) = F_A(U[x, 1/n]) \qquad \text{and} \qquad \alpha_n(x) = \alpha_A(U[x, 1/n])$$

for each $x \in A$ and $n = 1, 2, \ldots$. Corollary 3.4.2 and Proposition 3.2.2 imply that the function α has only countably many discontinuities (cf. [42, Theorem 4.30, p. 96]), and by Corollary 2.3.2, so does F. Since the same is clearly true for the functions α_n and G_n, these functions are α_A-measurable by Lemma 6.3.2. According to Lemma 5.4.3, there is an α_A-negligible set N such that $\alpha_n(x) > 0$ for each $x \in A - N$ and $n = 1, 2, \ldots$. Thus the extended real-valued function

$$g = \limsup \frac{G_n}{\alpha_n}$$

is defined on $A - N$, and it is α_A-measurable by Propositions 3.6.8 and 3.6.6. It follows from Theorem 5.4.9 that $f(x) = F_\alpha^{sym}(x) = g(x)$ for α_A-almost all $x \in A$. An application of Proposition 3.6.9 and Exercise 3.6.4, 2, 3 completes the argument. ∎

The next theorem gives a complete characterization of $\mathcal{R}(A, \alpha)$ in terms of $\mathcal{R}_*(A, \alpha)$. A different characterization is given by Theorem 6.3.9 below.

Theorem 6.3.4 *Let f be a function defined on a cell A. Then $f \in \mathcal{R}(A, \alpha)$ if and only if both f and $|f|$ belong to $\mathcal{R}_*(A, \alpha)$.*

PROOF. Let f and $|f|$ belong to $\mathcal{R}_*(A, \alpha)$. In view of Theorem 6.3.3, the functions $f^{\pm}$ belong to $\mathcal{R}(A, \alpha)$ by the Lebesgue test (Theorem 4.4.7), and we conclude that so does $f = f^+ - f^-$. As $\mathcal{R}(A, \alpha) \subset \mathcal{R}_*(A, \alpha)$, the converse follows from Proposition 2.2.3. ∎

Corollary 6.3.5 *A nonnegative function on a cell A is P-integrable if and only if it is integrable.* ∎

Exercise 6.3.6 Let A be a cell and let f and g be in $\mathcal{R}_*(A, \alpha)$.

1. Show that if there is an $h \in \mathcal{R}_*(A, \alpha)$ such that $f \leq h$ and $g \leq h$, then $\max\{f, g\}$ belongs to $\mathcal{R}_*(A, \alpha)$.
2. Show by example that, in general, the maximum of two P-integrable functions need not be P-integrable.

Proposition 6.3.7 *Let A be a cell and let $f \in \mathcal{R}_*(A, \alpha)$. If $\int_B f \, d\alpha = 0$ for every cell $B \subset A$, then $f = 0$ α_A-almost everywhere.*

PROOF. Henstock's lemma implies that $|f| \in \mathcal{R}_*(A, \alpha)$ and $\int_A |f| = 0$. By Corollary 6.3.5, we have $|f| \in \mathcal{R}(A, \alpha)$, and the proposition is a consequence of Theorem 4.1.7. ∎

Proposition 6.3.8 *Let A be a cell and let $\{f_n\}$ be a sequence in $\mathcal{R}_*(A, \alpha)$ that converges to a function f on A. Then $f \in \mathcal{R}_*(A, \alpha)$ and*

$$\int_A f \, d\alpha = \lim \int_A f_n \, d\alpha$$

whenever either of the following conditions is fulfilled:

1. *$f_n \leq f_{n+1}$ for $n = 1, 2, \ldots$ and $\lim \int_A f \, d\alpha < +\infty$;*
2. *$g \leq f_n \leq h$ for some $g, h \in \mathcal{R}_*(A, \alpha)$ and $n = 1, 2, \ldots$.*

PROOF. In view of Corollary 6.3.5, it suffices to apply the monotone and dominated convergence theorems for the McShane integral to the sequences $\{f_n - f_1\}$ and $\{f_n - g\}$, respectively. ∎

Theorem 6.3.9 *Let A be a cell, and let F be an indefinite integral of a* *function $f \in \mathcal{R}_*(A, \alpha)$. Then $f \in \mathcal{R}(A, \alpha)$ if and only if $V(F, A) < +\infty$, in which case VF is an indefinite integral of $|f|$.*

PROOF. If $f \in \mathcal{R}(A, \alpha)$ then $|f| \in \mathcal{R}(A, \alpha)$, and we have

$$\sum_{D \in \mathcal{D}} |F(D)| \leq \sum_{D \in \mathcal{D}} \int_D |f| = \int_A |f|$$

for each division $\mathcal{D}$ of A. It follows that $V(F, A) \leq \int_A |f| < +\infty$.

Conversely, let $V(F, A) < +\infty$. Choose an $\varepsilon > 0$ and find a division $\mathcal{D}$ of A so that

$$V(F, A) < \sum_{D \in \mathcal{D}} |F(D)| + \frac{\varepsilon}{2}.$$

By Henstock's lemma, there is a positive function δ on A such that

$$\sum_{i=1}^{p} |f(x_i)\alpha(A_i) - F(A_i)| < \frac{\varepsilon}{2}$$

for each δ-fine P-partition $\{(A_1, x_1), \ldots, (A_p, x_p)\}$ in A. Making δ smaller, we may assume that $D \cap U(x, \delta(x)) = \emptyset$ for each $D \in \mathcal{D}$ and $x \in A - D$. Now select a δ-fine P-partition $P = \{(A_1, x_1), \ldots, (A_p, x_p)\}$ of A. Due to the choice of δ, replacing each (A_i, x_i) by the collection

$$\{(A_i \cap D, x_i) : D \in \mathcal{D} \text{ and } A_i^\circ \cap D^\circ \neq \emptyset\},$$

we obtain a new δ-fine P-partition Q of A such that $\{B : (B, x) \in Q\}$ is a segmentation of $\mathcal{D}$ and $\sigma(g, Q) = \sigma(g, P)$ for any function g on A. Thus, with no loss of generality, we may assume that $\{A_1, \ldots, A_p\}$ is already a segmentation of $\mathcal{D}$. Since

$$|\sigma(|f|, P) - V(F, A)|$$

$$\leq \sum_{i=1}^{p} \left| |f(x_i)|\alpha(A_i) - |F(A_i)| \right| + V(F, A) - \sum_{D \in \mathcal{D}} \sum_{A_i \subset D} |F(A_i)|$$

$$\leq \sum_{i=1}^{p} |f(x_i)\alpha(A_i) - F(A_i)| + V(F, A) - \sum_{D \in \mathcal{D}} |F(D)| < \varepsilon,$$

we conclude that $|f| \in \mathcal{R}_*(A, \alpha)$ and $\int_A |f| = V(F, A)$. Thus $f \in \mathcal{R}(A, \alpha)$ by Theorem 6.3.4, and repeating the previous argument for a cell $B \subset A$ completes the proof. ∎

Exercise 6.3.10 Show that the statements of this section remain correct when $\mathcal{R}(A, \alpha)$ and $\mathcal{R}_*(A, \alpha)$ are replaced by $\overline{\mathcal{R}}(A, \alpha)$ and $\overline{\mathcal{R}}_*(A, \alpha)$, respectively.

6.4 AC_* functions

We generalize the concept of an AC function (Definition 5.1.6) and use it for obtaining a partial descriptive definition of the P-integral.

Definition 6.4.1 A function F on a cell A is called AC_* if, given a negligible set $E \subset A$ and $\varepsilon > 0$, there is a positive function δ on E such that $\sum_{i=1}^{p} |F(A_i)| < \varepsilon$ for each P-partition $\{(A_1, x_1), \ldots, (A_p, x_p)\}$ in A anchored in E that is δ-fine.

Note. Using the asterisk as a subscript to distinguish a concept defined by means of P-partitions, as opposed to partitions, is consistent with our previous notation (cf. $\mathcal{R}_*(A, \alpha)$ versus $\mathcal{R}(A, \alpha)$). Nonetheless, we warn the reader that the AC_* functions of Definition 6.4.1 may *not coincide* with the AC_* functions used by other authors (e.g., [44, Chapter VII, Section 8]).

Proposition 6.4.2 *On a cell A each AC function is AC_*, and each AC_* function is continuous.*

PROOF. The first statement follows from Proposition 5.1.9, 3. If F is an AC_* function on A and $x \in A$, then applying Definition 6.4.1 to the negligible set $E = \{x\}$ yields the continuity of F at x. ∎

Exercise 6.4.3 Show that the function F of Example 2.4.12 is AC_* but not AC on the cell $[0, 1]$.

The following theorem gives a *partial descriptive definition* of the Henstock–Kurzweil integral (cf. Remark 5.3.6). While a full descriptive definition also exists, we do not discuss it in this book, since it is highly technical and restricted only to dimension one. The interested reader is referred to [44, Chapters VII and VIII].

Theorem 6.4.4 *Let F be a function on a cell A that is differentiable almost everywhere in A. Then F is AC_* if and only if $F' \in \overline{\mathcal{R}}_*(A, \lambda)$ and F is an indefinite integral of F'.*

PROOF. Let $F' \in \overline{\mathcal{R}}_*(A, \lambda)$ and let F be an indefinite integral of F'. Choose a negligible set $E \subset A$ and $\varepsilon > 0$. Enlarging E, we may assume that F is differentiable at each $x \in A - E$ and set

$$f(x) = \begin{cases} F'(x) & \text{if } x \in A - E, \\ 0 & \text{if } x \in E. \end{cases}$$

It follows from Proposition 6.3.1 that $f \in \mathcal{R}_*(A, \lambda)$ and F is an indefinite integral of f. By Henstock's lemma, there is a positive function δ on A such that

$$\sum_{i=1}^{p} |f(x)\lambda(A_i) - F(A_i)| < \varepsilon$$

for each δ-fine P-partition $P = \{(A_1, x_1), \ldots, (A_p, x_p)\}$ in A. If such a P is anchored in E, then $\sum_{i=1}^{p} |F(A_i)| < \varepsilon$ and we see that F is AC$_*$.

Conversely, let F be AC$_*$. Denote by E the negligible set of all $x \in A$ for which F is not differentiable at x, and define a function f on A as above. It suffices to show that $f \in \mathcal{R}_*(A, \lambda)$ and F is an indefinite integral of f. To this end, choose an $\varepsilon > 0$ and find a positive function β on E associated with E and $\varepsilon/2$ according to Definition 6.4.1. Given $x \in A - E$, there is a $\gamma(x) > 0$ such that

$$|f(x)\lambda(C) - F(C)| < \frac{\varepsilon}{2\lambda(A)}\lambda(C)$$

for each cell $C \subset A \cap U(x, \gamma(x))$ with $x \in C$. Define a positive function δ on A by setting

$$\delta(x) = \begin{cases} \beta(x) & \text{if } x \in E, \\ \gamma(x) & \text{if } x \in A - E, \end{cases}$$

and choose a δ-fine P-partition $P = \{(B_1, y_1), \ldots, (B_q, y_q)\}$ of a cell $B \subset A$. We obtain

$$|\sigma(f, P; \lambda) - F(B)| \leq \sum_{j=1}^{q} |f(y_j)\lambda(B_j) - F(B_j)|$$

$$\leq \sum_{y_j \in E} |F(B_j)| + \frac{\varepsilon}{2\lambda(A)} \sum_{y_j \notin E} \lambda(B_j) < \varepsilon,$$

which implies that $f \in \mathcal{R}_*(B, \lambda)$ and $\int_B f\, d\lambda = F(B)$. $\blacksquare$

Proposition 6.4.5 *Let F be a function of bounded variation defined on a cell A. Then F is AC if and only if it is AC$_*$.*

PROOF. As the converse follows from Proposition 6.4.2, assume that F is AC$_*$. By Corollary 5.3.9, the function F is differentiable almost everywhere in A, and it follows from Theorem 6.4.4 that $F' \in \mathcal{R}_*(A, \lambda)$ and F is an indefinite integral of F'. Now Theorem 6.3.9 implies that $F' \in \mathcal{R}(A, \lambda)$, so F is AC by Corollary 5.1.10. $\blacksquare$

Let E be a subset of $\mathbf{R}$. A nonnegative function δ on E is called a *gage* on E if the *null set*

$$Z_\delta = \{x \in E : \delta(x) = 0\}$$

of δ is countable. If δ is a gage on E, we say that a partition $P = \{(A_1, x_1), \ldots, (A_p, x_p)\}$ anchored in E is *δ-fine* whenever $A_i \subset U(x_i, \delta(x_i))$ for $i = 1, \ldots, p$. While formally this is not different from Definition 1.2.2, the δ-fine partition P is more restricted if δ is a gage with $Z_\delta \neq \emptyset$; specifically, P is anchored in the set $E - Z_\delta$.

The following characterization of AC_* functions will be used in Section 6.5 to derive a substantial generalization of the fundamental theorem of calculus.

Proposition 6.4.6 *A continuous function F on a cell A is AC_* whenever for each negligible set $E \subset A$ and each $\varepsilon > 0$, there is a gage δ on E such that $\sum_{i=1}^{p} |F(A_i)| < \varepsilon$ for every P-partition $\{(A_1, x_1), \ldots, (A_p, x_p)\}$ in A anchored in E that is δ-fine.*

PROOF. Given $\varepsilon > 0$ and a negligible set $E \subset A$, find a gage δ on E so that $\sum_{i=1}^{p} |F(A_i)| < \varepsilon/2$ for each P-partition $\{(A_1, x_1), \ldots, (A_p, x_p)\}$ in A anchored in E that is δ-fine. Let $\{z_1, z_2, \ldots\}$ be an enumeration of the null set Z_δ of δ. As F is continuous, there are $\eta_n > 0$ such that $|F(B)| < \varepsilon 2^{-n-1}$ for each cell $B \subset A \cap U(z_n, \eta_n)$ and $n = 1, 2, \ldots$. Now define a positive function Δ on E by setting

$$\Delta(x) = \begin{cases} \delta(x) & \text{if } x \in E - Z_\delta, \\ \eta_n & \text{if } x = z_n \text{ and } n = 1, 2, \ldots, \end{cases}$$

and choose a P-partition $\{(A_1, x_1), \ldots, (A_p, x_p)\}$ in A anchored in E that is Δ-fine. Since

$$\sum_{i=1}^{p} |F(A_i)| < \sum_{x_i \notin Z_\delta} |F(A_i)| + \sum_{n=1}^{\infty} \sum_{x_i = z_n} |F(A_i)| < \frac{\varepsilon}{2} + \sum_{n=1}^{\infty} \varepsilon 2^{-n-1} < \varepsilon,$$

we see that F is AC_*. ∎

Exercise 6.4.7 Show that a function F on a cell A is AC_* whenever either of the following conditions is satisfied.

1. Given a negligible set $E \subset A$ and $\varepsilon > 0$, there is a positive function δ on E such that $|\sum_{i=1}^{p} F(A_i)| < \varepsilon$ for each P-partition $\{(A_1, x_1), \ldots, (A_p, x_p)\}$ in A anchored in E that is δ-fine.

2. The function F is continuous, and given a negligible set $E \subset A$ and $\varepsilon > 0$, there is a gage δ on E such that $|\sum_{i=1}^{p} F(A_i)| < \varepsilon$ for each P-partition $\{(A_1, x_1), \ldots, (A_p, x_p)\}$ in A anchored in E that is δ-fine.

6.5 Densities

For $E \subset \mathbf{R}$ and $x \in \mathbf{R}$, we call the numbers

$$\overline{\Theta}(E, x) = \limsup_{r \to 0+} \frac{\lambda(E \cap U[x, r])}{\lambda(U[x, r])} \quad \text{and} \quad \underline{\Theta}(E, x) = \liminf_{r \to 0+} \frac{\lambda(E \cap U[x, r])}{\lambda(U[x, r])}$$

the *upper* and *lower density* of E at x, respectively. Clearly,

$$0 \leq \underline{\Theta}(E, x) \leq \overline{\Theta}(E, x) \leq 1,$$

and when $\underline{\Theta}(E, x) = \overline{\Theta}(E, x)$, this common value is called the *density* of E at x, denoted by $\Theta(E, x)$. Intuitively, the density $\Theta(E, x)$ measures the fraction of $\mathbf{R}$ occupied by E in the immediate vicinity of x. Points x for which $\Theta(E, x) = 1$ or $\Theta(E, x) = 0$ have special importance; we call them *density* or *dispersion* points of E, respectively.

Exercise 6.5.1 Show that the following statements hold for a set $A \subset \mathbf{R}$.

1. Each dispersion point of $\mathbf{R} - A$ is a density point of A.
2. If A is measurable, then a point $x \in \mathbf{R}$ is a density point of A if and only if it is a dispersion point of $\mathbf{R} - A$.
3. If a set $B \subset \mathbf{R}$ differs from A by a negligible set, then both A and B have the same density and dispersion points.

Proposition 6.5.2 *Let E be a subset of $\mathbf{R}$. For almost all $x \in \mathbf{R}$, the set E has a density at x and*

$$\Theta(E, x) = \lim \frac{\lambda(E \cap K_n)}{\lambda(K_n)}$$

for any sequence $\{K_n\}$ of cells containing x such that $\lim \lambda(K_n) = 0$. If E is measurable, then $\Theta(E, x) = \chi_E(x)$ for almost all $x \in \mathbf{R}$.

PROOF. For each integer n define an increasing function E_n on the cell $[n, n+1]$ by letting

$$E_n(x) = \lambda(E \cap [n, x])$$

for every $x \in [n, n+1]$, and use Proposition 3.3.2, 4 to observe that $E_n(K) = \lambda(E \cap K)$ for each interval $K \subset [n, n+1]$. Thus, if E_n is differentiable at a point $x \in (n, n+1)$, then by Proposition 5.3.3 the set E has a density at x and

$$\Theta(E, x) = E'_n(x) = \lim \frac{\lambda(E \cap K_n)}{\lambda(K_n)}$$

for any sequence $\{K_n\}$ of cells containing x such that $\lim \lambda(K_n) = 0$. Now Theorem 5.3.8 and Proposition 3.5.3, 3 imply the first part of the proposition.

If E is measurable, then $\chi_E \in \mathcal{R}([n, n+1], \lambda)$ and E_n is an indefinite integral of χ_E by Theorem 4.1.1. In view of this and the first part of the proof, the proposition is a direct consequence of Theorem 5.3.5 and Proposition 3.5.3, 3. ∎

Corollary 6.5.3 *If E is a subset of $\mathbf{R}$, then almost all points of E are density points of E.*

PROOF. By Lemma 3.3.14, there is a measurable set A such that $E \subset A$ and $\lambda(E) = \lambda(A)$. If B is a measurable set, the Carathéodory test for measurability implies that

$$\lambda(E) = \lambda(E \cap B) + \lambda(E - B) \leq \lambda(A \cap B) + \lambda(A - B) = \lambda(A).$$

It follows that $\lambda(E \cap B) = \lambda(A \cap B)$ for each measurable set B, and we see that the upper and lower densities of the sets E and A coincide. As almost all $x \in E$ are density points of A by Proposition 6.5.2, the corollary is proved. ∎

Lemma 6.5.4 *Let $x \in \mathbf{R}$ be a density point of a set $E \subset \mathbf{R}$, and let $\{r_n\}$ be a sequence of positive numbers with $\lim r_n = 0$. If $A_n \subset U[x, r_n]$ is a measurable set and $\lambda(A_n) \geq c\lambda(U[x, r_n])$ for $n = 1, 2, \ldots$ and a fixed $c > 0$, then*

$$\lim \frac{\lambda(E \cap A_n)}{\lambda(A_n)} = 1.$$

PROOF. Proceeding towards a contradiction, assume that

$$\liminf \frac{\lambda(E \cap A_n)}{\lambda(A_n)} < a < 1,$$

and observe that there is a number $b < 1$ such that $1 - b < c(1 - a)$. If

$U_n = U[x, r_n]$ for $n = 1, 2, \ldots$, then $\lim[\lambda(E \cap U_n)/\lambda(U_n)] = 1$. Thus we can find an integer $k \geq 1$ so that

$$\lambda(E \cap A_k) < a\lambda(A_k) \qquad \text{and} \qquad \lambda(E \cap U_k) > b\lambda(U_k).$$

Since the set A_k is measurable, Proposition 3.3.2, 4 yields a contradiction:

$$\begin{aligned}
\lambda(E \cap [U_k - A_k]) &= \lambda(E \cap U_k) - \lambda(E \cap A_k) > b\lambda(U_k) - a\lambda(A_k) \\
&= \lambda(U_k) - \lambda(A_k) - (1 - b)\lambda(U_k) + (1 - a)\lambda(A_k) \\
&\geq \lambda(U_k - A_k) + [c(1 - a) - (1 - b)]\lambda(U_k) \\
&> \lambda(U_k - A_k) \geq \lambda(E \cap [U_k - A_k]) . \quad \blacksquare
\end{aligned}$$

Exercise 6.5.5　Let $x \in \mathbf{R}$ be a density point of a set $A \subset \mathbf{R}$ and a nondispersion point of a measurable set B. Show that x is a nondispersion point of $A \cap B$.

6.6 Almost differentiable functions

Definition 6.6.1　A function F defined on a set $S \subset \mathbf{R}$ is called *almost differentiable* at $z \in S^\circ$ if

$$\limsup_{x \to z} \frac{|F(x) - F(z)|}{|x - z|} < +\infty .$$

Exercise 6.6.2　Let F be a function defined on a cell A. Show that F is almost differentiable at $z \in A^\circ$ if and only if the extended real numbers $D_\lambda F(z)$ and $D^\lambda F(z)$ are both finite.

Proposition 6.6.3　*If F is a continuous function on a cell A that is almost differentiable at all but countably many points of A°, then F is AC_*.*

PROOF.　Choose a negligible set $E \subset A$ and an $\varepsilon > 0$. Let $C \subset A$ be a countable set such that $\partial A \subset C$ and F is almost differentiable at each $x \in A - C$. For $n = 1, 2, \ldots$, set

$$E_n = \left\{ z \in E - C : n - 1 \leq \limsup_{x \to z} \frac{|F(x) - F(z)|}{|x - z|} < n \right\}$$

and find open sets $U_n \subset A$ so that $E_n \subset U_n$ and $\lambda(U_n) < \varepsilon 2^{-n}/n$. If $y < z < x$ are points in A, then

$$\left| \frac{F(x) - F(y)}{x - y} \right| \leq \left| \frac{F(x) - F(z)}{x - z} \right| \cdot \frac{x - z}{x - y} + \left| \frac{F(z) - F(y)}{z - y} \right| \cdot \frac{z - y}{x - y}.$$

Thus, given $x \in E_n$, we can find a $\delta_n(x) > 0$ so that $U(x, \delta_n(x)) \subset U_n$ and $|F(B)| < n\lambda(B)$ for each cell $B \subset U(x, \delta_n(x))$ with $x \in B$. Since $E - C$ is the disjoint union of the sets $E_1, E_2, \ldots$, the formula

$$\delta(x) = \begin{cases} \delta_n(x) & \text{if } x \in E_n \text{ and } n = 1, 2, \ldots, \\ 0 & \text{if } x \in E \cap C, \end{cases}$$

defines a gage on E. Given a P-partition $\{(A_1, x_1), \ldots, (A_p, x_p)\}$ in A anchored in E that is δ-fine, we obtain

$$\sum_{i=1}^{p} |F(A_i)| = \sum_{n=1}^{\infty} \sum_{x_i \in E_n} |F(A_i)| < \sum_{n=1}^{\infty} \sum_{x_i \in E_n} n\lambda(A_i)$$

$$\leq \sum_{n=1}^{\infty} n\lambda(U_n) < \sum_{n=1}^{\infty} \varepsilon 2^{-n} = \varepsilon,$$

and F is AC_* by Proposition 6.4.6. ∎

Our next goal is to justify the term "almost differentiable" by showing that almost differentiable functions are differentiable almost everywhere (Theorem 6.6.8 below). This requires a delicate argument in which the densities introduced in Section 6.5 will be used.

A function F on a set $E \subset \mathbf{R}$ is called *Lipschitz* if there is a real number c such that

$$|F(y) - F(x)| \leq c|y - x|$$

for each $x, y \in E$ (cf. Exercise 2.2.7). It is easy to see that the least number $c \geq 0$ for which the above inequality holds exists; we call it the *Lipschitz constant* of F, denoted by $\mathrm{Lip}(F)$.

Remark 6.6.4 A direct verification shows that a Lipschitz function F on a cell A is AC and hence differentiable almost everywhere in A by Corollary 5.3.9. Moreover, $|F'(x)| \leq \mathrm{Lip}(F)$ for each $x \in A$ at which F is differentiable.

Proposition 6.6.5 *Each Lipschitz function F on a set $E \subset \mathbf{R}$ can be extended to a Lipschitz function G on $\mathbf{R}$ so that $\mathrm{Lip}(G) = \mathrm{Lip}(F)$.*

PROOF. Avoiding a triviality, assume that $E \neq \emptyset$ and let $c = \mathrm{Lip}(F)$. We employ an idea from the proof of Proposition 4.2.3 (cf. Exercise 4.2.4), and define a function G on $\mathbf{R}$ by letting

$$G(x) = \inf_{z \in E}[F(z) + c|x - z|]$$

for each $x \in \mathbf{R}$. Select $x \in \mathbf{R}$ and $t \in E$. Since

$$F(z) + c|x - z| \geq F(t) - c|t - z| + c|x - z| \geq F(t) - c|x - t|$$

for every $z \in E$, we have

$$-\infty < F(t) - c|x - t| \leq G(x) \leq F(t) + c|x - t| < +\infty.$$

Letting $x = t$ yields $G(t) = F(t)$. Next choose $x, y \in \mathbf{R}$ and $\varepsilon > 0$. As there is a $z_x \in E$ with

$$F(z_x) + c|x - z_x| < G(x) + \varepsilon,$$

we obtain

$$G(y) - G(x) \leq F(z_x) + c|y - z_x| - [F(z_x) + c|x - z_x| - \varepsilon] \leq c|y - x| + \varepsilon.$$

Now $G(y) - G(x) \leq c|y - x|$ by the arbitrariness of ε, and it follows from symmetry that G is a Lipschitz function with $\mathrm{Lip}(G) \leq \mathrm{Lip}(F)$. As the reverse inequality is obvious, the proposition is proved. ∎

Exercise 6.6.6 Perform the following tasks.

1. Make up a definition of a Lipschitz function on an arbitrary metric space.
2. Show that Proposition 6.6.5 and its proof remain correct for a Lipschitz function defined on a subset of a metric space.
3. Find an alternative proof of Proposition 6.6.5 in $\mathbf{R}$ using the facts that f has a unique extension to E^- and that $\mathbf{R} - E^-$ is a countable union of disjoint open segments (see [42, Chapter 2, Exercise 29, p. 45 and Chapter 4, Exercise 13, p. 99]).

Lemma 6.6.7 *Let a function F defined on a set $S \subset \mathbf{R}$ and a set $C \subset S$ satisfy the following conditions:*

1. *there are positive numbers c and η such that $U(y, \eta) \subset S$ and*

$$|F(x) - F(y)| \leq c|x - y|$$

whenever $y \in C$ and $x \in U(y, \eta)$;

2. *there is a Lipschitz function G on $\mathbf{R}$ such that $G(y) = F(y)$ for each $y \in C$.*

If $z \in C$ is a density point of C and G is differentiable at z, then so is F and $F'(z) = G'(z)$.

PROOF. Let $z \in C$ be a density point of C at which G is differentiable, and choose a positive $\varepsilon \leq 1$. Making c bigger and η smaller, we may assume that $\mathrm{Lip}(G) \leq c$, $\eta < \varepsilon$, and

$$|G(x) - G(z) - G'(z)(x - z)| < \varepsilon |x - z|$$

for every $x \in U(z, \eta)$. Now for each $x \in \mathbf{R} - \{z\}$, let $r_x = |x - z|$ and use Lemma 6.5.4 to deduce that

$$\lim_{x \to z} \frac{\lambda(C \cap U[x, \varepsilon r_x])}{\lambda(U[x, \varepsilon r_x])} = 1.$$

Thus, making η still smaller, we may assume that $\lambda[C \cap U(x, \varepsilon r_x)] > 0$ for each $x \in U(z, \eta)$; in particular, $C \cap U(x, \varepsilon r_x) \neq \emptyset$ whenever $x \in U(z, \eta)$. Now fix an $x \in U(z, \eta)$, and select a $y \in C \cap U[x, \varepsilon r_x]$. Since $F(y) = G(y)$ and $|y - x| \leq \varepsilon |x - z| < \eta$, we have

$$
\begin{aligned}
|F(x) &- F(z) - G'(z)(x - z)| \\
&\leq |G(x) - G(z) - G'(z)(x - z)| + |F(x) - F(y)| + |G(y) - G(x)| \\
&< \varepsilon |x - z| + 2c|x - y| \leq \varepsilon(1 + 2c)|x - z|,
\end{aligned}
$$

and the lemma is established. $\blacksquare$

We are now ready to prove the main result of this section, known as the *Stepanoff theorem*.

Theorem 6.6.8 *Let F be a function defined on a set $S \subset \mathbf{R}$ and let E be a subset of $S°$. If F is almost differentiable at all $x \in E$, then it is differentiable at almost all $x \in E$.*

PROOF. For $n = 1, 2, \ldots$, let E_n be the set of all $z \in E$ such that $U(z, 1/n)$ is a subset of S and

$$|F(x) - F(z)| \leq n|x - z|$$

whenever $x \in U(z, 1/n)$. By the assumption, $E = \bigcup_{n=1}^{\infty} E_n$. For an integer k and a positive integer n, let $C_{n,k} = E_n \cap [k/n, (k + 1)/n)$ and observe that for each $y \in C_{n,k}$ and $x \in U(y, 1/n)$, we have $x \in S$ and

$$|F(x) - F(y)| \leq n|x - y|.$$

In particular, the function F restricted to $C_{n,k}$ is Lipschitz, and it can be extended to a Lipschitz function G on $\mathbf{R}$ by Proposition 6.6.5. Now it follows from Lemma 6.6.7 that F is differentiable at each $z \in C_{n,k}$ that is a density point of $C_{n,k}$ at which G is differentiable. In view of Remark 6.6.4 and Corollary 6.5.3, these conditions are met almost everywhere in $C_{n,k}$. As

$$E = \bigcup_{n=1}^{\infty} \bigcup_{k=-\infty}^{\infty} C_{n,k},$$

the theorem is a consequence of Proposition 3.5.3, 3. ∎

Theorem 6.6.9 *Let C be a countable subset of a cell $[a, b]$, and let F be a continuous function on $[a, b]$ that is almost differentiable at each point $x \in (a, b) - C$. Then $F' \in \overline{\mathcal{R}}_*([a, b], \lambda)$ and*

$$\int_a^b F' \, d\lambda = F(b) - F(a).$$

PROOF. The theorem follows immediately from Proposition 6.6.3 and Theorems 6.6.8 and 6.4.4. ∎

Example 5.3.11 shows that in Theorem 6.6.9, the countable set C cannot be replaced by a negligible set; that the continuity of F is also essential follows from Example 6.6.10 below.

Example 6.6.10 If $F = \chi_{[0,1]}$ is the characteristic function of the cell $[0, 1]$, then F is differentiable at every $x \in (-1, 0) \cup (0, 1)$ with $F'(x) = 0$. Yet,

$$\int_{-1}^1 F' \, d\lambda = 0 \neq 1 = F(1) - F(-1).$$

6.7 Gages and calibers

Since gages proved useful, it is natural to investigate whether they can be employed directly in defining the Henstock–Kurzweil integral. A simpleminded approach of replacing positive functions δ by gages does not work. Indeed, if δ is a gage on a cell A, no δ-fine partition of A may exist; the gage $\delta : x \mapsto |x|$ on $[-1, 1]$ provides an easy example. However, the following observation paves the way: for any gage δ on a cell A, there is a δ-fine Perron partition $\{(A_1, x_1), \ldots, (A_p, x_p)\}$ in A such that the set $A - \bigcup_{i=1}^p A_i$ is "small."

it suffices to show that $f \in \mathcal{R}_*(A, \lambda)$ and $\int_A f = F(A)$. To this end, choose an $\varepsilon > 0$, and use Lemma 4.3.8 to find a nonnegative increasing function β on A so that $\beta(A) < \varepsilon$ and $\beta'(x) = +\infty$ for each $x \in D$.

Let $x \in D$. Since $(\beta \pm F)'(x) = +\infty$, there is a $\vartheta_x > 0$ such that $\beta(B) \pm F(B) \geq 0$ for every cell $B \subset A \cap U(x, \vartheta_x)$ with $x \in B$. In particular,

$$|f(x)\lambda(B) - F(B)| = |F(B)| \leq \beta(B)$$

whenever $B \subset A \cap U(x, \vartheta_x)$ is a cell containing x.

If $x \in A° - (C \cup D)$, then there is a $\theta_x > 0$ such that

$$|f(x)\lambda(B) - F(B)| < \varepsilon \lambda(B)$$

for each cell $B \subset A \cap U(x, \theta_x)$ with $x \in B$. Finally, by Proposition 6.7.3, there is a caliber η such that $|F(B)| < \varepsilon$ for each (ε, η)-small figure $B \subset A$.

Now define a gage δ on A by setting

$$\delta(x) = \begin{cases} \theta_x & \text{if } x \in A° - (C \cup D), \\ \vartheta_x & \text{if } x \in D, \\ 0 & \text{if } x \in C \cup \partial A, \end{cases}$$

and choose a δ-fine P-partition $P = \{(A_1, x_1), \ldots, (A_p, x_p)\}$ of A mod (ε, η). Since the figure $B = A \ominus \bigcup P$ is (ε, η)-small, we obtain

$$|\sigma(f, P) - F(A)| \leq \sum_{i=1}^{p} |f(x_i)\lambda(A_i) - F(A_i)| + |F(B)|$$

$$< \sum_{x_i \in D} \beta(A_i) + \sum_{x_i \in A° - (C \cup D)} \varepsilon \lambda(A_i) + \varepsilon$$

$$\leq \beta(A) + \varepsilon \lambda(A) + \varepsilon < \varepsilon[2 + \lambda(A)],$$

and an application of Theorem 6.7.5 completes the proof. ∎

Remark 6.7.7 Theorem 6.7.5 offers an alternative definition of the Henstock–Kurzweil integral. Together with Proposition 6.7.6 it suggests the multidimensional definition of the *gage integral* adopted in Chapter 11.

Part II

Multidimensional integration

Chapter 7

Preliminaries

Throughout the remainder of this book, m is a fixed positive integer. Our ambient space is the m-fold Cartesian product of $\mathbf{R}$, denoted by $\mathbf{R}^m$. The *origin* $(0, \ldots, 0)$ of $\mathbf{R}^m$ is denoted by $\mathbf{0}$. For points $x = (\xi_1, \ldots, \xi_m)$ and $y = (\eta_1, \ldots, \eta_m)$ in $\mathbf{R}^m$, we let

$$x \cdot y = \sum_{i=1}^{m} \xi_i \eta_i, \qquad \|x\| = \sqrt{x \cdot x}, \qquad \text{and} \qquad |x| = \max\{|\xi_1|, \ldots, |\xi_m|\}.$$

Let A and B be subsets of $\mathbf{R}^m$. The extended real numbers

$$d(A) = \sup\{|x - y| : x, y \in A\}$$

and

$$\operatorname{dist}(A, B) = \inf\{|x - y| : x \in A \text{ and } y \in B\}$$

are called, respectively, the *diameter* of A and the *distance* between A and B. For $x \in \mathbf{R}^m$, we write $\operatorname{dist}(x, A)$ instead of $\operatorname{dist}(\{x\}, A)$.

The function $(x, y) \mapsto |x - y|$ is a metric in $\mathbf{R}^m$ that is more convenient for our purposes than the usual *Euclidean metric* induced by the norm $\|\cdot\|$. Since $|x| \leq \|x\| \leq \sqrt{m}\,|x|$ for each $x \in \mathbf{R}^m$, both metrics produce the same topology in $\mathbf{R}^m$, and they coincide if $m = 1$. For $x \in \mathbf{R}^m$ and $r > 0$, we let

$$U(x, r) = \{y \in \mathbf{R}^m : |x - y| < r\} \quad \text{and} \quad U[x, r] = \{y \in \mathbf{R}^m : |x - y| \leq r\}.$$

The closure, interior, and boundary of a set $E \subset \mathbf{R}^m$ are denoted by E^-, E°, and ∂E, respectively.

7.1 Intervals

An *interval* in $\mathbf{R}^m$ is the Cartesian product

$$A = \prod_{i=1}^{m} A_i$$

where $A_i = [\xi_i, \eta_i]$ are intervals in $\mathbf{R}$ for $i = 1, \ldots, m$. If each A_i is a cell in $\mathbf{R}$, we say that A is a *cell* in $\mathbf{R}^m$; otherwise, we say that A is a *degenerate interval* in $\mathbf{R}^m$. Since $A^\circ = \prod_{i=1}^{m} A_i^\circ$, it is clear that A is a cell in $\mathbf{R}^m$ if and only if $A^\circ \neq \emptyset$. When $\eta_i - \xi_i = h > 0$ for all i, the cell A is called a *cube* of diameter h; in particular, $U[x,r]$ is a cube of diameter $2r$. *Dyadic cubes* are the cells

$$\prod_{i=1}^{m} [k_i 2^{-n}, (k_i + 1)2^{-n}]$$

where $k_1, \ldots, k_m$ and n are integers with $n \geq 0$.

Throughout Part II of this book, we agree that intervals and cells are always intervals and cells in $\mathbf{R}^m$, respectively. When a need arises to consider intervals and cells in $\mathbf{R}^k$ for a positive integer $k < m$, we refer to them, respectively, as *k-dimensional intervals* and *k-dimensional cells*, or simply as *k-intervals* and *k-cells*. Thus the intervals and cells of Part I are now called the one-dimensional intervals and one-dimensional cells, or one-intervals and one-cells, respectively.

A collection of intervals is called *nonoverlapping* if their interiors are disjoint. Clearly, a degenerate interval overlaps no interval (including itself). It is easy to see that two dyadic cubes are either nonoverlapping or one is a subset of the other. Consequently, any family $\mathcal{K}$ of dyadic cubes contains a nonoverlapping subfamily $\mathcal{C}$ such that $\bigcup \mathcal{K} = \bigcup \mathcal{C}$. This fact, which makes dyadic cubes extremely useful, is referred to as the *fundamental property of dyadic cubes*.

Although the essential properties of intervals in $\mathbf{R}$ and $\mathbf{R}^m$ are the same, the latter are far from obvious and require careful proofs. The straightforward verification of the next two lemmas is left to the reader.

Lemma 7.1.1 *Let A, X and B, Y be two pairs of abstract sets. Then*

$$(A \times X) \cup (B \times X) = (A \cup B) \times X,$$

$$(A \times X) \cap (B \times Y) = (A \cap B) \times (X \cap Y),$$

$$A \times X - B \times Y = [(A - B) \times X] \cup [(A \cap B) \times (X - Y)]. \quad \blacksquare$$

Lemma 7.1.2 *Let r and s be positive integers, and let $A \subset \mathbf{R}^r$ and $B \subset \mathbf{R}^s$. Then $(A \times B)^- = A^- \times B^-$, $(A \times B)^\circ = A^\circ \times B^\circ$, and*

$$\partial(A \times B) = (\partial A \times B^-) \cup (A^- \times \partial B). \quad \blacksquare$$

Proposition 7.1.3 *If K and L are cells, then the following statements hold.*

1. *The intersection $K \cap L$ is an interval, possibly degenerate.*
2. *There is a finite, possibly empty, collection $\mathcal{C}$ of nonoverlapping cells such that $(K - L)^- = \bigcup \mathcal{C}$.*

PROOF. The proposition is true for $m = 1$. Proceeding inductively, let $m = r+s$ where r and s are positive integers, and assume that the proposition holds in $\mathbf{R}^r$ and $\mathbf{R}^s$. Since $K = A \times X$ and $L = B \times Y$ where A, B are r-cells and X, Y are s-cells, the first statement follows from Lemma 7.1.1 and the induction hypothesis. By Lemmas 7.1.1 and 7.1.2, we have

$$(K - L)^- = [(A - B)^- \times X] \cup [(A \cap B) \times (X - Y)^-],$$

and the second statement follows from the induction hypothesis. $\blacksquare$

Exercise 7.1.4 If $E, C \subset \mathbf{R}^m$, prove that $(E - C)^- = (E^- - C)^-$ whenever C is a closed set. Show by example that the closeness of C is essential.

Corollary 7.1.5 *If K and $L_1, \ldots, L_p$ are cells, then there is a finite, possibly empty, family $\mathcal{C}$ of nonoverlapping cells such that $(K - \bigcup_{i=1}^p L_i)^- = \bigcup \mathcal{C}$.*

PROOF. By Proposition 7.1.3, the corollary is true for $p = 1$. Proceeding by induction, suppose that $(K - \bigcup_{i=1}^{p-1} L_i)^- = \bigcup_{j=1}^q C_j$ where $C_1, \ldots, C_q$ are nonoverlapping cells. Since L_p is a closed subset of $\mathbf{R}^m$, Exercise 7.1.4 yields

$$\left(K - \bigcup_{i=1}^p L_i\right)^- = \left[\left(K - \bigcup_{i=1}^{p-1} L_i\right)^- - L_p\right]^- = \bigcup_{j=1}^q (C_j - L_p)^-.$$

Another application of Proposition 7.1.3 completes the proof. $\blacksquare$

The concepts of division and segmentation introduced in Sections 5.1 and 3.2, respectively, carry over verbatim to cells in $\mathbf{R}^m$. We recall the definitions.

- A *division* of a cell A is a finite collection $\mathcal{D}$ of nonoverlapping cells such that $\bigcup \mathcal{D} = A$.

- A *segmentation* of a family $\mathcal{C}$ of cells is a finite collection $\mathcal{D}$ of nonoverlapping cells such that each $D \in \mathcal{D}$ is contained in a $C \in \mathcal{C}$ and for every $C \in \mathcal{C}$, the family $\{D \in \mathcal{D} : D \subset C\}$ is a division of C.

In view of Proposition 7.1.3 and Corollary 7.1.5, the proof of the next lemma is identical to that of Lemma 3.2.9.

Lemma 7.1.6 *Each finite family of cells has a segmentation.* ∎

Lemma 7.1.7 *Let $m = r + s$ where r and s are positive integers, and let A, B, C and X, Y, Z be r-cells and s-cells, respectively. The cells $A \times X$ and $B \times Y$ do not overlap and*

$$(A \times X) \cup (B \times Y) = C \times Z$$

if and only if either of the following symmetric conditions holds:

1. $X = Y = Z$, *A and B do not overlap, and $C = A \cup B$;*
2. $A = B = C$, *X and Y do not overlap, and $Z = X \cup Y$.*

PROOF. By Lemma 7.1.1, either of the symmetric conditions implies the statement of the lemma. Conversely, we have $A \cup B = C$ and $X \cup Y = Z$ and we may assume that A and B do not overlap. If there is an $\eta \in X - Y$, choose a $\xi \in B°$, and observe that the point $x = (\xi, \eta)$ lies in $C \times Z$ but not in $(A \times X) \cup (B \times Y)$, a contradiction. Consequently $X \subset Y$, and by symmetry, $X = Y = Z$. ∎

The reader should note that Lemma 7.1.7 is false if cells are replaced by intervals.

7.2 Volumes

An *additive function* in a set $E \subset \mathbf{R}^m$ is a function F defined on the family of all subintervals of E such that

$$F(B \cup C) = F(B) + F(C)$$

for each pair of nonoverlapping intervals $B, C \subset E$ for which $B \cup C$ is an interval.

Exercise 7.2.1 If F is an additive function in a set $E \subset \mathbf{R}^m$, show that $F(D) = 0$ for each degenerate interval $D \subset E$.

Proposition 7.2.2 *Let F be an additive function in a cell K. Then*

$$F(K) = \sum_{C \in \mathcal{C}} F(C)$$

for each division $\mathcal{C}$ of K.

PROOF. The proposition is true for $m = 1$. Proceeding by induction, let $m = r+s$ where r and s are positive integers, and assume that the proposition holds in $\mathbf{R}^r$ and $\mathbf{R}^s$. There are r-cells A, A_C and s-cells B, B_C such that $K = A \times B$ and $C = A_C \times B_C$ for each $C \in \mathcal{C}$. By Lemma 7.1.6, there are segmentations $\{A_1, \ldots, A_p\}$ and $\{B_1, \ldots, B_q\}$ of the families $\{A_C : C \in \mathcal{C}\}$ and $\{B_C : C \in \mathcal{C}\}$, respectively. Letting $C_{i,j} = A_i \times B_j$, it is easy to verify that the collection

$$\{C_{i,j} : i = 1, \ldots, p; \ j = 1, \ldots, q\}$$

is a segmentation of $\mathcal{C}$. It follows from Lemma 7.1.7 that for each r-interval $X \subset A$ and each s-interval $Y \subset B$ the functions

$$U \mapsto F(U \times Y) \qquad \text{and} \qquad V \mapsto F(X \times V)$$

defined on r-subintervals of A and s-subintervals of B are additive functions in A and B, respectively. Using this and the induction hypothesis, we obtain

$$F(K) = \sum_{i=1}^{p} F(A_i \times B) = \sum_{i=1}^{p} \sum_{j=1}^{q} F(C_{i,j}) = \sum_{C \in \mathcal{C}} \sum_{C_{i,j} \subset C} F(C_{i,j})$$

$$= \sum_{C \in \mathcal{C}} \sum_{A_i \subset A_C} F(A_i \times B_C) = \sum_{C \in \mathcal{C}} F(C) . \quad \blacksquare$$

Let $m = r + s$ where r and s are positive integers, and let F and G be additive functions in sets $R \subset \mathbf{R}^r$ and $S \subset \mathbf{R}^s$, respectively. We define a function $F \otimes G$ on the family of all subintervals of $R \times S$ by letting

$$(F \otimes G)(A \times X) = F(A)G(X)$$

for every r-interval $A \subset R$ and s-interval $X \subset S$.

Proposition 7.2.3 *The function $F \otimes G$ is additive.*

PROOF. Let $H = F \otimes G$. Choose r-cells A, B, C in $\mathcal{R}$ and s-cells X, Y, Z in $\mathcal{S}$ so that the cells $A \times X$ and $B \times Y$ do not overlap and

$$(A \times X) \cup (B \times Y) = C \times Z.$$

In view of Lemma 7.1.7, we may assume that $X = Y = Z$, A and B do not overlap, and $C = A \cup B$. Consequently

$$H(C \times Z) = [F(A) + F(B)]G(Z) = F(A)G(X) + F(B)G(Y)$$
$$= H(A \times X) + H(B \times Y),$$

and the additivity of H follows from Exercise 7.2.1. $\blacksquare$

Corollary 7.2.4 *For $i = 1, \ldots, m$, let F_i be a function on a set $S_i \subset \mathbf{R}$, and let $S = \prod_{i=1}^{m} S_i$. The function $H = \bigotimes_{i=1}^{m} F_i$ defined by*

$$H(A) = \prod_{i=1}^{m} F_i(A_i)$$

for each subinterval $A = \prod_{i=1}^{m} A_i$ of S is an additive function in S.

PROOF. For $i = 1, \ldots, m$, it follows from the discussion in Section 1.1 that the associated interval function F_i is an additive function in S_i. Since

$$\bigotimes_{i=1}^{m} F_i = \left(\bigotimes_{i=1}^{r} F_i \right) \otimes \left(\bigotimes_{i=r+1}^{m} F_i \right)$$

for each positive integer $r < m$, the corollary follows from Proposition 7.2.3 by induction. $\blacksquare$

Definition 7.2.5 A nonnegative additive function α in a set $E \subset \mathbf{R}^m$ is called a *volume* in E. If A is a subinterval of E, the number $\alpha(A)$ is called the α-*volume* of A.

Convention. Throughout Part II of this book, we always tacitly assume that a volume α is defined on each cell in question.

Note. In Section 1.1, a length in $\mathbf{R}$, which is a volume in $\mathbf{R}^1 = \mathbf{R}$, was induced by an increasing function on $\mathbf{R}$. It is possible to show that a volume in $\mathbf{R}^m$ is also induced by a function on $\mathbf{R}^m$ that "increases" in a well-defined sense. The interested reader is referred to [3, Chapter 2, Section 12].

Remark 7.2.6 In view of Lemma 7.1.6, it is easy to see that Proposition 3.2.10 and Lemma 6.7.1 together with their proofs hold, respectively, for volumes and additive functions in a set $E \subset \mathbf{R}^m$.

If $\alpha_i = \lambda$ for $i = 1, \ldots, m$, then the additive function

$$\lambda_m = \bigotimes_{i=1}^{m} \alpha_i$$

in $\mathbf{R}^m$ is called the *Lebesgue volume* in $\mathbf{R}^m$. Note that for $m = 1, 2$, and 3, the Lebesgue volume of a cell A is the *usual length, area,* and *volume* of A, respectively.

The *shape* of a cell A in $\mathbf{R}^m$ is the number

$$s(A) = \frac{\lambda_m(A)}{[d(A)]^m},$$

which measures how much A differs from a cube. We have $0 < s(A) \leq 1$, and the equality is attained if and only if A is a cube.

7.3 Partitions

A *Lebesgue partition*, or simply a *partition*, in $\mathbf{R}^m$ is a collection, possibly empty,

$$P = \{(A_1, x_1), \ldots, (A_p, x_p)\}$$

where $A_1, \ldots, A_p$ are nonoverlapping cells in $\mathbf{R}^m$, and $x_1, \ldots, x_p$ are points of $\mathbf{R}^m$. The set $\bigcup_{i=1}^{p} A_i$ is called the *body* of P, denoted by $\bigcup P$. Clearly, a partition of Definition 1.2.1 is a partition in $\mathbf{R}^1 = \mathbf{R}$. Unless stated differently, in Part II of this book a partition is always a partition in $\mathbf{R}^m$.

Definition 7.3.1 Let $P = \{(A_1, x_1), \ldots, (A_p, x_p)\}$ be a partition in $\mathbf{R}^m$.

1. If $x_i \in A_i$ for $i = 1, \ldots, p$, we say that P is a *Perron partition* (abbreviated as P-partition).
2. If $E \subset \mathbf{R}^m$ and $\{x_1, \ldots, x_p\} \subset E$, we say that P is *anchored* in E.
3. If δ is a positive function on a set $E \subset \mathbf{R}^m$ in which P is anchored and $A_i \subset U(x_i, \delta(x_i))$ for $i = 1, \ldots, p$, we say that P is a *δ-fine partition*.
4. If $\varepsilon > 0$ and $s(A_i) > \varepsilon$ for $i = 1, \ldots, p$, we say that P is an *ε-shapely partition*.
5. If P is anchored in a set $E \subset \mathbf{R}^m$ and $\bigcup P \subset E$ or $\bigcup P = E$, we say that P is a partition *in E* or *of E*, respectively.

It is obvious that for $m = 1$, the Perron and δ-fine partitions defined above coincide with those introduced in Section 1.2. On the other hand, ε-shapely partitions, which play an important role in multidimensional integration, are new. As the shape of each one-dimensional cell equals one, using ε-shapely partitions in $\mathbf{R}$ is of no value.

Lemma 7.3.2 *Let δ be a positive function on a cell $K = \prod_{j=1}^{m}[r_j, s_j]$ where $r_j < s_j$ are integers for $j = 1, \ldots, m$. Then there is a δ-fine Perron partition $\{(K_1, x_1), \ldots, (K_p, x_p)\}$ of K such that $K_1, \ldots, K_p$ are dyadic cubes.*

PROOF. The proof is similar to that of Proposition 1.2.4. To shorten our expressions, we call a partition $\{(K_1, x_1), \ldots, (K_p, x_p)\}$ *dyadic* if $K_1, \ldots, K_p$ are dyadic cubes. For $n = 0, 1, \ldots$, denote by $\mathcal{K}_n$ the family of all dyadic cubes $L \subset K$ such that $d(L) = 2^{-n}$ and there is no δ-fine dyadic P-partition of L. If $\mathcal{K}_0 = \emptyset$, then for each dyadic cube $L \subset K$ with $d(L) = 1$ we can find a δ-fine dyadic P-partition P_L of L, and the union of all these partitions P_L is a δ-fine dyadic P-partition of K. Using this observation, we proceed by contradiction. Assuming that the lemma is false, we construct inductively a sequence $\{L_n\}_{n=0}^{\infty}$ of dyadic cubes such that $L_{n+1} \subset L_n$ and $L_n \in \mathcal{K}_n$ for $n = 0, 1, \ldots$.. The nested intervals theorem ([42, Theorem 2.38, p. 38]), implies that

$$\bigcap_{n=0}^{\infty} L_n = \{z\}$$

for a point $z \in K$. Since $\delta(z) > 0$, we can find an integer $k \geq 0$ so that $d(L_k) < \delta(z)$. Thus $\{(L_k, z)\}$ is a δ-fine dyadic P-partition of L_k, contrary to our assumption. $\blacksquare$

The next proposition is the m-dimensional version of *Cousin's lemma* (cf. Proposition 1.2.4).

Proposition 7.3.3 *For each positive function δ on a cell A and each positive $\varepsilon < 1$ there is a δ-fine ε-shapely Perron partition of A.*

PROOF. Choose a positive $\varepsilon < 1$. Translating A, we may assume that $A = \prod_{i=1}^{m}[-h_i, h_i]$.

Claim. There are real numbers $c_i > 0$ such that $c_i h_i$ is an integer for $i = 1, \ldots, p$ and

$$\frac{\min\{c_1, \ldots, c_m\}}{\max\{c_1, \ldots, c_m\}} > \sqrt[m]{\varepsilon}\,.$$

Assume that the claim is true, and let $c = \min\{c_1, \ldots, c_m\}$. The linear map

$$\Phi : (\xi_1, \ldots, \xi_m) \mapsto \left(\frac{\xi_1}{c_1}, \ldots, \frac{\xi_m}{c_m}\right)$$

maps bijectively the cell $B = \prod_{i=1}^{m}[-c_i h_i, c_i h_i]$ onto A. By Lemma 7.3.2, there is a $(c\delta \circ \Phi)$-fine P-partition $\{(B_1, y_1), \ldots, (B_q, y_q)\}$ of B where the cells $B_1, \ldots, B_q$ are dyadic cubes. It follows that

$$\{(\Phi(B_1), \Phi(y_1)), \ldots, (\Phi(B_q), \Phi(y_q))\}$$

is a δ-fine P-partition of A with

$$s(\Phi(B_j)) = \frac{c^m}{c_1 \cdots c_m} \geq \left[\frac{\min\{c_1, \ldots, c_m\}}{\max\{c_1, \ldots, c_m\}}\right]^m > \varepsilon$$

for $j = 1, \ldots, q$.

Thus it remains to establish the claim. For this purpose, find integers $S > s > 0$ with $s/S > \sqrt[m]{\varepsilon}$, and rationals $r_i > 0$ so that $s < r_i h_i < S$ for $i = 1, \ldots, m$. If $t_i = s/(r_i h_i)$, then $s/S < t_i < 1$ for all i, and consequently

$$\frac{\min\{t_1, \ldots, t_m\}}{\max\{t_1, \ldots, t_m\}} > \frac{s}{S} > \sqrt[m]{\varepsilon}.$$

If $r_i = p_i/q_i$ where p_i and q_i are positive integers, then $p_i t_i h_i = q_i s$. Hence, letting $p = \prod_{i=1}^{m} p_i$, we see that $p t_i h_i = q_i s \prod_{j \neq i} p_j$ is an integer, and it suffices to let $c_i = p t_i$ for $i = 1, \ldots, m$. ∎

Using Proposition 7.3.3 and Corollary 7.1.5, the proof of the following corollary is identical to that of Corollary 1.2.5.

Corollary 7.3.4 *Let δ be a positive function on a cell A. Each δ-fine partition P in A is a subset of a δ-fine partition Q of A. If P is, respectively, Perron or ε-shapely for a positive $\varepsilon < 1$, then so is Q.* ∎

7.4 Stieltjes sums

Let α be a volume in a cell A, and let $P = \{(A_1, x_1), \ldots, (A_p, x_p)\}$ be a partition in A. In complete analogy with Section 1.3, for any function f on $\{x_1, \ldots, x_p\}$, we set

$$\sigma(f, P; \alpha) = \sum_{i=1}^{p} f(x_i)\alpha(A_i)$$

and call this number the α-*Stieltjes sum*, or simply the *Stieltjes sum*, of f associated with P. Often we write $\sigma(f, P)$ or $\sigma(P)$ instead of $\sigma(f, P; \alpha)$. We note that $\sigma(f, P; \alpha) = 0$ whenever any of the following cases occurs: $f = 0$, $P = \emptyset$, or $\alpha(A) = 0$.

As in dimension one, if f is a function defined on a cell A, then an integral of f over A with respect to α is a real number approximated by α-Stieltjes sums of f.

Chapter 8

The McShane integral

We define the generalized Riemann integral of McShane in $\mathbf{R}^m$ and show that a large portion of its development parallels that of the one-dimensional case. Nonetheless, there are aspects of multidimensional integration connected with the divergence and Fubini theorems, which require new ideas and techniques.

Convention. Many one-dimensional results established in Part I translate easily, together with their proofs, into $\mathbf{R}^m$. In such cases, when no confusion is possible, we simply quote the one-dimensional result and leave it to the reader to apply the appropriate multidimensional version.

8.1 The integral

The definition of the McShane integral in $\mathbf{R}^m$ is formally the same as that in $\mathbf{R}$ (cf. Definition 2.1.1).

Definition 8.1.1 Let α be a volume in a cell A. A function f on A is called *integrable* over A with respect to α if there is a real number I having the following property: given $\varepsilon > 0$, we can find a positive function δ on A such that

$$|\sigma(f, P; \alpha) - I| < \varepsilon$$

for each δ-fine partition P of A.

Using Cousin's lemma in $\mathbf{R}^m$ (Proposition 7.3.3) and arguing as in Remark 2.1.2, we see that the number I from Definition 8.1.1 is determined uniquely by the integrable function f. We call it the *integral* of f over A

with respect to α, denoted by $\int_A f\, d\alpha$. When the volume α with respect to which we integrate is clearly understood from the context, we may write $\int_A f$ instead of $\int_A f\, d\alpha$. By $\mathcal{R}(A, \alpha)$ we denote the family of all functions integrable over A with respect to α.

Convention. As in Part I, we let $\int_A f\, d\alpha = 0$ whenever A is a degenerate interval.

Because of the formal similarity between the Definitions 2.1.1 and 8.1.1, Proposition 2.1.3 and Corollary 2.1.4 together with their proofs hold for the integral in $\mathbf{R}^m$. As the sum of two volumes and a nonnegative multiple of a volume are again volumes, the proof of the next proposition is similar to that of Proposition 2.1.5, and we shall omit it.

Proposition 8.1.2 *Let α and β be volumes on a cell A, let $c \geq 0$, and let $f \in \mathcal{R}(A, \alpha) \cap \mathcal{R}(A, \beta)$. Then $f \in \mathcal{R}(A, \alpha + \beta) \cap \mathcal{R}(A, c\alpha)$ and*

$$\int_A f\, d(\alpha + \beta) = \int_A f\, d\alpha + \int_A f\, d\beta \qquad and \qquad \int_A f\, d(c\alpha) = c \int_A f\, d\alpha\,.$$

If $\alpha = 0$, then any function g on A belongs to $\mathcal{R}(A, \alpha)$ and $\int_A g\, d\alpha = 0$. ∎

Cauchy's test for integrability and its proof translate verbatim to the m-dimensional situation. Consequently, the proof of the next proposition is nearly identical to that of Proposition 2.1.9.

Proposition 8.1.3 *If A is a cell and $f \in \mathcal{R}(A, \alpha)$, then $f \in \mathcal{R}(B, \alpha)$ for each cell $B \subset A$.*

PROOF. Choose an $\varepsilon > 0$ and use Cauchy's test to find a positive function δ on A so that $|\sigma(P) - \sigma(Q)| < \varepsilon$ for each pair of δ-fine partitions P and Q of A. If B is a subcell of A, then by Proposition 7.1.3, 2, there is a collection $\mathcal{C}$ of nonoverlapping cells such that $(A - B)^- = \bigcup \mathcal{C}$. According to the m-dimensional version of Cousin's lemma, there is a δ-fine partition P_C of C for each $C \in \mathcal{C}$. If P_B and Q_B are δ-fine partitions of B, then

$$P = P_B \cup \bigcup_{C \in \mathcal{C}} P_C \qquad and \qquad Q = Q_B \cup \bigcup_{C \in \mathcal{C}} P_C$$

are δ-fine partitions of A and

$$\sigma(P) = \sigma(P_B) + \sum_{C \in \mathcal{C}} \sigma(P_C) \qquad and \qquad \sigma(Q) = \sigma(Q_B) + \sum_{C \in \mathcal{C}} \sigma(P_C)\,.$$

Therefore

$$\varepsilon > |\sigma(P) - \sigma(Q)| = |\sigma(P_B) - \sigma(Q_B)|,$$

and the proposition follows from Cauchy's test for integrability. ∎

Proposition 8.1.4 *Let f be a function on a cell A. If $\mathcal{C}$ is a division of A and $f \in \mathcal{R}(C, \alpha)$ for each $C \in \mathcal{C}$, then $f \in \mathcal{R}(A, \alpha)$ and*

$$\int_A f \, d\alpha = \sum_{C \in \mathcal{C}} \int_C f \, d\alpha.$$

PROOF. Let $I = \sum_{C \in \mathcal{C}} \int_C f$. Choose an $\varepsilon > 0$, and suppose that the division $\mathcal{C}$ has n elements. Given $C \in \mathcal{C}$, find a positive function δ_C on C so that

$$\left| \sigma(f, Q) - \int_C f \right| < \frac{\varepsilon}{n}$$

for each δ_C-fine partition Q of C. Making δ_C smaller, we may assume that $B \cap U(x, \delta_C(x)) = \emptyset$ for every $B \in \mathcal{C}$ and every $x \in C - B$. Define a positive function δ on A by setting

$$\delta(x) = \min\{\delta_C(x) : C \in \mathcal{C} \text{ and } x \in C\}$$

for each $x \in A$, and select a δ-fine partition $P = \{(A_1, x_1), \ldots, (A_p, x_p)\}$ of A. If $\mathcal{D}$ is a segmentation of the family $\mathcal{C} \cup \{A_1, \ldots, A_p\}$, it is easy to see that the collection

$$Q_C = \{(D, x_i) : x_i \in C, \ D \in \mathcal{D}, \text{ and } D \subset C \cap A_i\}$$

is a δ_C-fine partition of $C \in \mathcal{C}$, and that $\sigma(f, P) = \sum_{C \in \mathcal{C}} \sigma(f, Q_C)$. Thus we have

$$|\sigma(f, P) - I| \leq \sum_{C \in \mathcal{C}} \left| \sigma(f, Q_C) - \int_C f \right| < \varepsilon,$$

and the proposition is proved. ∎

Corollary 8.1.5 *Let A be a cell and $f \in \mathcal{R}(A, \alpha)$. The map*

$$F : B \mapsto \int_B f \, d\alpha$$

is an additive function in A. In particular, F is a volume in A whenever f is nonnegative. ∎

With the exception of Corollary 2.3.2, and Theorems 2.3.10 and 2.3.11, all results of Sections 2.2 and 2.3 together with their proofs hold for the McShane integral in $\mathbf{R}^m$. The reader can easily construct a multidimensional analog of Example 2.2.9, and in view of Corollary 8.1.5, also Theorem 2.4.1 and its proof translate to $\mathbf{R}^m$.

Corollary 2.3.2 and Theorem 2.3.11 have no obvious extension to higher dimensions. On the other hand, Theorem 2.3.10 is still valid in $\mathbf{R}^m$ but requires a more sophisticated proof: it follows from the m-dimensional versions of Lemma 6.3.2 and Corollary 4.4.8.

8.2 Dirac volumes

The length associated with an increasing gap function having a single gap of size one (such as the function γ_n of Section 2.5) is often called a *Dirac length*. We use Dirac lengths to define a Dirac volume and show how to evaluate the integral with respect to it.

Let $s = (\sigma_1, \ldots, \sigma_m)$ and $t = (\tau_1, \ldots, \tau_m)$ be points of $\mathbf{R}^m$ and $[0,1]^m$, respectively, and for $i = 1, \ldots, m$, set

$$\alpha_i(\xi) = \begin{cases} 0 & \text{if } \xi < \sigma_i, \\ \tau_i & \text{if } \xi = \sigma_i, \\ 1 & \text{if } \xi > \sigma_i. \end{cases}$$

The interval function associated with α_i according to Section 1.1, still denoted by α_i, is a length in $\mathbf{R}$. The volume $\alpha = \bigotimes_{i=1}^m \alpha_i$ in $\mathbf{R}^m$ is called the *Dirac volume* at s determined by t.

Exercise 8.2.1 Let α be a Dirac volume at $s \in \mathbf{R}^m$, and let A be a cell. Show that $\alpha(A) = 1$ when $s \in A^\circ$ and $\alpha(A) = 0$ when $x \notin A$. If α is determined by $t = (1/2, \ldots, 1/2)$, then prove that $\alpha(A) \geq 2^{-m}$ for each cell A containing s.

Now we fix a cell A and a sequence $\{s_n\}$ of distinct points in A°. For $n = 1, 2, \ldots$, we select Dirac volumes γ_n at s_n, and real numbers $c_n \geq 0$ such that $\sum_{n=1}^\infty c_n < +\infty$. It is easy to verify that

$$\gamma = \sum_{n=1}^\infty c_n \gamma_n$$

is a volume in $\mathbf{R}^m$.

Lemma 8.2.2 *Let f be a bounded function on A. Then $f \in \mathcal{R}(A, \gamma)$ and*

$$\int_A f \, d\gamma = \sum_{n=1}^{\infty} c_n f(s_n) \, .$$

PROOF. The proof is similar to that of Lemma 2.5.1. Since there is a positive constant c such that $|f| \leq c$, the series $\sum_{n=1}^{\infty} c_n f(s_n)$ converges. Choose an $\varepsilon > 0$ and find an integer $k \geq 1$ so that $\sum_{n>k} c_n < \varepsilon/(2c)$. Let

$$\mu = \sum_{n=1}^{k} c_n \gamma_n \qquad \text{and} \qquad \nu = \sum_{n=k+1}^{\infty} c_n \gamma_n \, ,$$

and define a positive function δ on A by setting

$$\delta(x) = \min\{|x - s_n| : 1 \leq n \leq k, \ s_n \neq x\}$$

for each $x \in A$. Let $P = \{(A_1, x_1), \ldots, (A_p, x_p)\}$ be a δ-fine partition of A. By the choice of δ, each A_i contains at most one point s_n with $1 \leq n \leq k$, and if such an s_n is in A_i, then $s_n = x_i$. Hence for $n = 1, \ldots, k$, the set $B_n = \bigcup_{x_i = s_n} A_i$ contains s_n in the interior, and contains no s_j with $1 \leq j \leq k$ and $j \neq n$. Since $U[s_n, r] \subset B_n$ for a sufficiently small $r > 0$, it follows from Exercise 8.2.1 and Proposition 3.2.10 that

$$1 = \gamma_n(U[s_n, r]) \leq \sum_{x_i = s_n} \gamma_n(A_i) \leq \gamma_n(A) = 1 \, .$$

As $\gamma_n(A_i) = 0$ for $n = 1, \ldots, k$ and $x_i \neq s_n$, we have

$$\sum_{i=1}^{p} f(x_i)\mu(A_i) = \sum_{n=1}^{k} \sum_{x_i = s_n} f(x_i)\mu(A_i) = \sum_{n=1}^{k} c_n f(s_n) \, .$$

From the last inequality we conclude that

$$\left| \sigma(f, P; \gamma) - \sum_{n=1}^{\infty} c_n f(s_n) \right| = \left| \sum_{i=1}^{p} f(x_i)\nu(A_i) - \sum_{n=k+1}^{\infty} c_n f(s_n) \right|$$

$$\leq c\nu(A) + c \sum_{n=k+1}^{\infty} c_n < \varepsilon$$

and the lemma is proved. ∎

With Lemma 8.2.2 at hand, it is easy to see that Proposition 2.5.2 and its proof translate verbatim into the present situation. This shows, in particular, that neither the existence nor the value of $\int_A f \, d\gamma$ depends on how the Dirac volumes γ_n are determined.

8.3 The divergence theorem

Throughout this section, we assume that $m \geq 2$. Our goal is to establish the m-dimensional analog of the fundamental theorem of calculus. As this task involves integrating over the boundary of a cell, we begin by introducing the necessary notation and terminology.

Let $A = \prod_{i=1}^{m}[a_-^i, a_+^i]$ be a cell in $\mathbf{R}^m$, and let f be a continuous function defined on ∂A. For $i = 1, \ldots, m$, set

$$A_{(i)} = [a_-^1, a_+^1] \times \cdots \times [a_-^{i-1}, a_+^{i-1}] \times [a_-^{i+1}, a_+^{i+1}] \times \cdots \times [a_-^m, a_+^m],$$

$$A_{\pm}^i = [a_-^1, a_+^1] \times \cdots \times [a_-^{i-1}, a_+^{i-1}] \times \{a_{\pm}^i\} \times [a_-^{i-1}, a_+^{i+1}] \times \cdots \times [a_-^m, a_+^m],$$

and call the sets $A_{\pm}^i$ the *faces* of A. Each $A_{(i)}$ is the common projection of A and the faces $A_{\pm}^i$ into $\mathbf{R}^{m-1}$ in the direction of the ith coordinate axis, and we have

$$\lambda_m(A) = (a_+^i - a_-^i)\lambda_{m-1}(A_{(i)})$$

and $\partial A = (\bigcup_{i=1}^{m} A_-^i) \cup (\bigcup_{i=1}^{m} A_+^i)$. The number

$$\|A\| = 2\sum_{i=1}^{m} \lambda_{m-1}(A_{(i)})$$

is called the *perimeter* of A. If $m = 2$, the perimeter of a cell A is the *usual* perimeter of A; if $m = 3$, the perimeter is the *surface area* of A. By Theorem 2.2.8, the functions

$$f_{\pm}^i : (\xi_1, \ldots, \xi_{i-1}, \xi_{i+1}, \ldots, \xi_m) \mapsto f(\xi_1, \ldots, \xi_{i-1}, a_{\pm}^i, \xi_{i+1}, \ldots, \xi_m)$$

defined on $A_{(i)}$ belong to $\mathcal{R}(A_{(i)}, \lambda_{m-1})$ and we set

$$\int_{A_{\pm}^i} f\, d\lambda_{m-1} = \int_{A_{(i)}} f_{\pm}^i\, d\lambda_{m-1}\,.$$

Observe that the symbols $\int_{A_{\pm}^i} f\, d\lambda_{m-1}$ are genuine integrals, provided we make the obvious identification between $A_{(i)} \subset \mathbf{R}^{m-1}$ and the faces $A_{\pm}^i$.

For $i = 1, \ldots, m$, let

$$n_{\pm}^i = (0, \ldots, \pm 1, \ldots, 0)$$

be the unit vectors whose ith coordinates equal ± 1. If A is a cell, the vectors $n_{\pm}^i$ are perpendicular to the hyperplanes spanned by the faces $A_{\pm}^i$, and when placed at a point of $A^i_\pm$, they point outward from A; they are referred to as the *exterior normals* of $A^i_\pm$.

A *vector field* on a set $E \subset \mathbf{R}^m$ is a map $v : E \to \mathbf{R}^m$. If A is a cell in $\mathbf{R}^m$ and v is a continuous vector field on ∂A, then the integrals $\int_{A^i_\pm} v \cdot n^i_\pm \, d\lambda_{m-1}$ exist for $i = 1, \ldots, m$. We set

$$\int_{\partial A} v \cdot n \, d\lambda_{m-1} = \sum_{i=1}^m \left(\int_{A^i_+} v \cdot n^i_+ \, d\lambda_{m-1} + \int_{A^i_-} v \cdot n^i_- \, d\lambda_{m-1} \right)$$

and call this number the *flux* of v *from* A. We shall not attempt to interpret the symbol $\int_{\partial A} v \cdot n \, d\lambda_{m-1}$ as an integral; the reader should view it as mere notation, which is intuitively suggestive.

Note. The term "flux" is justified by the following physical interpretation. If A is a cell in $\mathbf{R}^3$ and v is the vector field of velocities of water flowing in $\mathbf{R}^3$, then it is easy to calculate that $\int_{\partial A} v \cdot n \, d\lambda_{m-1}$ is equal to the volume of water that flows from A into $\mathbf{R}^3 - A$ in a unit of time.

Proposition 8.3.1 *For a cell A the following statements are true.*

1. *The map $v \mapsto \int_{\partial A} v \cdot n \, d\lambda_{m-1}$ is a linear functional on the linear space of all continuous vector fields v on ∂A.*
2. *If v is a continuous vector field on ∂A, then*

$$\left| \int_{\partial A} v \cdot n \, d\lambda_{m-1} \right| \leq \|A\| \sup_{x \in \partial A} |v(x)| .$$

PROOF. The first statement follows from Proposition 2.1.3. Suppose that $v = (f_1, \ldots, f_m)$ and $c = \sup_{x \in \partial A} |v(x)|$. Then $|v(x) \cdot n^i_\pm| = |f_i(x)| \leq c$ for all $x \in \partial A$, so

$$\left| \int_{\partial A} v \cdot n \, d\lambda_{m-1} \right| \leq 2 \sum_{i=1}^m c\lambda_{m-1}(A_{(i)}) = c\|A\| . \quad \blacksquare$$

If v is a continuous vector field on a set $E \subset \mathbf{R}^m$, we set

$$F(C) = \begin{cases} \int_{\partial C} v \cdot n \, d\lambda_{m-1} & \text{if } C \subset E \text{ is a cell,} \\ 0 & \text{if } C \subset E \text{ is a degenerate interval,} \end{cases}$$

and call the function F the *flux* of v *in* E.

Proposition 8.3.2 *If v is a continuous vector field on a set $E \subset \mathbf{R}^m$, then the flux F of v in E is an additive function in E.*

PROOF. Let $A = \prod_{i=1}^m [a_-^i, a_+^i]$ and $B = \prod_{i=1}^m [b_-^i, b_+^i]$ be nonoverlapping subcells of E such that $C = A \cup B$ is a cell. Extending Lemma 7.1.7 by induction, we may assume that $a_-^1 < a_+^1 = b_-^1 < b_+^1$, and $[a_-^i, a_+^i] = [b_-^i, b_+^i]$ for $i = 2, \ldots, m$. It follows that

$$C_-^1 = A_-^1, \quad C_+^1 = B_+^1, \quad \text{and} \quad A_+^1 = B_-^1,$$

and that $C_{(i)}$ is the nonoverlapping union of $A_{(i)}$ and $B_{(i)}$ for $i = 2, \ldots, m$. As

$$\int_{A_+^1} v \cdot n_+^1 \, d\lambda_{m-1} = - \int_{B_-^1} v \cdot n_-^1 \, d\lambda_{m-1},$$

we have

$$F(A) + F(B) = \int_{B_+^1} v \cdot n_+^1 \, d\lambda_{m-1} + \int_{A_-^1} v \cdot n_-^1 \, d\lambda_{m-1}$$

$$+ \sum_{i=2}^m \left(\int_{A_+^i} v \cdot n_+^i \, d\lambda_{m-1} + \int_{B_+^i} v \cdot n_+^i \, d\lambda_{m-1} \right)$$

$$+ \sum_{i=2}^m \left(\int_{A_-^i} v \cdot n_-^i \, d\lambda_{m-1} + \int_{B_-^i} v \cdot n_-^i \, d\lambda_{m-1} \right) = F(C). \quad \blacksquare$$

We summarize briefly a few elementary facts about differentiable maps from $\mathbf{R}^r$ to $\mathbf{R}^s$ where r and s are positive integers. For proofs and additional properties we refer to [42, Chapter 9].

Let $E \subset \mathbf{R}^r$ and $z \in E^\circ$. We say that a map $\Phi : E \to \mathbf{R}^s$ is *differentiable* at z if there is a linear map $L : \mathbf{R}^r \to \mathbf{R}^s$ such that

$$\lim_{x \to z} \frac{|\Phi(x) - \Phi(z) - L(x - z)|}{|x - z|} = 0.$$

The linear map L, which is uniquely determined by Φ, is called the *differential* of Φ at z, denoted by $D\Phi(z)$. If $\Phi = (f_1, \ldots, f_s)$, then Φ is differentiable at z if and only if the functions $f_1, \ldots, f_s$ viewed as maps from E to $\mathbf{R}$ are differentiable at z, in which case the *partial derivatives* $\partial f_j(z)/\partial \xi_i$ exist for $i = 1, \ldots, r$ and $j = 1, \ldots, s$. Letting

$$\text{grad} \, f_j(z) = \left(\frac{\partial f_j(z)}{\partial \xi_1}, \ldots, \frac{\partial f_j(z)}{\partial \xi_r} \right)$$

for $j = 1, \ldots, s$, we have

$$[D\Phi(z)](x) = ([\text{grad} \, f_1(z)] \cdot x, \ldots, [\text{grad} \, f_s(z)] \cdot x)$$

for each $x \in \mathbf{R}^r$; in particular, $[Df_j(z)](x) = [\operatorname{grad} f_j] \cdot x$. The vector $\operatorname{grad} f_j(z)$ is called the *gradient* of f_j at z.

If $v = (f_1, \ldots, f_m)$ is a vector field on a set $E \subset \mathbf{R}^m$ that is differentiable at $z \in E^\circ$, we set

$$\operatorname{div} v(z) = \sum_{i=1}^{m} \frac{\partial f_i(z)}{\partial \xi_i}$$

and call this number the *divergence* of v at z. A simple calculation reveals that

$$\operatorname{div}[Dv(z)](x) = \operatorname{div} v(z)$$

for every $x \in \mathbf{R}^m$. This is an important identity, which will be used in the proof of Lemma 8.3.4 below.

Lemma 8.3.3 *If w is a linear vector field on $\mathbf{R}^m$, then*

$$\int_A \operatorname{div} w \, d\lambda_m = \int_{\partial A} w \cdot n \, d\lambda_{m-1}$$

for each cell A in $\mathbf{R}^m$.

PROOF. Using symmetry and Propositions 8.3.1, 1 and 2.1.3, it suffices to prove the lemma for $w = (f, 0, \ldots, 0)$ where $f(\xi_1, \ldots, \xi_m) = \xi_j$ for each $(\xi_1, \ldots, \xi_m)$ in $\mathbf{R}^m$. Let $A = \prod_{i=1}^{m}[a_-^i, a_+^i]$ and observe that

$$\int_{\partial A} w \cdot n \, d\lambda_{m-1} = \int_{A_+^1} f \, d\lambda_{m-1} - \int_{A_-^1} f \, d\lambda_{m-1} = \int_{A_{(1)}} (f_+^1 - f_-^1) \, d\lambda_{m-1}.$$

Now if $j = 1$, then

$$\int_{A_{(1)}} (f_+^1 - f_-^1) \, d\lambda_{m-1} = (a_+^1 - a_-^1)\lambda_{m-1}(A_{(1)}) = \lambda_m(A) = \int_A \operatorname{div} w \, d\lambda_m.$$

If $j = 2, \ldots, m$, then $f_+^1 = f_-^1$ and $\operatorname{div} w = 0$, which completes the proof. ∎

Lemma 8.3.4 *Let v be a continuous vector field on a set $E \subset \mathbf{R}^m$ that is differentiable at $z \in E^\circ$. Given $\varepsilon > 0$, we can find a $\delta > 0$ so that*

$$\left| \operatorname{div} v(z)\lambda_m(B) - \int_{\partial B} v \cdot n \, d\lambda_{m-1} \right| < \varepsilon d(B)\|B\|$$

for each cell $B \subset E \cap U(z, \delta)$ containing z.

PROOF.   Choose an $\varepsilon > 0$, and let $w(x) = v(z) + [Dv(z)](x - z)$ for every $x \in \mathbf{R}^m$. Then $\operatorname{div} w(x) = \operatorname{div} v(z)$ for each $x \in \mathbf{R}^m$, and there is a $\delta > 0$ such that $|v(x) - w(x)| < \varepsilon|x - z|$ for all $x \in E \cap U(z, \delta)$. If $B \subset E \cap U(z, \delta)$ contains z, then $|v(x) - w(x)| < \varepsilon d(B)$ for each $x \in B$. Applying Lemma 8.3.3 and Proposition 8.3.1, we obtain

$$
\left| \operatorname{div} v(z) \lambda_m(B) - \int_{\partial B} v \cdot n \, d\lambda_{m-1} \right|
$$

$$
= \left| \int_B \operatorname{div} w \, d\lambda_m - \int_{\partial B} v \cdot n \, d\lambda_{m-1} \right|
$$

$$
= \left| \int_{\partial B} w \cdot n \, d\lambda_{m-1} - \int_{\partial B} v \cdot n \, d\lambda_{m-1} \right|
$$

$$
= \left| \int_{\partial B} (w - v) \cdot n \, d\lambda_{m-1} \right| \leq \varepsilon d(B) \|B\| . \quad \blacksquare
$$

The following result is called the *divergence theorem*. Its proof is based on the same idea that was used to prove the fundamental theorem of calculus (Theorem 2.4.8).

Theorem 8.3.5 *Let v be a vector field defined on an open set $U \subset \mathbf{R}^m$ and let $A \subset U$ be a cell. If v is differentiable at each $x \in A$ and $\operatorname{div} v$ belongs to $\mathcal{R}(A, \lambda_m)$, then*

$$
\int_A \operatorname{div} v \, d\lambda_m = \int_{\partial A} v \cdot n \, d\lambda_{m-1} .
$$

PROOF.   Since v is continuous on A, the flux F of v in A is defined. Choose an $\varepsilon > 0$ and find a positive function δ on A so that

$$
\left| \sigma(\operatorname{div} v, P; \lambda_m) - \int_A \operatorname{div} v \, d\lambda_m \right| < \varepsilon
$$

for every δ-fine partition P of A. Lemma 8.3.4 implies that making δ smaller, we may assume that

$$
|\operatorname{div} v(x) \lambda_m(B) - F(B)| < \varepsilon d(B) \|B\|
$$

for each $x \in A$ and each cell $B \subset A \cap U(x, \delta(x))$ containing x. According to Proposition 7.3.3, there is a δ-fine $(1/2)$-shapely P-partition $P = \{(A_1, x_1), \ldots, (A_p, x_p)\}$ of A. Observe that

$$
d(A_i) \|A_i\| \leq 2m[d(A_i)]^m = 2m\lambda_m(A_i) \frac{1}{s(A_i)} < 4m\lambda_m(A_i)
$$

for $i = 1, \ldots, p$. Using Proposition 8.3.2 and Lemma 8.3.4, we obtain

$$\left| \int_A \operatorname{div} v \, d\lambda_m - F(A) \right|$$

$$\leq \left| \int_A \operatorname{div} v \, d\lambda_m - \sigma(\operatorname{div} v, P; \lambda_m) \right| + \sum_{i=1}^p |\operatorname{div} v(x_i)\lambda_m(A_i) - F(A_i)|$$

$$< \varepsilon + \sum_{i=1}^p \varepsilon d(A_i)\|A_i\| \leq \varepsilon + 4m\varepsilon \sum_{i=1}^p \lambda_m(A_i) = \varepsilon[1 + 4m\lambda_m(A)],$$

and the theorem follows from the arbitrariness of ε. ∎

8.4 Measures and measurability

Throughout this section, we assume that α is a fixed volume in $\mathbf{R}^m$.

The α-measure

If δ is a positive function on a set $E \subset \mathbf{R}^m$, we let

$$\alpha^\delta(E) = \sup \sum_{i=1}^p \alpha(A_i)$$

where the supremum is taken over all partitions $\{(A_1, x_1), \ldots, (A_p, x_p)\}$ anchored in E that are δ-fine. The α-*measure* of a set $E \subset \mathbf{R}^m$ is the number

$$\alpha^*(E) = \inf \alpha^\delta(E)$$

where the infimum is taken over all positive functions δ defined on E.

As the definition of α^* is a repetition of Definition 3.2.1, all statements and proofs of Section 3.2 remain valid in $\mathbf{R}^m$. The only exceptions are Proposition 3.2.2 and Exercise 3.2.3, which are replaced by the next proposition (cf. Remark 3.2.16).

Proposition 8.4.1 *The following statements are true.*

1. *If A is an interval, then*

$$\alpha^*(A) = \inf \alpha(C)$$

where the infimum is taken over all cells C with $A \subset C^\circ$.

2. *If A is a cell, then*

$$\alpha^*(A^\circ) = \sup \alpha(C)$$

where the supremum is taken over all cells $C \subset A^\circ$.

PROOF. Since a cell has divisions consisting of arbitrarily large numbers of cells, there are cells whose α-volumes are arbitrarily small. It follows that the first statement is true for $A = \emptyset$. If A is a nonempty interval and U is an open set containing A, then $\text{dist}(A, \mathbf{R}^m - U) > 0$. Thus there is a cell $C \subset U$ with $A \subset C^\circ$. By Proposition 3.2.4, 2 and Corollary 3.2.15, we have

$$\alpha^*(A) \le \alpha^*(C^\circ) \le \alpha(C) \le \alpha^*(C) \le \alpha^*(U),$$

and the first statement follows from Proposition 3.2.8.

Now let A be a cell and let $K \subset A^\circ$ be a nonempty compact set. Since $\text{dist}(K, \mathbf{R}^m - A^\circ) > 0$, there is a cell $C \subset A^\circ$ with $K \subset C^\circ$. As above, we conclude that

$$\alpha^*(K) \le \alpha^*(C^\circ) \le \alpha(C) \le \alpha^*(C) \le \alpha^*(A^\circ)$$

and the second statement follows from Proposition 3.2.14. ∎

Exercise 8.4.2 Prove Proposition 8.4.1 directly from the definition of α^*.

α-measurable sets

A set $E \subset \mathbf{R}^m$ is called *α-measurable* if, given $\varepsilon > 0$, there is a positive function δ on $\mathbf{R}^m$ such that

$$\sum_{i=1}^{p} \sum_{j=1}^{q} \alpha(A_i \cap B_j) < \varepsilon$$

for all δ-fine partitions $\{(A_1, x_1), \ldots, (A_p, x_p)\}$ and $\{(B_1, y_1), \ldots, (B_q, y_q)\}$ anchored in E and $\mathbf{R}^m - E$, respectively.

Since the above definition is formally the same as Definition 3.3.1, all properties of α-measurable sets, together with their proofs, stated in Section 3.3 carry over to $\mathbf{R}^m$. As an example, we prove the m-dimensional version of Proposition 3.3.4.

Proposition 8.4.3 *The union of countably many α-measurable sets is α-measurable.*

PROOF. Let E be the union of countably many α-measurable sets. If E is a subset of a cell, then proving its α-measurability amounts to a verbatim repetition of the corresponding one-dimensional argument. Thus we may assume that the sets

$$C_N = E \cap \prod_{k=1}^{m} [n_k, n_k + 1]$$

are α-measurable for each $N = (n_1, \ldots, n_m)$ where $n_1, \ldots, n_m$ are integers. Given such an N, let $|N| = \sum_{i=1}^{m} |n_i|$ and find a positive function δ_N on $\mathbf{R}^m$ such that

$$\sum_{i=1}^{p} \sum_{j=1}^{q} \alpha(A_i \cap B_j) < \varepsilon 2^{-|N|}$$

for all δ_N-fine partitions $\{(A_1, x_1), \ldots, (A_p, x_p)\}$ and $\{(B_1, y_1), \ldots, (B_q, y_q)\}$ anchored in C_N and $\mathbf{R}^m - C_N$, respectively. We may assume that $\delta_N \leq 1$ and consequently, we may also assume that $\delta_N(x) = 1$ whenever x lies outside $\prod_{k=1}^{m} [n_k - 2, n_k + 3]$. It follows that $\delta = \inf \delta_N$ is a positive function on $\mathbf{R}^m$. Let $\{(A_1, x_1), \ldots, (A_p, x_p)\}$ and $\{(B_1, y_1), \ldots, (B_q, y_q)\}$ be δ-fine partitions anchored in E and $\mathbf{R}^m - E$, respectively. Since $E = \bigcup_N C_N$, we have

$$\sum_{i=1}^{p} \sum_{j=1}^{q} \alpha(A_i \cap B_j) = \sum_N \sum_{x_i \in C_N} \sum_{j=1}^{q} \alpha(A_i \cap B_j) < \sum_N \varepsilon 2^{-|N|}$$

$$\leq \sum_{n_1=-\infty}^{\infty} \cdots \sum_{n_m=-\infty}^{\infty} \varepsilon 2^{-|n_1|} \ldots 2^{-|n_m|}$$

$$= \varepsilon \left(\sum_{n=-\infty}^{\infty} 2^{-|n|} \right)^m = \varepsilon 3^m,$$

and the α-measurability of E is established. ∎

Calculating α-measures

A *hyperplane* of $\mathbf{R}^m$ is a set

$$H = \{(\xi_1, \ldots, \xi_m) \in \mathbf{R}^m : \xi_i = c\}$$

where i is an integer with $1 \leq i \leq m$ and c is a real number. Thus hyperplanes are translates of $(m-1)$-dimensional subspaces of $\mathbf{R}^m$ that are parallel to the coordinate axes.

Remark 8.4.4 We want to stress that "slanted hyperplanes," such as the set $\{(\xi_1, \ldots, \xi_m) \in \mathbf{R}^m : \xi_1 = \xi_2\}$, are *not* hyperplanes according to our terminology.

As Proposition 3.4.1 and its proof hold in $\mathbf{R}^m$, we obtain the following m-dimensional analog of Corollary 3.4.2.

Proposition 8.4.5   *For all but countably many hyperplanes H of $\mathbf{R}^m$, we have $\alpha^*(H) = 0$.* ∎

Lemma 8.4.6   *Each nonempty open set $U \subset \mathbf{R}^m$ is the union of nonoverlapping cells $K_1, K_2, \ldots$ such that $\alpha^*(\partial K_n) = 0$ for all n. In particular,*

$$\alpha^*(U) = \sum_{n=1}^{\infty} \alpha(K_n) \,.$$

PROOF. If $E \subset \mathbf{R}^m$ and $z \in \mathbf{R}^m$, we let $E + z = \{x + z : x \in E\}$. Since the boundaries of all dyadic cubes are contained in a countable union of hyperplanes, it follows from Proposition 8.4.5 and Lemma 3.4.3 that there is a $z \in \mathbf{R}^m$ such that $\alpha^*(\partial(K + z)) = 0$ for each dyadic cube K. The rest of the proof is identical to that of Lemma 3.4.4. ∎

In view of Lemma 8.4.6, it is easy to see that Proposition 3.4.5 and its proof hold in $\mathbf{R}^m$.

Exercise 8.4.7   If $E \subset \mathbf{R}^m$, $z \in \mathbf{R}^m$, and $c \in \mathbf{R}$, prove the following claims:

1. $\lambda^*(E + z) = \lambda^*(E)$ and $\lambda^*(cE) = c^m \lambda^*(E)$;
2. If E is λ-measurable, then so are $E + z$ and cE.

α-negligible sets and α-measurable functions

The α-*negligible sets* and the corresponding expressions of α-*almost everywhere* and α-*almost all* are defined exactly as in the one-dimensional case (Section 3.5). The same is true for α-*measurable functions* and α-*simple functions* (Section 3.6). The properties listed in Sections 3.5 and 3.6 together with their proofs remain valid in $\mathbf{R}^m$.

Note that the *characteristic function* χ_E of a set $E \subset \mathbf{R}^m$ is defined by the formula

$$\chi_E(x) = \begin{cases} 1 \text{ if } x \in E, \\ 0 \text{ if } x \in \mathbf{R}^m - E. \end{cases}$$

Exercise 8.4.8   Show that every hyperplane of $\mathbf{R}^m$ is λ_m-negligible.

The α_A-measure

Let A be a cell. If F is a function on the family of all subintervals of A, we define a function F_A on the family of all intervals in $\mathbf{R}^m$ by setting

$$F_A(B) = F(A \cap B)$$

for every interval B in $\mathbf{R}^m$. In particular, α_A is a volume in $\mathbf{R}^m$ and

$$\alpha_A(B) = \alpha(A \cap B) \leq \alpha(B)$$

for every interval B. Consequently, $\alpha_A^*(E) \leq \alpha^*(E)$ for each set $E \subset \mathbf{R}^m$. We note that the volume α_A in $\mathbf{R}^m$ can be defined whenever α is a volume in A; the fact that α is a volume in the whole of $\mathbf{R}^m$ is irrelevant (cf. Remark 3.7.1).

The proof of the next proposition, which is similar to that of Proposition 3.7.2, is left to the reader.

Proposition 8.4.9 *If $\mathcal{D}$ is a division of a cell A, then*

$$\alpha_A^* = \sum_{D \in \mathcal{D}} \alpha_D^* . \quad \blacksquare$$

It is easy to see that Propositions 3.7.3 and 3.7.4 together with their proofs remain valid in $\mathbf{R}^m$. On the other hand, the higher-dimensional counterpart of Proposition 3.7.5 is quite complicated, and we shall not bother with its formulation. Assuming that $\{A_n\}$ is a sequence of cubes in $\mathbf{R}^m$, the reader may work out an appropriately modified Exercise 3.7.6.

Exercise 8.4.10 Choose a set $E \subset \mathbf{R}^m$, and define a volume β in $\mathbf{R}^m$ by setting $\beta(A) = \alpha_A^*(E)$ for each cell A (see Proposition 8.4.9). Prove the following facts.

1. $\alpha^*(E \cap S) \leq \beta^*(S)$ for each set $S \subset \mathbf{R}^m$. *Hint.* If $A_1, A_2, \ldots$ are cells with $S \subset \bigcup_{n=1}^\infty A_n^\circ$, then

$$\alpha^*(E \cap S) \leq \sum_{n=1}^\infty \alpha^*(E \cap A_n^\circ) \leq \sum_{n=1}^\infty \beta^*(A_n) .$$

2. $\alpha^*(E \cap S) = \beta^*(S)$ for each set S that is β-measurable. *Hint.* Using Lemma 8.4.6, prove it first for an open set S. Then apply 1 and Theorem 3.3.13.

3. $\alpha^*(E \cap S) = \beta^*(E \cap S)$ for each set $S \subset \mathbf{R}^m$. *Hint.* Use Proposition 3.2.8 together with 1 and 2.

4. If E is α-measurable, then $\alpha^*(E \cap S) = \beta^*(S)$ for each set $S \subset \mathbf{R}^m$. *Hint.* Find a G_δ set U containing S so that $\alpha^*(S) = \alpha^*(U)$ and $\beta^*(S) = \beta^*(U)$, and apply 2.

Exercise 8.4.11 Let E be a subset of a cell A such that $\alpha^*(E \cap \partial A) = 0$. Show that E is α-measurable if and only if it is α_A-measurable.

8.5 The Perron test

All results of Section 4.1 and their proofs translate verbatim to m dimensions. In particular, for each cell A, the McShane integral can be extended to the family $\overline{\mathcal{R}}(A, \alpha)$ (Definition 4.1.9). If A is a cell and $f \in \overline{\mathcal{R}}(A, \alpha)$, then it follows from Corollary 8.1.5 that the map

$$B \mapsto \int_B f \, d\alpha$$

is an additive function in A, called the *indefinite integral* of f.

The results of Section 4.4 (i.e., the Vitali–Carathéodory approximation theorem and its consequences) and their proofs translate to $\mathbf{R}^m$ as well, provided we establish the m-dimensional version of Perron's test for integrability (Theorem 4.3.12). This can be accomplished by systematically reformulating the one-dimensional concepts of Section 4.3 in terms of interval functions.

Note first that the definition of semicontinuous functions presented in Section 4.2 is meaningful when $\mathbf{R}$ is replaced by $\mathbf{R}^m$, and their properties given in Section 4.2 remain valid. The straightforward verification of these facts is left to the reader.

If F is a function defined on all subcells of a cell A and α is a volume in A, then the *lower* and *upper Lebesgue derivates* $\mathcal{D}_\alpha F$ and $\mathcal{D}^\alpha F$ are defined as in Section 4.3. We omit the proof of the next proposition, which is the same as in dimension one (cf. Proposition 4.3.4 and Lemmas 4.3.6 and 4.3.8).

Proposition 8.5.1 *Let A be a cell.*

1. *If F is a function defined on all subcells of A, then $\mathcal{D}_\alpha F$ is a lower semicontinuous function on A.*
2. *If F is an additive function in A and $\mathcal{D}_\alpha F \geq 0$, then $F \geq 0$.*
3. *If $\varepsilon > 0$ and $E \subset A$ is α-negligible, then there is a volume β in A such that $\beta(A) < \varepsilon$ and $\mathcal{D}_\alpha \beta(x) = +\infty$ for each $x \in E$.* ∎

Definition 8.5.2 Let α be a volume in a cell A, and let f be an extended real-valued function on A. We say that an additive function F in A is an *α-majorant* or *α-minorant* of f in A whenever

$$-\infty \neq \mathcal{D}_\alpha F \geq f \qquad \text{or} \qquad +\infty \neq \mathcal{D}^\alpha F \leq f,$$

respectively.

Defining the extended real numbers $U(f, A; \alpha)$ and $L(f, A; \alpha)$ as in Section 4.3, it is easy to verify that they have the properties stated in Corollary 4.3.7 and Proposition 4.3.9. The proof of the following lemma is identical to that of Lemma 4.4.1.

Lemma 8.5.3 *Let A be a cell and let $h \in \overline{\mathcal{R}}(A, \alpha)$. If h is lower semicontinuous and $h > -\infty$, then the indefinite integral H of h is an α-majorant of h in A.* ∎

Lemma 8.5.4 *Let S be a finite function defined on subcells of a cell A, and let $\mathcal{D}_\alpha S > -\infty$ and $\sum_{D \in \mathcal{D}} S(D) \leq S(A)$ for each division $\mathcal{D}$ of A. Then $\mathcal{D}_\alpha S \in \overline{\mathcal{R}}(A, \alpha)$ and $\int_A \mathcal{D}_\alpha S \, d\alpha \leq S(A)$; in particular, $\mathcal{D}_\alpha S < +\infty$ α_A-almost everywhere.*

PROOF. By Propositions 8.5.1, 1 and 4.2.3, we can find a sequence $\{f_n\}$ of finite continuous functions on A so that $f_n \nearrow \mathcal{D}_\alpha S$. Choose an integer $n \geq 1$ and $\varepsilon > 0$. There is a positive function δ on A such that $|\sigma(f_n, P) - \int_A f_n| < \varepsilon$ for each δ-fine partition P of A. Making δ smaller, we may assume that

$$S(B) \geq [f_n(x) - \varepsilon]\alpha(B)$$

for each $x \in A$ and each cell $B \subset A \cap U(x, \delta(x))$. Thus, if $\{(A_1, x_1), \ldots, (A_p, x_p)\}$ is a δ-fine partition of A, then

$$\int_A f_n - \varepsilon < \sum_{i=1}^p f_n(x_i)\alpha(A_i) < \sum_{i=1}^p [S(A_i) + \varepsilon\alpha(A_i)] \leq S(A) + \varepsilon\alpha(A).$$

By the arbitrariness of ε, we have $\int_A f_n \leq S(A)$, and the lemma follows from the monotone convergence theorem (Proposition 4.1.10). ∎

Remark 8.5.5 Note that the assumptions of Lemma 8.5.4 are satisfied whenever S is an α-majorant of any extended real-valued function on A.

Now we are ready to prove the *Perron test* for integrability.

Theorem 8.5.6   *An extended real-valued function f on a cell A belongs to $\overline{\mathcal{R}}(A, \alpha)$ if and only if*

$$-U(-f, A; \alpha) = U(f, A; \alpha) \neq \pm\infty,$$

in which case this common value equals $\int_A f \, d\alpha$.

PROOF.   Let $I = L(f, A) = U(f, A)$ be a finite number. Choose an $\varepsilon > 0$ and find an α-majorant M and an α-minorant m of f in A so that

$$M(A) < I + \varepsilon \qquad \text{and} \qquad m(A) > I - \varepsilon.$$

It follows from Lemma 8.5.4 and Remark 8.5.5 that there is a finite function g on A equal to f α_A-almost everywhere such that M is an α-majorant and m is an α-minorant of g in A. Now proceeding as in the proof of Lemma 4.3.10, we can show that $g \in \mathcal{R}(A, \alpha)$ and $\int_A g = I$. Thus, by definition, $f \in \overline{\mathcal{R}}(A, \alpha)$ and $\int_A f = I$.

Conversely, assume that $f \in \overline{\mathcal{R}}(A, \alpha)$ and find a $g \in \mathcal{R}(A, \alpha)$ that equals f α_A-almost everywhere. Choose an $\varepsilon > 0$ and use Henstock's lemma to find a positive function δ on A so that

$$\sum_{i=1}^{p} \left| g(x_i)\alpha(A_i) - \int_{A_i} g \right| < \varepsilon$$

for each δ-fine partition $\{(A_1, x_1), \ldots, (A_p, x_p)\}$ in A. If B is a subcell of A, let Π_B be the family of all δ-fine partitions Q in A for which $\bigcup Q = B$, and let $S(B) = \sup_{Q \in \Pi_B} \sigma(g, Q)$. Since

$$\int_B g - \varepsilon \leq S(B) \leq \int_B g + \varepsilon$$

by the choice of δ, the function S is finite. Let $\mathcal{D}$ be a division of A. If $Q_D \in \Pi_D$ for each $D \in \mathcal{D}$, then $Q = \bigcup_{D \in \mathcal{D}} Q_D$ belongs to Π_A and

$$\sum_{D \in \mathcal{D}} \sigma(g, Q_D) = \sigma(g, Q) \leq S(A).$$

As the partitions Q_D are arbitrary, we have $\sum_{D \in \mathcal{D}} S(D) \leq S(A)$. Now choose an $x \in A$ and a cell $B \subset A \cap U(x, \delta(x))$. Since $Q = \{(B, x)\}$ belongs to Π_B, we obtain

$$S(B) \geq \sigma(g, Q) = g(x)\alpha(B)$$

and hence $\mathcal{D}_\alpha S(x) \geq g(x) > -\infty$. By Lemma 8.5.4, the function $\mathcal{D}_\alpha S$ belongs to $\overline{\mathcal{R}}(A, \alpha)$ and $\int_A \mathcal{D}_\alpha S \leq S(A)$. If M is the indefinite integral of

$\mathcal{D}_\alpha S$, then M is an α-majorant of $\mathcal{D}_\alpha S$ according to Proposition 8.5.1, 1 and Lemma 8.5.3. It follows that M is an α-majorant of g, and Proposition 4.3.9 implies that

$$U(f, A) = U(g, A) \leq M(A) \leq S(A) \leq \int_A f + \varepsilon.$$

From the arbitrariness of ε, we conclude that $U(f, A) \leq \int_A f$. Applying this result to $-f$ and using Corollary 4.3.7 completes the proof. ∎

8.6 The Fubini theorem

Throughout this section, we fix positive integers r and s so that $r + s = m$. We also fix an r-cell A with an r-dimensional volume α in A and an s-cell B with an s-dimensional volume β in B. In the m-cell $C = A \times B$ we consider the m-dimensional volume $\gamma = \alpha \otimes \beta$ (see Proposition 7.2.3). It follows from Lemma 7.1.1 that $\gamma_C = \alpha_A \otimes \beta_B$.

Let f be an extended real-valued function on C. If $x \in A$ and $y \in B$, we define extended real-valued functions f_x on B and f^y on A by setting

$$f_x(v) = f(x, v) \qquad \text{and} \qquad f^y(u) = f(u, y)$$

for each $v \in B$ and $u \in A$. Our goal is to give a precise meaning to the formula

$$\int_C f \, d\gamma = \int_A \left[\int_B f_x(y) \, d\beta(y) \right] d\alpha(x) = \int_B \left[\int_A f^y(x) \, d\alpha(x) \right] d\beta(y),$$

which relates the *double integral* of f to the *iterated integrals* of f.

Proposition 8.6.1 *If $f \in \mathcal{R}(C, \gamma)$, then $f_x \in \mathcal{R}(B, \beta)$ for α_A-almost all $x \in A$ and $f^y \in \mathcal{R}(A, \alpha)$ for β_B-almost all $y \in B$.*

PROOF. Let $N = \{x \in A : f_x \notin \mathcal{R}(B, \beta)\}$. By Cauchy's test for integrability, given $x \in N$, there is an $\varepsilon_x > 0$ such that for each positive function θ on B, we can find θ-fine partitions Q_x and T_x of B so that

$$\sigma(f_x, Q_x; \beta) - \sigma(f_x, T_x; \beta) \geq \varepsilon_x;$$

this is achieved by selecting Q_x and T_x so that $\sigma(f_x, Q_x; \beta) \geq \sigma(f_x, T_x; \beta)$. Define a nonnegative function g on A by setting

$$g(x) = \begin{cases} \varepsilon_x & \text{if } x \in N, \\ 0 & \text{if } x \in A - N, \end{cases}$$

and observe that in view of Theorem 4.1.7, it suffices to show that g belongs to $\mathcal{R}(A, \alpha)$ and $\int_A g \, d\alpha = 0$.

To this end, choose an $\varepsilon > 0$ and using Cauchy's test again, find a positive function Δ on C such that

$$|\sigma(f, Q; \gamma) - \sigma(f, T; \gamma)| < \varepsilon$$

for all Δ-fine partitions Q and T of C. If $x \in N$, select Δ_x-fine partitions

$$Q_x = \{(K_j^x, y_j^x) : j = 1, \ldots, q_x\} \quad \text{and} \quad T_x = \{(L_k^x, z_k^x) : k = 1, \ldots, t_x\}$$

of B so that $\sigma(f_x, Q_x; \beta) - \sigma(f_x, T_x; \beta) \geq \varepsilon_x$. If $x \in A - N$, let $Q_x = T_x$ be any Δ_x-fine partition of B. Define a positive function δ on A by setting

$$\delta(x) = \min\{\Delta_x(y_1^x), \ldots, \Delta_x(y_{q_x}^x), \Delta_x(z_1^x), \ldots, \Delta_x(z_{t_x}^x)\}$$

for each $x \in A$. If $P = \{(A_1, x_1), \ldots, (A_p, x_p)\}$ is a δ-fine partition of A, it is easy to verify that

$$Q = \{(A_i \times K_j^{x_i}, (x_i, y_j^{x_i})) : j = 1, \ldots, q_{x_i}; \; i = 1, \ldots, p\},$$

$$T = \{(A_i \times L_k^{x_i}, (x_i, z_k^{x_i})) : k = 1, \ldots, t_{x_i}; \; i = 1, \ldots, p\}$$

are Δ-fine partitions of C (draw a picture for $r = s = 1$!). Therefore

$$
\begin{aligned}
|\sigma(g, P; \alpha)| &= \sum_{i=1}^{p} g(x_i)\alpha(A_i) \\
&\leq \sum_{i=1}^{p} [\sigma(f_{x_i}, Q_{x_i}; \beta) - \sigma(f_{x_i}, T_{x_i}; \beta)]\alpha(A_i) \\
&= \sum_{i=1}^{p} \left[\sum_{j=1}^{q_{x_i}} f_{x_i}(y_j^{x_i})\beta(K_j^{x_i}) - \sum_{k=1}^{t_{x_i}} f_{x_i}(z_k^{x_i})\beta(L_k^{x_i}) \right] \alpha(A_i) \\
&= \sigma(f, Q; \gamma) - \sigma(f, T; \gamma) < \varepsilon,
\end{aligned}
$$

which implies that $g \in \mathcal{R}(A, \alpha)$ and $\int_A g \, d\alpha = 0$. The proposition follows by symmetry. ∎

Lemma 8.6.2 *If $N \subset \mathbf{R}^r$ is α_A-negligible, then $N \times \mathbf{R}^s$ is γ_C-negligible.*

PROOF. Choose an $\varepsilon > 0$ and an s-cell K. By Proposition 3.5.2, there is a sequence $\{A_n\}$ of r-cells such that $N \subset \bigcup_{n=1}^{\infty} A_n^\circ$ and $\sum_{n=1}^{\infty} \alpha_A(A_n) < \varepsilon$. If L is an s-cell with $K \subset L^\circ$, then $N \times K \subset \bigcup_{n=1}^{\infty} (A_n \times L)^\circ$ and

$$\sum_{n=1}^{\infty} \gamma_C(A_n \times L) = \beta_B(L) \sum_{n=1}^{\infty} \alpha_A(A_n) < \varepsilon \beta(B) \, .$$

Thus $N \times K$ is γ_A-negligible according to Proposition 3.5.2. Since $\mathbf{R}^s$ is a countable union of s-cells, the lemma follows from Proposition 3.5.3, 3. ∎

Proposition 8.6.3 *If $f \in \mathcal{R}(C, \gamma)$, then the functions*

$$F : x \mapsto \int_B f_x \, d\beta \quad and \quad G : y \mapsto \int_A f^y \, d\alpha$$

belong to $\overline{\mathcal{R}}(A, \alpha)$ and $\overline{\mathcal{R}}(B, \beta)$, respectively, and

$$\int_C f \, d\gamma = \int_A F \, d\alpha = \int_B G \, d\beta \, .$$

PROOF. Using Proposition 8.6.1 and Lemma 8.6.2, it is easy to construct a function g on C so that $g = f$ γ_C-almost everywhere, $g_x = f_x$ whenever $f_x \in \mathcal{R}(B, \beta)$, and $g_x \in \mathcal{R}(B, \beta)$ for all $x \in A$. Thus, with no loss of generality, we may assume that $f_x \in \mathcal{R}(B, \beta)$ for all $x \in A$, which implies that the function F is defined on all of A. Choose an $\varepsilon > 0$ and find a positive function Δ on C such that

$$\left| \sigma(f, S; \gamma) - \int_C f \, d\gamma \right| < \varepsilon$$

for every Δ-fine partition S of C. For each $x \in A$, select a Δ_x-fine partition $Q_x = \{(B_j^x, y_j^x) : j = 1, \ldots, q_x\}$ of B so that

$$|\sigma(f_x, Q_x, \beta) - F(x)| < \varepsilon \, ,$$

and define a positive function δ on A by setting

$$\delta(x) = \min\{\Delta_x(y_1^x), \ldots, \Delta_x(y_{q_x}^x)\}$$

for each $x \in A$. If $P = \{(A_1, x_1), \ldots, (A_p, x_p)\}$ is a δ-fine partition of A, it is easy to see that

$$S = \{(A_i \times B_j^{x_i}, (x_i, y_j^{x_i})) : j = 1, \ldots, q_{x_i}; \ i = 1, \ldots, p\}$$

is a Δ-fine partition of C (draw a picture for $r = s = 1$!). Therefore

$$\left| \sigma(F, P; \alpha) - \int_C f \, d\gamma \right| \leq |\sigma(F, P; \alpha) - \sigma(f, S; \gamma)| + \left| \sigma(f, S; \gamma) - \int_C f \, d\gamma \right|$$

$$< \varepsilon + \left| \sum_{i=1}^{p} \left[F(x_i) - \sum_{j=1}^{q_{x_i}} f_{x_i}(y_j^{x_i}) \beta(B_j^{x_i}) \right] \alpha(A_i) \right|$$

$$\leq \varepsilon + \sum_{i=1}^{p} |F(x_i) - \sigma(f_{x_i}, Q_{x_i}; \beta)| \alpha(A_i)$$

$$< \varepsilon + \sum_{i=1}^{p} \varepsilon \alpha(A_i) = \varepsilon[1 + \alpha(A)] ,$$

which implies that $F \in \mathcal{R}(A, \alpha)$ and $\int_A F \, d\alpha = \int_C f \, d\gamma$. The proposition follows by symmetry. ∎

Corollary 8.6.4 Let $E \subset \mathbf{R}^m$ be a γ_C-measurable set. The sets

$$E_x = \{v \in \mathbf{R}^s : (x, v) \in E\} \qquad \text{and} \qquad E^y = \{u \in \mathbf{R}^r : (u, y) \in E\}$$

are, respectively, β_B-measurable for α_A-almost all $x \in \mathbf{R}^r$ and α_A-measurable for β_B-almost all $y \in \mathbf{R}^s$. The functions $x \mapsto \beta_B^*(E_x)$ and $y \mapsto \alpha_A^*(E^y)$ belong to $\mathcal{R}(A, \alpha)$ and $\mathcal{R}(B, \beta)$, respectively, and

$$\gamma_C^*(E) = \int_A \beta_B^*(E_x) \, d\alpha(x) = \int_B \alpha_A^*(E^y) \, d\beta(y) .$$

In particular, E is γ_C-negligible if and only if E_x is β_B-negligible for α_A-almost all $x \in \mathbf{R}^r$, or if and only if E^y is α_A-negligible for β_B-almost all $y \in \mathbf{R}^s$.

PROOF. Since $(\chi_E)_x = \chi_{E_x}$ and $(\chi_E)^y = \chi_{E^y}$ for each $x \in \mathbf{R}^r$ and each $y \in \mathbf{R}^s$, the corollary follows from Proposition 8.6.3 and Theorems 4.1.1 and 4.1.7. ∎

The following example shows that the existence and equality of the iterated integrals of a function f on C need not imply that f is integrable over C. For more striking examples, we refer to [33, Exercises (16–13) and (16–14), pp. 206 and 207].

Example 8.6.5 Assume that $r = s = 1$ and $A = B = [-1, 1]$. For each $(x, y) \in C$, let

$$f(x, y) = \begin{cases} xy(x^2 + y^2)^{-2} & \text{if } x^2 + y^2 > 0, \\ 0 & \text{if } x^2 + y^2 = 0. \end{cases}$$

If $\alpha = \beta = \lambda$, it is easy to see that $f_x \in \mathcal{R}(B, \beta)$ for all $x \in A$ and $f^y \in \mathcal{R}(A, \alpha)$ for all $y \in B$, and that

$$\int_{-1}^{1} \left(\int_{-1}^{1} f_x(y) \, dy \right) dx = \int_{-1}^{1} \left(\int_{-1}^{1} f^y(x) \, dx \right) dy = 0$$

where dx and dy stand for $d\lambda(x)$ and $d\lambda(y)$, respectively. On the other hand, $\gamma = \lambda_2$ and it follows from Proposition 8.6.3 that $f \notin \mathcal{R}([0,1]^2, \gamma)$. In view of this and Proposition 8.1.3, we conclude that $f \notin \mathcal{R}(C, \gamma)$.

Combining Propositions 8.6.1 and 8.6.3 with Corollary 8.6.4, we obtain the well-known *Fubini theorem*.

Theorem 8.6.6 *Let f be an extended real-valued function defined on C. If $f \in \overline{\mathcal{R}}(C, \gamma)$, then $f_x \in \overline{\mathcal{R}}(B, \beta)$ for α_A-almost all $x \in A$ and $f^y \in \overline{\mathcal{R}}(A, \alpha)$ for β_B-almost all $y \in B$. Moreover, the functions*

$$F : x \mapsto \int_B f_x \, d\beta \qquad and \qquad G : y \mapsto \int_A f^y \, d\alpha$$

belong to $\overline{\mathcal{R}}(A, \alpha)$ and $\overline{\mathcal{R}}(B, \beta)$, respectively, and

$$\int_C f \, d\gamma = \int_A F \, d\alpha = \int_B G \, d\beta . \quad \blacksquare$$

For an extended real-valued function f on C, the Fubini theorem has two principal applications.

1. It tells us that the double integral of f *does not exist* if the two iterated integrals of f have different values.
2. If the double integral of f *does exist*, it facilitates its evaluation.

The next proposition, called the *Tonelli theorem*, is a test for the existence of the double integral of f. In practice, performing the test amounts to evaluating the double integral by means of Fubini's theorem without an a priori verification that it exists.

Proposition 8.6.7 *Let $f \geq 0$ be a γ_C-measurable extended real-valued function on C such that either $x \mapsto \int_B f_x \, d\beta$ belongs to $\overline{\mathcal{R}}(A, \alpha)$ or $y \mapsto \int_A f^y \, d\alpha$ belongs to $\overline{\mathcal{R}}(B, \beta)$. Then $f \in \overline{\mathcal{R}}(C, \gamma)$.*

PROOF. By symmetry, we may assume that $F : x \mapsto \int_B f_x \, d\beta$ belongs to $\overline{\mathcal{R}}(A, \alpha)$. According to Corollary 4.4.8, the functions $f_n = \min\{f, n\}$ belong to $\mathcal{R}(C, \gamma)$ for $n = 1, 2 \ldots$, and clearly $f_n \nearrow f$. From Proposition 8.6.3

we infer that the function $F_n : x \mapsto \int_B (f_n)_x \, d\beta$ belongs to $\overline{\mathcal{R}}(A, \alpha)$ and $\int_A F_n \, d\alpha = \int_C f_n \, d\gamma$. Since $(f_n)_x \nearrow f_x$, the monotone convergence theorem implies that $F_n(x) \nearrow F(x)$ whenever $F(x)$ is defined, which is α_A-almost everywhere by the assumption. Moreover,

$$\lim \int_C f_n \, d\gamma = \lim \int_A F_n \, d\alpha \leq \int_A F \, d\alpha < +\infty \,,$$

and another application of the monotone convergence theorem shows that f belongs to $\overline{\mathcal{R}}(C, \gamma)$. ∎

If f and g are extended real-valued functions on A and B, respectively, we define an extended real-valued function $f \otimes g$ on C by letting

$$(f \otimes g)(x, y) = f(x)g(y)$$

for each $(x, y) \in C$.

Proposition 8.6.8 *If $f \in \overline{\mathcal{R}}(A, \alpha)$ and $g \in \overline{\mathcal{R}}(B, \beta)$, then $f \otimes g \in \overline{\mathcal{R}}(C, \gamma)$ and*

$$\int_C f \otimes g \, d\gamma = \left(\int_A f \, d\alpha \right) \cdot \left(\int_B g \, d\beta \right).$$

PROOF. In view of Lemma 8.6.2, we may assume that $f \in \mathcal{R}(A, \alpha)$ and $g \in \mathcal{R}(B, \beta)$. Suppose first that f and g are nonnegative and choose an $\varepsilon > 0$. By the Vitali–Carathéodory theorem and Exercise 4.2.2, 5, there are f_1, f_2 in $\overline{\mathcal{R}}(A, \alpha)$ and g_1, g_2 in $\overline{\mathcal{R}}(B, \beta)$ such that f_1, g_1 are upper semicontinuous, f_2, g_2 are lower semicontinuous, and

$$0 \leq f_1 \leq f \leq f_2 \,, \qquad 0 \leq g_1 \leq g \leq g_2 \,, \qquad \max\{f_1, g_1\} < +\infty \,,$$

$$\max\left\{ \int_A (f_2 - f_1) \, d\alpha, \int_B (g_2 - g_1) \, d\beta \right\} < \varepsilon \,.$$

In addition, we have

$$\int_A f_2 \, d\alpha = \int_A (f_2 - f_1) \, d\alpha + \int_A f_1 \, d\alpha < \varepsilon + \int_A f \, d\alpha$$

and similarly $\int_B g_2 \, d\beta < \varepsilon + \int_A g \, d\beta$. It is easy to verify that $f_1 \otimes g_1$ and $f_2 \otimes g_2$ are, respectively, upper and lower semicontinuous extended real-valued functions on C such that

$$0 \leq f_1 \otimes g_1 \leq f \otimes g \leq f_2 \otimes g_2 \qquad \text{and} \qquad f_1 \otimes g_1 < +\infty \,.$$

By Exercise 4.2.2, 1, lower semicontinuous extended real-valued functions are γ_C-measurable, and the Tonelli and Fubini theorems yield

$$\int_C (f_2 \otimes g_2 - f_1 \otimes g_1)\, d\gamma$$

$$\leq \int_C f_2 \otimes [g_2 - g_1]\, d\gamma + \int_C [f_2 - f_1] \otimes g_2\, d\gamma$$

$$= \left(\int_A f_2\, d\alpha \right) \cdot \left(\int_B [g_2 - g_1]\, d\beta \right) + \left(\int_A [f_2 - f_1]\, d\alpha \right) \cdot \left(\int_B g_2\, d\beta \right)$$

$$< \varepsilon \left(2\varepsilon + \int_A f\, d\alpha + \int_B g\, d\beta \right).$$

Thus $f \otimes g \in \mathcal{R}(C, \gamma)$ by the Vitali–Carathéodory theorem. If f and g are arbitrary, then

$$f \otimes g = f^+ \otimes g^+ + f^- \otimes g^- - f^+ \otimes g^- - f^- \otimes g^+$$

belongs to $\mathcal{R}(C, \gamma)$ by the first part of the proof. The proposition follows from Fubini's theorem. ∎

Corollary 8.6.9 *Let $X \subset \mathbf{R}^r$ and $Y \subset \mathbf{R}^s$ be, respectively, α_A-measurable and β_B-measurable. Then $X \times Y$ is γ_C-measurable and*

$$\gamma_C^*(X \times Y) = \alpha_A^*(X)\beta_B^*(Y).$$

PROOF. In view of Theorem 4.1.1 and Proposition 8.6.8, it suffices to observe that $\chi_{X \times Y} = \chi_X \otimes \chi_Y$. ∎

Chapter 9
Descriptive definition

The λ_m-measure λ_m^* induced by the Lebesgue volume λ_m in $\mathbf{R}^m$ is called the *m-dimensional Lebesgue measure* or simply the *Lebesgue measure* in $\mathbf{R}^m$. Since the Lebesgue measure in $\mathbf{R}^m$ plays a very special role in our presentation, we agree that throughout the remainder of this book the words "measure," "measurable," and "negligible" always refer to this measure. The same applies to the expressions "almost all" and "almost everywhere." We also agree to write $\lambda_m(E)$ or $|E|$ instead of $\lambda_m^*(E)$ for every set $E \subset \mathbf{R}^m$. As $\lambda_m^*(A) = \lambda_m(A)$ for each interval A in $\mathbf{R}^m$ (an easy consequence of Proposition 8.4.1, 1), this will cause no confusion.

Employing the ideas of Chapter 5, we shall present a full descriptive definition of the McShane integral with respect to the Lebesgue measure in $\mathbf{R}^m$.

9.1 AC functions

The definitions of functions of bounded variation and AC functions in $\mathbf{R}^m$ and $\mathbf{R}$ are formally the same.

Definition 9.1.1 Let F be an additive function in a cell A. The *variation* of F on a cell $B \subset A$ is the extended real number

$$V(F, B) = \sup \sum_{D \in \mathcal{D}} |F(D)|$$

where the supremum is taken over all divisions $\mathcal{D}$ of B. If $V(F, A) < +\infty$, we say that F is of *bounded variation*.

As in dimension one, it is convenient to let $V(F, B) = 0$ whenever B is a degenerate interval.

Definition 9.1.2 An additive function F in a set $E \subset \mathbf{R}^m$ is called AC if, given $\varepsilon > 0$, there is an $\eta > 0$ such that

$$\sum_{k=1}^{n} |F(B_k)| < \varepsilon$$

whenever $\{B_1, \ldots, B_n\}$ is a collection of nonoverlapping subcells of E with $\sum_{k=1}^{n} |B_k| < \eta$.

The proofs of the next proposition and its corollaries are similar to those of Proposition 5.1.2 and Corollaries 5.1.3 and 5.1.5. The reader can do them as an exercise.

Proposition 9.1.3 *Let F be an additive function in a cell A. If C is a division of A, then*

$$V(F, A) = \sum_{C \in \mathcal{C}} V(F, C). \quad \blacksquare$$

Corollary 9.1.4 *Let F be an additive function in a cell A, and let*

$$VF(B) = V(F, B)$$

for each interval $B \subset A$. If F is of bounded variation, then VF is an additive function in A such that $|F| \leq VF$. In particular, VF and $VF - F$ are volumes in A. $\blacksquare$

Corollary 9.1.5 *An additive function F in a cell A is of bounded variation if and only if it is the difference of two volumes in A.* $\blacksquare$

We omit the proof of the following proposition, since it is a verbatim repetition of the proof of Proposition 5.1.8.

Proposition 9.1.6 *A volume α in a cell A is AC if and only if each negligible set E is α_A-negligible.* $\blacksquare$

Remark 9.1.7 Propositions 5.1.7 and 5.1.9 as well as Corollary 5.1.10 hold in $\mathbf{R}^m$ together with their proofs.

9.2 Covering theorems

We establish the m-dimensional analogs of the covering results given in Section 5.2. The reader should keep in mind that every family of nonoverlapping, in particular disjoint, cells is countable.

Let $A = \prod_{i=1}^{m}[a_i, b_i]$ be a cell, and let $c_i = (a_i + b_i)/2$ and $r_i = (b_i - a_i)/2$ for $i = 1, \ldots, m$. The point $c_A = (c_1, \ldots, c_m)$ and the numbers

$$r_A = \min\{r_1, \ldots, r_m\} \qquad \text{and} \qquad R_A = \max\{r_1, \ldots, r_m\}$$

are called, respectively, the *center* and the *inner* and *outer radius* of A. With this notation it is clear that

$$d(A) = 2R_A \qquad \text{and} \qquad U[c_A, r_A] \subset A \subset U[c_A, R_A].$$

If $s = s(A)$ is the shape of A (see Section 7.2), let

$$A^{\bullet} = \prod_{i=1}^{m}\left[c_i - \frac{5}{s}r_i, c_i + \frac{5}{s}r_i\right]$$

and observe that

$$r_{A^{\bullet}} = \frac{5}{s}r_A = 5\frac{(2R_A)^m}{(2r_1)\cdots(2r_m)}r_A \geq 5\frac{(R_A)^m}{r_A(R_A)^{m-1}}r_A = 5R_A$$

and $|A^{\bullet}| = (5/s)^m|A|$; in particular, $s(A^{\bullet}) = s$.

Note. Since the shape of a one-cell equals one, the present notation is consistent with that of Section 5.2.

Lemma 9.2.1 *Let $\mathcal{E}$ be a family of subcells of a cell A. There is a disjoint family $\mathcal{F} \subset \mathcal{E}$ such that for each $B \in \mathcal{E}$ we can find a $C \in \mathcal{F}$ with $B \cap C \neq \emptyset$ and $B \subset C^{\bullet}$. In particular,*

$$\bigcup \mathcal{E} \subset \bigcup\{C^{\bullet} : C \in \mathcal{F}\}.$$

PROOF. The proof is similar to that of Lemma 5.2.1. If $R = R_A$ is the outer radius of A, set $\mathcal{E}_n = \{B \in \mathcal{E} : R2^{-n} < R_B \leq R2^{-n+1}\}$ for $n = 1, 2, \ldots$. Unlike in the one-dimensional case, the family $\mathcal{E}_n$ may contain infinitely many disjoint cells. Nonetheless, a simple application of the *Hausdorff maximal principle* ([41, Chapter 1, Section 8, p. 24]) shows that each family $\mathcal{C} \subset \mathcal{E}_n$ still contains a maximal disjoint subfamily.

Let $\mathcal{F}_1$ be a maximal disjoint subfamily of $\mathcal{E}_1$, and assuming the families $\mathcal{F}_1 \subset \mathcal{E}_1, \ldots, \mathcal{F}_{n-1} \subset \mathcal{E}_{n-1}$ have been defined so that $\bigcup_{i=1}^{n-1} \mathcal{F}_i$ is disjoint, let $\mathcal{F}_n$ be a maximal disjoint subfamily of

$$\left\{ B \in \mathcal{E}_n : B \cap \left[\bigcup_{i=1}^{n-1} \left(\bigcup \mathcal{F}_i \right) \right] = \emptyset \right\}.$$

We claim that $\mathcal{F} = \bigcup_{n=1}^{\infty} \mathcal{F}_n$ is the desired family. Indeed, $\mathcal{F}$ is disjoint by construction. If $B \in \mathcal{E}$, then $B \in \mathcal{E}_n$ for a unique integer $n \geq 1$, and it follows from the maximality of $\mathcal{F}_n$ that B meets a cell $C \in \bigcup_{i=1}^{n} \mathcal{F}_i$. Since $R_B \leq 2^{-n+1} < 2R_C$, it is easy to verify that

$$B \subset U[c_B, R_B] \subset U[c_C, 5R_C] \subset U[c_{C^{\bullet}}, r_{C^{\bullet}}] \subset C^{\bullet}. \quad \blacksquare$$

Exercise 9.2.2 To each cell A associate a cube $A^{\star} = U[c_A, 5R_A]$. Show that $|A^{\star}| = [5/s(A)]^m |A|$, and that Lemma 9.2.1 remains valid when $C^{\bullet}$ is replaced by $C^{\star}$.

Definition 9.2.3 A family $\mathcal{E}$ of cells is called a *Vitali cover* of a set $E \subset \mathbf{R}^m$ if there is a positive function γ on E such that for each $x \in E$ and each $\eta > 0$ we can find a $B \in \mathcal{E}$ with $x \in B$, $B \subset U(x, \eta)$, and $s(B) > \gamma(x)$.

The *shape condition* in Definition 9.2.3, which is automatically satisfied when $m = 1$, plays a critical role in proving the m-dimensional version of the *Vitali covering theorem*.

Proposition 9.2.4 *If a family $\mathcal{E}$ of cells is a Vitali cover of a set $E \subset \mathbf{R}^m$, then there is a disjoint family $\mathcal{F} \subset \mathcal{E}$ such that $E - \bigcup \mathcal{F}$ is negligible.*

PROOF. Assume first that there is an integer $p \geq 2$ such that $B \subset U(0, p)$ and $s(B) > 1/p$ for every $B \in \mathcal{E}$. Let $\{C_1, C_2, \ldots\}$ be an enumeration of a family $\mathcal{F} \subset \mathcal{E}$ that satisfies the conditions of Lemma 9.2.1. Exactly as in the proof of Proposition 5.2.3, we show that $E - \bigcup_{i=1}^{n} C_i \subset \bigcup_{i>n} C_i^{\bullet}$ for $n = 1, 2, \ldots$. Thus $E - \bigcup \mathcal{F} = \emptyset$ if $\mathcal{F}$ is finite, and

$$\left| E - \bigcup \mathcal{F} \right| \leq \left| E - \bigcup_{i=1}^{n} C_i \right| \leq \left| \bigcup_{i=n+1}^{\infty} C_i^{\bullet} \right| \leq (5p)^m \sum_{i=n+1}^{\infty} |C_i|$$

for $n = 1, 2, \ldots$ if $\mathcal{F}$ is infinite. As $\sum_{C \in \mathcal{F}} |C| \leq (2p)^m$, we conclude that the set $E - \bigcup \mathcal{F}$ is always negligible.

In the general case, let γ be the positive function on E associated with $\mathcal{E}$ according to Definition 9.2.3, and let $E_p = \{x \in E \cap U(\mathbf{0}, p) : \gamma(x) > 1/p\}$ for $p = 2, 3, \ldots$. As the family

$$\mathcal{E}_2 = \left\{ B \in \mathcal{E} : B \subset U(\mathbf{0}, 2) \text{ and } s(B) > \frac{1}{2} \right\}$$

is a Vitali cover of the set E_2, it follows from the first part of the proof that there is a finite disjoint family $\mathcal{F}_2 \subset \mathcal{E}_2$ such that $|E_2 - \bigcup \mathcal{F}_2| < 1/2$. Since $\bigcup \mathcal{F}_2$ is a closed set, the family

$$\mathcal{E}_3 = \left\{ B \in \mathcal{E} : B \subset U(\mathbf{0}, 3) - \bigcup \mathcal{F}_2 \text{ and } s(B) > \frac{1}{3} \right\}$$

is a Vitali cover of the set $E_3 - \bigcup \mathcal{F}_2$. Again by the first part of the proof, there is a finite disjoint family $\mathcal{F}_3 \subset \mathcal{E}_3$ such that

$$\left| E_3 - \bigcup (\mathcal{F}_2 \cup \mathcal{F}_3) \right| < \frac{1}{3}.$$

In particular, $\mathcal{F}_2 \cup \mathcal{F}_3$ is a disjoint family. Proceeding inductively, we construct finite families $\mathcal{F}_p \subset \mathcal{E}$ so that the family $\mathcal{F} = \bigcup_{p=2}^{\infty} \mathcal{F}_p$ is disjoint and

$$\left| E_p - \bigcup \left(\bigcup_{n=2}^{p} \mathcal{F}_n \right) \right| < \frac{1}{p}$$

for $p = 2, 3, \ldots$. Since $E_2 \subset E_3 \subset \cdots$ and $E = \bigcup_{p=2}^{\infty} E_p$, it follows from Proposition 3.3.15 that

$$\left| E - \bigcup \mathcal{F} \right| = \lim \left| E_p - \bigcup \mathcal{F} \right| \leq \lim \left| E_p - \bigcup \left(\bigcup_{n=2}^{p} \mathcal{F}_n \right) \right| = 0. \quad \blacksquare$$

Corollary 9.2.5　*The union of any family $\mathcal{C}$ of cells is measurable.*

PROOF.　The family $\mathcal{K}$ of all subcubes of cells from $\mathcal{C}$ is a Vitali cover of $\bigcup \mathcal{C}$. By Proposition 9.2.4, there is a countable family $\mathcal{F} \subset \mathcal{K}$ such that the set $\bigcup \mathcal{C} - \bigcup \mathcal{F}$ is negligible. Since $\bigcup \mathcal{F}$ is a measurable subset of $\bigcup \mathcal{C}$, the corollary follows. $\blacksquare$

Warning. Corollary 9.2.5 is false if cells are replaced by intervals. Indeed, as singletons are intervals, this follows from the existence of a nonmeasurable set ([41, Chapter 3, Theorem 17, p. 64]).

Lemma 9.2.6 *Let r be a bounded positive function defined on a bounded set $E \subset \mathbf{R}^m$. There is an integer $q \geq 1$, depending only on the dimension m, and subsets $E_1, \ldots, E_q$ of E such that for $k = 1, \ldots, q$, the family $\{U[x, r(x)] : x \in E_k\}$ is disjoint and*

$$E \subset \bigcup_{k=1}^{q} \bigcup_{x \in E_k} U(x, r(x)).$$

PROOF. Up to some technical complications, the proof is the same as that of Lemma 5.2.4. Choose an $x_1 \in E$ so that $r(x_1) \geq \frac{3}{4}\sup\{r(x) : x \in E\}$. Proceeding inductively, assume that $x_1, \ldots, x_n$ have been chosen in E so that $x_i \notin \bigcup_{j=1}^{i-1} U(x_j, r(x_j))$ and

$$r(x_i) \geq \frac{3}{4}\sup\left\{r(x) : x \in E - \bigcup_{j=1}^{i-1} U(x_i, r(x_i))\right\}$$

for $i = 1, \ldots, n$. If $E \subset \bigcup_{i=1}^{n} U(x_i, r(x_i))$, the induction stops. Otherwise, select an $x_{n+1} \in E - \bigcup_{i=1}^{n} U(x_i, r(x_i))$ so that

$$r(x_{n+1}) \geq \frac{3}{4}\sup\left\{r(x) : x \in E - \bigcup_{i=1}^{n} U(x_i, r(x_i))\right\}.$$

The induction produces a finite or infinite sequence $\{x_1, x_2, \ldots\}$ of distinct points of E, and we let $r_n = r(x_n)$.

Claim 1. If $i \neq j$, then the cells $U[x_i, r_i/3]$ and $U[x_j, r_j/3]$ are disjoint.

Claim 2. $E \subset \bigcup_n U(x_n, r_n)$.

Claim 3. Each $U[x_n, r_n]$ intersects fewer than 26^m of the $U[x_i, r_i]$ with $i < n$ and $r_i \leq 9r_n$.

The proofs of Claims 1–3 are left to the reader; they are similar to those of the corresponding claims in the proof of Lemma 5.2.4.

Claim 4. Each $U[x_n, r_n]$ intersects no more than $6^m m^{m/2}$ of the $U[x_i, r_i]$ with $i < n$ and $r_i > 9r_n$.

Proof. Let $i < j < n$ be such that both $U[x_i, r_i]$ and $U[x_j, r_j]$ intersect $U[x_n, r_n]$ and $\min\{r_i, r_j\} > 9r_n$. Then $r_i \geq 3r_j/4$, and setting

$$a_i = \|x_i - x_n\|, \qquad a_j = \|x_j - x_n\|, \qquad a = \|x_i - x_j\|,$$

we have $a_i \leq r_i + r_n$ and $a_j \leq r_j + r_n$. Since x_n lies outside $U(x_i, r_i) \cup U(x_j, r_j)$ and x_j lies outside $U(x_i, r_i)$, we also have $r_i \leq a_i$, $r_j \leq a_j$, and $r_i \leq a$. From the identity

$$a^2 = \|(x_i - x_n) - (x_j - x_n)\|^2 = a_i^2 + a_j^2 - 2(x_i - x_n) \cdot (x_j - x_n),$$

we obtain

$$\frac{(x_i - x_n) \cdot (x_j - x_n)}{a_i a_j} = \frac{a_i^2 + a_j^2 - a^2}{2 a_i a_j} \leq \frac{(r_i + r_n)^2 + (r_j + r_n)^2 - r_i^2}{2 r_i r_j}$$

$$= \frac{r_n^2}{r_i r_j} + \frac{r_n}{r_i} + \frac{r_n}{r_j} + \frac{r_j}{2 r_i} < \frac{1}{9^2} + \frac{2}{9} + \frac{2}{3} = \frac{73}{81},$$

and hence

$$\left\| \frac{x_i - x_n}{a_i} - \frac{x_j - x_n}{a_j} \right\| = \left[2 - 2\frac{(x_i - x_n) \cdot (x_j - x_n)}{a_i a_j} \right]^{\frac{1}{2}} > \frac{4}{9}.$$

Since the unit sphere $S = \{x \in \mathbf{R}^m : \|x\| = 1\}$ is contained in $U[0, 1]$, an easy calculation reveals that S can be covered by $6^m m^{m/2}$ cubes whose diagonals are $1/3$ long. It follows from the pigeonhole principle that the last inequality is satisfied by no more than $6^m m^{m/2}$ points x_i, and the claim is proved.

Now letting $q = 26^m + 6^m m^{m/2}$ and using Claims 3 and 4, the proof is completed in exactly the same way as the proof of Lemma 5.2.4. $\blacksquare$

Remark 9.2.7 Definition 5.2.5 of a *centered Vitali cover*, as well as Proposition 5.2.6 and its proof, translate verbatim to the m-dimensional situation.

9.3 Derivability

Let α be a volume in a set $E \subset \mathbf{R}^m$, let F be any function defined on the family of all subcells of E, and let $x \in E$. If $\delta > 0$ and $\eta > 0$, we denote by $E_{\delta,\eta}$ the family of all cells $B \subset E \cap U(x, \delta)$ for which $x \in B$, $s(B) > \eta$, and the ratio $F(B)/\alpha(B)$ is defined. The extended real numbers

$$D_\alpha F(x) = \inf_{\eta > 0} \sup_{\delta > 0} \left[\inf_{B \in E_{\delta,\eta}} \frac{F(B)}{\alpha(B)} \right]$$

and $D^\alpha F(x) = -D_\alpha(-F)(x)$ are called, respectively, the *lower* and *upper* α-*derivates* of F at x. When $D_\alpha F(x) = D^\alpha F(x) \neq \pm\infty$, we denote this common value by $F'_\alpha(x)$ and say that F is α-*derivable* at x. If $m = 1$, it is clear that the above definitions of $D_\alpha F(x)$, $D^\alpha F(x)$, and $F'_\alpha(x)$ coincide with those introduced in Section 5.3.

Convention. We agree that the words *lower derivate*, *upper derivate*, and *derivable* always refer to λ_m-lower derivate, λ_m-upper derivate, and λ_m-derivable, respectively. We also agree to write $F'(x)$ instead of $F'_{\lambda_m}(x)$.

Exercise 9.3.1 Let F be a function defined on the family of all subcells of a set $E \subset \mathbf{R}^m$. Show that F is α-derivable at $x \in E$ if and only if the following conditions are satisfied.

1. There is a sequence $\{B_n\}$ of subcells of E such that
   - $x \in B_n$ and $F(B_n)/\alpha(B_n)$ is defined for $n = 1, 2, \ldots$;
   - $\lim d(B_n) = 0$ and $\inf s(B_n) > 0$.

2. For each sequence $\{B_n\}$ specified in the first condition, a finite limit $\lim[F(B_n)/\alpha(B_n)]$ exists.

Hint. If sequences $\{B_n\}$ and $\{C_n\}$ satisfy the first condition, then so does the sequence $\{B_1, C_1, B_2, C_2, \ldots\}$. Thus all limits in the second condition have the same value, equal to $F'_\alpha(x)$.

Exercise 9.3.2 Let F be a function defined on the family of all subcells of a set $E \subset \mathbf{R}^m$, and let α be a volume in E such that $0 < \alpha'(x) < +\infty$ for a point $x \in E$. Show that F is α-derivable at x if and only if it is derivable at x, in which case $F'_\alpha(x) = F'(x)/\alpha'(x)$.

Proposition 9.3.3 *Let α be a volume in a set $E \subset \mathbf{R}^m$, and let F be a function defined on the family of all subcells of E. For each $c \in \mathbf{R}$, the set $\{x \in E : D_\alpha F(x) < c\}$ is measurable. In particular, the extended real-valued function $x \mapsto D_\alpha F(x)$ on E is measurable whenever E is measurable.*

PROOF. Choose a $c \in \mathbf{R}$ and let $S = \{x \in E : D_\alpha F(x) < c\}$. It follows from the definition of $D_\alpha F(x)$ that $x \in S$ if and only if there is an integer $p \geq 1$ such that for each integer $q \geq 1$ we can find a cell $B \subset E$ containing x and having the following properties:

1. $d(B) < 1/q$ and $s(B) > 1/p$;
2. the ratio $F(B)/\alpha(B)$ is defined and smaller than $c - 1/p$.

Thus, if $S_{p,q}$ is the union of all cells $B \subset E$ having properties 1 and 2, then

$$S = \bigcup_{p=1}^{\infty} \bigcap_{q=1}^{\infty} S_{p,q}.$$

As $S_{p,q}$ is measurable by Corollary 9.2.5, the proposition follows from Theorem 3.3.7 and Lemma 3.6.3. ∎

Theorem 9.3.4 *Let A be a cell and $f \in \overline{\mathcal{R}}(A, \lambda_m)$. If F is the indefinite integral of f, then for almost all $x \in A$ the function F is derivable at x and $F'(x) = f(x)$.*

PROOF. The proof is similar to that of Theorem 5.3.5. We may assume that $f \in \mathcal{R}(A, \lambda_m)$. Let E be the set of all $x \in A$ for which either F is not derivable at x or $F'(x) \neq f(x)$. Given $x \in E$, we can find a $\gamma(x) > 0$ so that for each $\beta > 0$ there is a cell $B \subset A \cap U(x, \beta)$ with $x \in B$, $s(B) > \gamma(x)$, and

$$\left| f(x)|B| - F(B) \right| \geq \gamma(x)|B|.$$

Fix a positive integer n, let $E_n = \{x \in E : \gamma(x) \geq 1/n\}$, and choose an $\varepsilon > 0$. Using Henstock's lemma, find a positive function δ on A so that

$$\sum_{i=1}^{p} \left| f(x_i)|A_i| - F(A_i) \right| < \frac{\varepsilon}{n^{m+1}}$$

for each δ-fine partition $\{(A_1, x_1), \ldots, (A_p, x_p)\}$ in A. Let $\mathcal{E}$ be the collection of all cells $B \subset A$ such that $B \subset U(x_B, \delta(x_B))$ for a point $x_B \in B \cap E_n$, $s(B) > 1/n$, and

$$\left| f(x_B)|B| - F(B) \right| \geq \frac{|B|}{n}.$$

By Lemma 9.2.1, there is a disjoint subfamily $\mathcal{F}$ of $\mathcal{E}$ for which

$$E_n \subset \bigcup \{C^{\bullet} : C \in \mathcal{F}\}.$$

For each finite family $\mathcal{T} \subset \mathcal{F}$, the collection $\{(C, x_C) : C \in \mathcal{T}\}$ is a δ-fine $(1/n)$-shapely Perron partition in A. Hence

$$\sum_{C \in \mathcal{T}} |C| \leq n \sum_{C \in \mathcal{T}} \left| f(x_C)|C| - F(C) \right| < \frac{\varepsilon}{n^m}.$$

and consequently

$$|E_n| \le \sum_{C \in \mathcal{F}} |C^\bullet| < (5n)^m \sum_{C \in \mathcal{F}} |C| \le 5^m \varepsilon.$$

It follows from the arbitrariness of ε that E_n is negligible, and according to Proposition 3.5.3, 3, so is $E = \bigcup_{n=1}^{\infty} E_n$. ∎

In view of Proposition 9.2.4, the proof of the next theorem is identical to that of Theorem 5.3.8.

Theorem 9.3.5 *If α is a volume in a cell A, then α is derivable almost everywhere in A.* ∎

The immediate corollary of Theorem 9.3.5 is that each additive function of bounded variation, in particular each AC function, in a cell A is derivable almost everywhere in A (cf. Corollary 5.3.9).

The proof of the following lemma is the same as that of Lemma 5.3.13.

Lemma 9.3.6 *If F is an AC function on a cell A and $F' \ge 0$ almost everywhere in A, then $F \ge 0$.* ∎

Theorem 9.3.7 *An additive function F in a cell A is AC if and only if the following conditions are satisfied:*

1. *F is derivable almost everywhere in A;*
2. *$F' \in \overline{\mathcal{R}}(A, \lambda_m)$ and F is the indefinite integral of F'.*

PROOF. As the converse follows from Corollary 5.1.10 (see Remark 9.1.7), assume that F is an AC function. In view of Corollary 9.1.5 and Remark 9.1.7, it suffices to consider the case when F is a volume. By Theorem 9.3.5, the function F is derivable almost everywhere in A, and F' is a nonnegative measurable function by Proposition 9.3.3. According to Corollary 4.4.8, the functions $f_n = \min\{F', n\}$ belong to $\overline{\mathcal{R}}(A, \lambda_m)$ for $n = 1, 2, \ldots$, and we denote by F_n the indefinite integral of f_n. Now $F - F_n$ is an AC function and $(F - F_n)' = F' - f_n$ is nonnegative almost everywhere in A. Thus, by Lemma 9.3.6, we have

$$\lim \int_A f_n \, d\lambda_m = \lim F_n(A) \le F(A) < +\infty,$$

and the monotone convergence theorem implies that $F' \in \overline{\mathcal{R}}(A, \lambda_m)$. If G is the indefinite integral of F', then $F - G$ is AC and $(F - G)' = 0$ almost everywhere in A. Hence $F = G$ by Lemma 9.3.6, and the theorem is established. ∎

Definition 9.3.8 A volume α in a cell A is called *singular* whenever $\alpha' = 0$ almost everywhere in A.

A Dirac volume (Section 8.2) is an example of a singular volume. Since results analogous to Lemmas 5.4.2 and 5.4.3 are true for volumes (the proofs are verbatim translations of the one-dimensional ones), we have the following characterization of singular volumes (cf. Proposition 5.4.5).

Proposition 9.3.9 *A volume α in a cell A is singular if and only if there is an α_A-negligible set $E \subset A$ such that $|A - E| = 0$.* ∎

The *Lebesgue decomposition theorem* holds also for volumes, but it requires a different proof.

Theorem 9.3.10 *Let α be a volume in a cell A. There is a unique AC volume β and a unique singular volume γ such that $\alpha = \beta + \gamma$.*

PROOF. Let $\mathcal{E}$ be the family of all α_A-measurable sets $E \subset A$ such that $\alpha_A^*(E) > 0$ and $|E| = 0$. An application of the *Hausdorff maximal principle* ([41, Chapter 1, Section 8, p. 24]) shows that there is a maximal disjoint subfamily $\mathcal{C}$ of $\mathcal{E}$, and we let $C = \bigcup \mathcal{C}$. By Proposition 3.4.1, the family $\mathcal{C}$ is countable. Hence C is α_A-measurable and $|C| = 0$. Using Theorem 4.1.1, we define volumes β and γ in A by setting

$$\beta(B) = \int_B \chi_{A-C} \, d\alpha \qquad \text{and} \qquad \gamma(B) = \int_B \chi_C \, d\alpha$$

for each cell $B \subset A$. Clearly, $\alpha = \beta + \gamma$ and we show that β is AC and γ is singular.

Let $E \subset A$ be a G_δ set with $|E| = 0$. Then E is α_A-measurable as well as β_A-measurable, and Theorems 4.1.1 and 2.4.1 yield

$$\beta_A^*(E) = \int_A \chi_E \, d\beta = \int_A \chi_E \chi_{A-C} \, d\alpha = \int_A \chi_{E-C} \, d\alpha = \alpha_A^*(E - C) = 0 ;$$

the last equality follows from the maximality of $\mathcal{C}$. Since each negligible subset of A is contained in a G_δ negligible subset of A, we see that β is AC. Another application of Theorems 4.1.1 and 2.4.1 shows that

$$\gamma_A^*(A - C) = \int_A \chi_{A-C} \, d\gamma = \int_A \chi_{A-C} \chi_C \, d\alpha = 0 .$$

As $|C| = 0$, the volume γ is singular by Proposition 9.3.9.

If μ and ν are, respectively, AC and singular volumes in A such that $\alpha = \mu + \nu$, then $F = \beta - \mu = \nu - \gamma$ is an AC function in A and $F' = 0$ almost everywhere in A. It follows from Lemma 9.3.6 that $F = 0$, which establishes the uniqueness. $\blacksquare$

Corollary 9.3.11 *Let α be a volume in a cell A. Then $\alpha' \in \overline{\mathcal{R}}(A, \lambda_m)$ and*

$$\int_A \alpha' \, d\lambda_m \leq \alpha(A).$$

PROOF. By the Lebesgue decomposition theorem, $\alpha = \beta + \gamma$ where β is an AC volume and γ is a singular volume. Thus $\alpha' = \beta'$ almost everywhere in A and the corollary follows from Theorem 9.3.7. $\blacksquare$

Theorem 5.4.7 and its proof translate easily to $\mathbf{R}^m$. Defining the symmetric α-derivate F_α^{sym} in the obvious way, Theorem 5.4.9 together with its proof can be translated to higher dimensions as well.

Chapter 10
Change of variables

Throughout this chapter, the words "integral" and "integrable" always refer to integration with respect to the Lebesgue measure λ_m. If E is a measurable subset of $\mathbf{R}^m$ and $\Phi : E \to \mathbf{R}^m$ is a sufficiently nice map, we show how to calculate the measure of $\Phi(E)$. Our main result (Theorem 10.6.9) is a vast generalization of Corollary 2.4.10.

10.1 Integrating over a set

The reader should keep in mind that the boundary of each cell is negligible (Exercise 8.4.8). For a function f defined on a set $E \subset \mathbf{R}^m$, let

$$f^E(x) = \begin{cases} f(x) & \text{if } x \in E, \\ 0 & \text{if } x \in \mathbf{R}^m - E, \end{cases}$$

and observe that if E and f are measurable, then so is f^E. Using this notation we define the integral over bounded measurable sets as follows.

Let f be a function defined on a bounded measurable set E. We say that f is *integrable* over E whenever f^E belongs to $\mathcal{R}(A, \lambda_m)$ for a cell A containing E. In such a case, we set

$$\int_E f \, d\lambda_m = \int_A f^E \, d\lambda_m$$

and call this number the *integral* of f over E. As before, when no confusion is possible we write only $\int_E f$ instead of $\int_E f \, d\lambda_m$.

The family of all functions f that are integrable over a bounded measurable set E is denoted by $\mathcal{R}(E, \lambda_m)$ or simply by $\mathcal{R}(E)$. It follows from

Propositions 8.1.3 and 8.1.4 and Corollary 4.1.6 that neither the integrability of f nor the value of $\int_E f$ depends on the choice of a cell A containing E. In particular, if E is a cell, the present definition and notation are consistent with those of Section 8.1.

Note. While in principle the integral can be defined over arbitrary bounded sets, its good behavior is limited to the measurable sets only. In other words, we lose little by requiring that the integration domains be measurable.

Lemma 10.1.1 *Let E be a bounded measurable set and let $f \in \mathcal{R}(E, \lambda_m)$. Then $|f| \in \mathcal{R}(E, \lambda_m)$, and $f \in \mathcal{R}(S, \lambda_m)$ for each measurable set $S \subset E$.*

PROOF. Since $|f^E| = |f|^E$ and $f^S = f^E \chi_S$ for each set $S \subset E$, the lemma follows from Proposition 2.2.3 and Theorems 4.1.1 and 2.3.12. ∎

Lemma 10.1.2 *Let a bounded set E be the union of disjoint measurable sets $E_1, E_2, \ldots$, and let $f \geq 0$ be a function on E. Then $f \in \mathcal{R}(E, \lambda_m)$ if and only if $f \in \mathcal{R}(E_n, \lambda_m)$ for $n = 1, 2, \ldots$ and $\sum_{n=1}^{\infty} \int_{E_n} f \, d\lambda_m < +\infty$. In this case,*

$$\int_E f \, d\lambda_m = \sum_{n=1}^{\infty} \int_{E_n} f \, d\lambda_m \, .$$

PROOF. Observe that $f^E = \sum_{n=1}^{\infty} f^{E_n}$, and choose a cell A containing E. If $f \in \mathcal{R}(E_n)$ for $n = 1, 2, \ldots$ and $\sum_{n=1}^{\infty} \int_{E_n} f < +\infty$, then it follows from Theorem 2.3.13 that $f \in \mathcal{R}(E)$ and

$$\int_E f = \int_A f^E = \sum_{n=1}^{\infty} \int_A f^{E_n} = \sum_{n=1}^{\infty} \int_{E_n} f \, .$$

Conversely, if $f \in \mathcal{R}(E)$, then $f \in \mathcal{R}(E_n)$ for $n = 1, 2, \ldots$ by Lemma 10.1.1. Since

$$\sum_{n=1}^{k} \int_{E_n} f = \sum_{n=1}^{k} \int_A f^{E_n} = \int_A \left(\sum_{n=1}^{k} f^{E_n} \right)$$

$$\leq \int_A f^E = \int_E f < +\infty$$

for $k = 1, 2, \ldots$, we see that $\sum_{n=1}^{\infty} \int_{E_n} f < +\infty$. ∎

Exercise 10.1.3 Use Example 2.4.12 to show that the assumption $f \geq 0$ in Lemma 10.1.2 is essential.

Lemma 10.1.4   *Let C and $\mathcal{D}$ be countable families of disjoint bounded measurable sets, and let $E = \bigcup C = \bigcup \mathcal{D}$. If $f \geq 0$ is a function on E that is integrable over each $B \in C \cup \mathcal{D}$, then*

$$\sum_{C \in C} \int_C f \, d\lambda_m = \sum_{D \in \mathcal{D}} \int_D f \, d\lambda_m .$$

In particular, both series converge or diverge simultaneously.

PROOF.   Lemma 10.1.2 and the usual theorem about switching summations in double series with nonnegative terms ([33, Exercise (5–9), p. 65]) yield

$$\sum_{C \in C} \int_C f = \sum_{C \in C} \left(\sum_{D \in \mathcal{D}} \int_{C \cap D} f \right) = \sum_{D \in \mathcal{D}} \left(\sum_{C \in C} \int_{C \cap D} f \right) = \sum_{D \in \mathcal{D}} \int_D f . \ \blacksquare$$

Definition 10.1.5   Let f be a function defined on a measurable set E. We say that f is *integrable* over E whenever E is the union of disjoint bounded measurable sets $E_1, E_2, \ldots$ such that $f \in \mathcal{R}(E_n, \lambda_m)$ for $n = 1, 2, \ldots$ and $\sum_{n=1}^{\infty} \int_{E_n} |f| \, d\lambda_m < +\infty$. In such a case, we set

$$\int_E f \, d\lambda_m = \sum_{n=1}^{\infty} \int_{E_n} f \, d\lambda_m$$

and call this number the *integral* of f over E.

It follows from Lemma 10.1.4 that neither the integrability of f nor the value of $\int_E f \, d\lambda_m$ depends on the choice of the sets $E_1, E_2, \ldots$. Moreover, Lemma 10.1.2 shows that Definition 10.1.5 is consistent with that given at the beginning of this section. Thus, with no danger of confusion, we denote by $\mathcal{R}(E, \lambda_m)$ or $\mathcal{R}(E)$ the family of all integrable functions on any measurable set E, bounded or not.

Virtually all properties we established for the integral over cells are still true for the integral over measurable sets. Proving these properties is easy in most cases. The following exercises should give the reader some feeling for what is involved.

Exercise 10.1.6   Denote by $\mathcal{K}$ the family of all dyadic cubes of the same fixed diameter. For a function f on a measurable set E prove the following facts.

1. The function f belongs to $\mathcal{R}(E, \lambda_m)$ if and only if $f^E \in \mathcal{R}(K, \lambda_m)$ for each $K \in \mathcal{K}$ and $\sum_{K \in \mathcal{K}} \int_K |f^E| \, d\lambda_m < +\infty$.

2. If $f \in \mathcal{R}(E, \lambda_m)$, then $\int_E f \, d\lambda_m = \sum_{K \in \mathcal{K}} \int_K f^E \, d\lambda_m$.

Exercise 10.1.7 If E is a measurable set, prove that the family $\mathcal{R}(E, \lambda_m)$ is a linear space and the map $f \mapsto \int_E f \, d\lambda_m$ is a nonnegative linear functional on $\mathcal{R}(E, \lambda_m)$.

Exercise 10.1.8 Show that Lemmas 10.1.1 and 10.1.2 hold for any set E, whether it is bounded or not.

The straightforward proofs of the next two propositions are left to the reader.

Proposition 10.1.9 *Let f be a function on a measurable set E. Then f belongs to $\mathcal{R}(E, \lambda_m)$ if and only if f^+ and f^- do, in which case*

$$\int_E f \, d\lambda_m = \int_E f^+ \, d\lambda_m - \int_E f^- \, d\lambda_m . \quad \blacksquare$$

Proposition 10.1.10 *Let f and g be functions on a measurable set E, and let $f = g$ almost everywhere in E. Then f belongs to $\mathcal{R}(E, \lambda_m)$ if and only if g does, in which case $\int_E f \, d\lambda_m = \int_E g \, d\lambda_m$.* $\blacksquare$

Proposition 10.1.10 allows us to introduce, in the obvious way, the family $\overline{\mathcal{R}}(E, \lambda_m) = \overline{\mathcal{R}}(E)$ of all integrable extended real-valued functions defined almost everywhere in a measurable set E.

Exercise 10.1.11 Show that the monotone convergence stated in Proposition 4.1.10 holds for functions defined on a measurable set. *Hint.* Prove it first for a bounded measurable set, then extend it using Proposition 10.1.9 and Exercise 10.1.6.

Exercise 10.1.12 Let E be a measurable set and let $f \in \overline{\mathcal{R}}(E, \lambda_m)$ be nonnegative. From Remark 3.3.6 recall the definition of a measure μ on a σ-algebra $\mathcal{M}$ in a set X, and prove the following facts.

1. The family $\mathcal{M}$ of all measurable subsets of E is a σ-algebra in E.
2. The function $\mu : B \mapsto \int_B f \, d\lambda_m$ is a measure on $\mathcal{M}$.
3. If $B \in \mathcal{M}$ and $\mu(B) = 0$, then $f = 0$ λ_m-almost everywhere in B.
4. Given $\varepsilon > 0$, there is an $\eta > 0$ such that $\mu(B) < \varepsilon$ for each $B \in \mathcal{M}$ with $|B| < \eta$. *Hint.* Observe that $\mu(B) = 0$ for each $B \in \mathcal{M}$ with $|B| = 0$, and adapt the argument used in the proof of Proposition 5.1.8.

Let $m = r + s$ where r and s are positive integers, and let f be an extended real-valued function on a set $E \subset \mathbf{R}^m$. If $x \in \mathbf{R}^r$, we let $E_x = \{y \in \mathbf{R}^s : (x, y) \in E\}$ and define f_x on E_x by setting $f_x(y) = f(x, y)$ for each $y \in E_x$. Note that for some $x \in \mathbf{R}^r$, the set E_x may be empty. In this case, f_x is the empty function and $\int_{E_x} f_x \, d\lambda_r = 0$, since $(f_x)^{E_x}$ is the zero function on $\mathbf{R}^r$. Employing the notation of this paragraph, we generalize the Fubini theorem.

Theorem 10.1.13 *Let f be an extended real-valued function defined on a measurable set $E \subset \mathbf{R}^m$. If $f \in \overline{\mathcal{R}}(E, \lambda_m)$, then for λ_r-almost all $x \in \mathbf{R}^r$, the set E_x is λ_s-measurable and $f_x \in \overline{\mathcal{R}}(E_x, \lambda_s)$. Moreover, the function $F : x \mapsto \int_{E_x} f_x \, d\lambda_s$ belongs to $\overline{\mathcal{R}}(\mathbf{R}^r, \lambda_r)$ and*

$$\int_E f \, d\lambda_m = \int_{\mathbf{R}^r} F \, d\lambda_r .$$

PROOF. In view of Proposition 10.1.9, we may assume that f is nonnegative. Let $A^1, A^2, \ldots$ be nonoverlapping r-cells whose union is $\mathbf{R}^r$, and let $B^1, B^2, \ldots$ be nonoverlapping s-cells whose union is $\mathbf{R}^s$. For $i, j = 1, 2, \ldots$, let $E^{i,j} = E \cap (A^i \times B^j)$ and $f^{i,j} = f^{E^{i,j}}$. By Corollary 8.6.4 and Exercise 8.4.10, the set $E_x^{i,j}$ is a λ_s-measurable subset of B^j for λ_r-almost all $x \in A^i$. Since $E_x = \bigcup_{j=1}^\infty E_x^{i,j}$ for all $x \in A^i$, we conclude that E_x is λ_s-measurable for λ_r-almost all $x \in \mathbf{R}^r$. The Fubini theorem (Theorem 8.6.6) implies the following facts:

1. $f_x^{i,j} \in \overline{\mathcal{R}}(B^j, \lambda_s)$ for λ_r-almost all $x \in A^i$;
2. the function $F^{i,j} : x \mapsto \int_{B^j} f_x^{i,j} \, d\lambda_s$ belongs to $\overline{\mathcal{R}}(A^i, \lambda_r)$;
3. $\int_{A^i} F^{i,j} \, d\lambda_r = \int_{A^i \times B^j} f^{i,j} \, d\lambda_m = \int_{E^{i,j}} f \, d\lambda_m$.

Fix an integer $i \geq 1$, and let $E^i = \bigcup_{j=1}^\infty E^{i,j}$. By Exercise 10.1.8 and Theorem 2.3.13, the extended real-valued function $F^i = \sum_{j=1}^\infty F^{i,j}$ belongs to $\overline{\mathcal{R}}(A^i, \lambda_r)$ and $\int_{A^i} F^i \, d\lambda_r = \int_{E^i} f \, d\lambda_m$. Since $F^i(x) < +\infty$ and

$$F^{i,j}(x) = \int_{B^j} f_x^{i,j} \, d\lambda_s = \int_{E_x \cap B^j} f_x \, d\lambda_s$$

for λ_r-almost all $x \in A^i$, it follows from Exercise 10.1.8 that $f_x \in \overline{\mathcal{R}}(E_x, \lambda_s)$ and $F^i(x) = \int_{E_x} f_x \, d\lambda_s$ for λ_r-almost all $x \in A^i$. Thus $F^i(x) = F(x)$ for λ_r-almost all $x \in A^i$, and another application of Exercise 10.1.8 yields

$$\int_{\mathbf{R}^r} F \, d\lambda_r = \sum_{i=1}^\infty \int_{A^i} F^i \, d\lambda_r = \sum_{i=1}^\infty \int_{E^i} f \, d\lambda_m = \int_E f \, d\lambda_m . \quad \blacksquare$$

Exercise 10.1.14 Let $E \subset \mathbf{R}^m$ be a measurable set, and let $f \geq 0$ be a measurable extended real-valued function on E such that $x \mapsto \int_{E_x} f_x \, d\lambda_s$ belongs to $\overline{\mathcal{R}}(\mathbf{R}^r, \lambda_r)$. Prove that $f \in \overline{\mathcal{R}}(E, \lambda_m)$ and compare this result with Tonelli's theorem (Proposition 8.6.7).

Exercise 10.1.15 Let $E \subset \mathbf{R}^m$ be a measurable set. Show that $|E| = 0$ if and only if E_x is λ_s-negligible for λ_r-almost all $x \in \mathbf{R}^r$.

10.2 Luzin maps

Let $E \subset \mathbf{R}^m$. A continuous map $\Phi : E \to \mathbf{R}^m$ is called *Luzin* if it satisfies the following conditions:

1. $|\Phi(N)| = 0$ for each negligible set $N \subset E$;
2. $|\Phi(B)| < +\infty$ for each bounded set $B \subset E$.

Remark 10.2.1 The essence of Luzin's map is condition 1, known as the *Luzin condition* (cf. [44, Chapter VII, Section 6]). Condition 2 has been introduced merely for convenience. It simplifies the presentation and it is satisfied by Lipschitz maps (Proposition 10.3.4 below). Some indication on how to get by without condition 2 is given in Section 10.6 (see also the proof of Theorem 10.2.6 below).

Proposition 10.2.2 *Let E be a measurable set. If $\Phi : E \to \mathbf{R}^m$ is a Luzin map, then $\Phi(E)$ is measurable.*

PROOF. It follows from Theorem 3.3.13 and Exercise 3.3.12, 5 that E is the union of a sequence $\{K_n\}$ of compact sets and a negligible set N. Hence $\Phi(E)$ is the union of the sequence $\{\Phi(K_n)\}$ and the negligible set $\Phi(N)$. As Φ is continuous, the sets $\Phi(K_n)$ are compact ([42, Theorem 4.14, p. 89]) and the proposition follows. ∎

Throughout the remainder of this section, we assume that E is a measurable set and that $\Phi : E \to \mathbf{R}^m$ is an injective Luzin map. For each cell A, we let

$$\varphi(A) = |\Phi(E \cap A)|.$$

Proposition 10.2.3 *The function φ is a volume in $\mathbf{R}^m$. If S is a measurable set, then S is φ-measurable and*

$$\varphi^*(S) = |\Phi(E \cap S)|.$$

PROOF. If A is a cell, then $|\Phi(E \cap \partial A)| = 0$. Thus $\varphi(A) = |\Phi(E \cap A^\circ)|$, and since Φ is injective, it is easy to deduce that φ is a volume in $\mathbf{R}^m$. Let A be a cell, and find cells $B_1 \subset B_2 \subset \cdots$ and $C_1 \supset C_2 \supset \cdots$ so that

$$A^\circ = \bigcup_{n=1}^{\infty} B_n \quad \text{and} \quad A = \bigcap_{n=1}^{\infty} C_n^\circ.$$

Propositions 8.4.1 and 3.3.15 together with Exercise 3.3.16 yield

$$\varphi^*(A) \le \inf \varphi(C_n) = \varphi(A) = \sup \varphi(B_n) \le \varphi^*(A^\circ) \le \varphi^*(A),$$

and consequently $\varphi^*(A) = \varphi^*(A^\circ) = \varphi(A)$. As each open set U is the union of a sequence $\{K_n\}$ of nonoverlapping dyadic cubes, we obtain

$$\varphi^*(U) \le \sum_{i=1}^{\infty} \varphi^*(K_i) = \sum_{i=1}^{\infty} \varphi(K_i) = \sum_{i=1}^{\infty} \varphi^*(K_i^\circ) \le \varphi^*(U),$$

$$|\Phi(E \cap U)| \le \sum_{i=1}^{\infty} |\Phi(E \cap K_i)| = \sum_{i=1}^{\infty} \varphi(K_i) = \sum_{i=1}^{\infty} |\Phi(E \cap K_i^\circ)| \le |\Phi(E \cap U)|,$$

and hence $\varphi^*(U) = |\Phi(E \cap U)|$. This and Exercise 3.3.16 imply that

$$\varphi^*(D) = |\Phi(E \cap D)|$$

for each bounded G_δ set D. Since $D = \bigcup_{n=1}^{\infty}[D \cap U(\mathbf{0}, n)]$, the equality still holds if D is an arbitrary G_δ set (see Proposition 3.3.15). From this and Lemma 3.3.14 we infer that each negligible set is φ-negligible. If S is a measurable set, then $S = D - N$ where D is a G_δ set and $N \subset D$ is a negligible set (Theorem 3.3.13). Thus S is φ-measurable and

$$\varphi^*(S) = \varphi^*(D) - \varphi^*(N) = |\Phi(E \cap D)| - |\Phi(E \cap N)| = |\Phi(E \cap S)|. \quad \blacksquare$$

Corollary 10.2.4 *The derivate φ' exists almost everywhere in $\mathbf{R}^m$. If S is a bounded measurable set, then $\varphi' \in \overline{\mathcal{R}}(S, \lambda_m)$ and*

$$|\Phi(E \cap S)| = \int_S \varphi' \, d\lambda_m.$$

PROOF. By Theorem 9.3.5, the derivate φ' exists almost everywhere in each cell, and as $\mathbf{R}^m$ is a countable union of cells, it exists almost everywhere in $\mathbf{R}^m$. Let S be a bounded measurable set, and let A be a cell containing S.

Since $\varphi_A^* \leq \varphi^*$, Propositions 10.2.3 and 9.1.6 imply that φ is an AC volume in A. Using Theorem 4.1.1, Proposition 3.7.3, and Remark 5.3.16, we obtain

$$|\Phi(E \cap S)| = \varphi^*(S) = \int_A \chi_S \, d\varphi = \int_A \chi_S \cdot \varphi' \, d\lambda_m = \int_S \varphi' \, d\lambda_m \, ;$$

for $(\varphi')^S = \chi_S \cdot \varphi'$ almost everywhere in $\mathbf{R}^m$. ∎

Recall that a set $U \subset E$ is *relatively* open in E if and only if there is an open set $V \subset \mathbf{R}^m$ such that $U = E \cap V$ ([42, Theorem 2.30, p. 36]). Consequently, $G \subset E$ is a relative G_δ set in E if and only if there is a G_δ set $H \subset \mathbf{R}^m$ such that $G = E \cap H$. In particular, each relative G_δ subset of E is measurable. It follows from the continuity of Φ that $\Phi^{-1}(T)$ is a relative G_δ subset of E whenever T is a relative G_δ subset of $\Phi(E)$.

Lemma 10.2.5 *Assume that E is bounded. If $S \subset E$ is such that $\Phi(S)$ is measurable, then $\chi_S \cdot \varphi' \in \overline{\mathcal{R}}(E, \lambda_m)$ and*

$$|\Phi(S)| = \int_E \chi_S \cdot \varphi' \, d\lambda_m \, .$$

PROOF. If $T = \Phi(S)$ is relatively G_δ, then $S = \Phi^{-1}(T)$ is measurable. Since $\int_E \chi_S \cdot \varphi' = \int_S \varphi'$, the lemma follows from Corollary 10.2.4. If T is negligible, it is contained in a negligible set $V \subset E$ that is relatively G_δ. By Corollary 10.2.4 and Exercise 10.1.12, 3, we have $\varphi' = 0$ almost everywhere in $U = \Phi^{-1}(V)$, and the lemma is again true. If T is an arbitrary measurable set, then $T = V - N$ where V is a G_δ set and N is a negligible subset of V. Thus, letting $U = \Phi^{-1}(V)$ and $M = \Phi^{-1}(N)$, we obtain

$$|T| = |V| - |N| = \int_E \chi_U \cdot \varphi' - \int_E \chi_M \cdot \varphi' = \int_E \chi_S \cdot \varphi' \, . \quad ∎$$

Note. In Lemma 10.2.5, the set S need not be measurable; nonetheless, the zeros of φ' make the function $\chi_S \cdot \varphi'$ integrable.

Theorem 10.2.6 *If $f \in \overline{\mathcal{R}}(\Phi(E), \lambda_m)$, then $(f \circ \Phi)\varphi'$ belongs to $\overline{\mathcal{R}}(E, \lambda_m)$ and*

$$\int_E (f \circ \Phi)\varphi' \, d\lambda_m = \int_{\Phi(E)} f \, d\lambda_m \, .$$

PROOF. Considering Proposition 10.1.9, we may assume that f is nonnegative. If $A = \prod_{i=1}^m [a_i, b_i]$ is a cell, let $A^\circ = \prod_{i=1}^m [a_i, b_i)$ and observe that A° is a G_δ set. Let $K_1, K_2, \ldots$ be an enumeration of all dyadic cubes whose diameters are equal to one. The sets $E_{i,j} = (E \cap K_i^\diamond) \cap \Phi^{-1}(K_j^\diamond)$ are disjoint bounded and measurable, and so are the sets $\Phi(E_{i,j}) = \Phi(E \cap K_i^\diamond) \cap K_j^\diamond$. Since $E = \bigcup_{i,j=1}^\infty E_{i,j}$, it suffices to prove the theorem under the assumption that both sets E and $\Phi(E)$ are bounded.

Given a measurable set $T \subset \Phi(E)$, let $S = \Phi^{-1}(T)$ and observe that $\chi_S(x) = \chi_T \circ \Phi(x)$ for each $x \in E$. By Lemma 10.2.5, the function $(\chi_T \circ \Phi)\varphi'$ belongs to $\overline{\mathcal{R}}(E)$ and

$$\int_{\Phi(E)} \chi_T = |T| = \int_E (\chi_T \circ \Phi)\varphi' .$$

Now let $\mathcal{S}$ be the family of all nonnegative simple functions s with $s(x) = 0$ for each $x \in \mathbf{R}^m - \Phi(E)$. According to Exercise 10.1.7, for each $s \in \mathcal{S}$ the function $(s \circ \Phi)\varphi'$ belongs to $\overline{\mathcal{R}}(E)$ and

$$\int_{\Phi(E)} s = \int_E (s \circ \Phi)\varphi' .$$

It follows from Proposition 3.6.14 that there is a sequence $\{s_n\}$ in $\mathcal{S}$ such that $s_n \nearrow f$ in $\Phi(E)$, and hence $(s_n \circ \Phi)\varphi' \nearrow (f \circ \Phi)\varphi'$ almost everywhere in E (recall that φ' is defined only almost everywhere in E). By the monotone convergence theorem (Exercise 10.1.11), $(f \circ \Phi)\varphi'$ belongs to $\overline{\mathcal{R}}(E)$ and

$$\int_{\Phi(E)} f = \lim \int_{\Phi(E)} s_n = \lim \int_E (s_n \circ \Phi)\varphi' = \int_E (f \circ \Phi)\varphi' . \ \blacksquare$$

Exercise 10.2.7 Prove that Lemma 10.2.5 holds without the assumption that the set E is bounded.

Our next goal is to show that for certain Luzin maps $\Phi = (f_1, \ldots, f_m)$ the derivate φ' can be expressed in terms of the partial derivatives of the functions $f_1, \ldots, f_m$.

10.3 Lipschitz maps

We present first a *Hausdorff-type evaluation* of the Lebesgue measure λ_m (cf. Section 12.4 below, in particular Remark 12.5.3).

Theorem 10.3.1 *Let $E \subset \mathbf{R}^m$ and $0 < \eta \leq +\infty$. Then*

$$|E| = \inf \sum_{i=1}^\infty [d(C_i)]^m$$

where the infimum is taken over all sequences $\{C_i\}$ of nonempty subsets of $\mathbf{R}^m$ such that $E \subset \bigcup_{i=1}^\infty C_i$ and $d(C_i) < \eta$ for $i = 1, 2, \ldots$.

PROOF. Let t denote the right side of the desired equality. If C is a nonempty subset of $\mathbf{R}^m$ and $d(C) < +\infty$, then C is contained in a cube K of diameter $d(C)$ and we see that

$$|C| \leq |K| = [d(C)]^m .$$

Thus, if $\{C_i\}$ is a sequence of nonempty subsets of $\mathbf{R}^m$ with $E \subset \bigcup_{i=1}^{\infty} C_i$ and $d(C_i) < \eta$ for $i = 1, 2, \ldots$, then

$$|E| \leq \sum_{i=1}^{\infty} |C_i| \leq \sum_{i=1}^{\infty} [d(C_i)]^m .$$

The inequality $|E| \leq t$ follows.

Proceeding towards a contradiction, suppose that $|E| < t$ and use Proposition 3.2.8 to find an open set $U \subset \mathbf{R}^m$ so that $E \subset U$ and $|U| < t$. It is easy to verify that U is the union of all dyadic cubes contained in U whose diameters are less than η. By the fundamental property of dyadic cubes, the set U is the union of a sequence $\{K_i\}$ of nonoverlapping dyadic cubes such that $d(K_i) < \eta$ for $i = 1, 2, \ldots$. Hence $E \subset \bigcup_{i=1}^{\infty} K_i$ and

$$\sum_{i=1}^{\infty} [d(K_i)]^m = \sum_{i=1}^{\infty} |K_i| = |U| < t,$$

which is a contradiction. ∎

Remark 10.3.2 Theorem 10.3.1 depends on our definition of the diameter introduced at the beginning of Chapter 7. If $d(C_i)$ is replaced by the usual *Euclidean diameter*

$$d_e(C_i) = \sup\{\|x - y\| : x, y \in C_i\}$$

of C_i, the theorem still holds, provided the right side of the equality is multiplied by the number

$$\omega_m = \left| \left\{ x \in \mathbf{R}^m : \|x\| < \frac{1}{2} \right\} \right| .$$

This is a deeper result, which we shall not need; the interested reader is referred to [9, Section 2.2]. Note that ω_m is the m-dimensional Lebesgue measure of the m-dimensional ball whose Euclidean diameter equals one.

The concept of a Lipschitz function defined in Section 6.5 generalizes easily to higher dimensions (cf. Exercise 6.6.6, 1).

Let $E \subset \mathbf{R}^m$, and let $n \geq 1$ be an integer. A map $\Phi : E \to \mathbf{R}^n$ is called *Lipschitz* if there is a real number c such that

$$|\Phi(y) - \Phi(x)| \leq c|y - x|$$

for each $x, y \in E$. The least number $c \geq 0$ for which the above inequality holds exists and it is called the *Lipschitz constant* of Φ, denoted by $\mathrm{Lip}(\Phi)$. If $\Phi = (f_1, \ldots, f_n)$, then Φ is Lipschitz if and only if the functions $f_1, \ldots, f_n$ viewed as maps from E to $\mathbf{R}$ are Lipschitz, in which case

$$\mathrm{Lip}(\Phi) = \max\{\mathrm{Lip}(f_1), \ldots, \mathrm{Lip}(f_n)\}.$$

The following useful result, called the *Kirszbraun theorem*, generalizes Proposition 6.6.5.

Theorem 10.3.3 *If $E \subset \mathbf{R}^m$, then each Lipschitz map $\Phi : E \to \mathbf{R}^n$ can be extended to a Lipschitz map $\Psi : \mathbf{R}^m \to \mathbf{R}^n$ so that $\mathrm{Lip}(\Psi) = \mathrm{Lip}(\Phi)$.*

PROOF. Let $\Phi = (f_1, \ldots, f_n)$. Proceeding exactly as in the proof of Proposition 6.6.5, we show that each Lipschitz function f_i on E can be extended to a Lipschitz function g_i on $\mathbf{R}^m$ so that $\mathrm{Lip}(g_i) = \mathrm{Lip}(f_i)$ (cf. Exercise 6.6.6, 1 and 2). Thus $\Psi = (g_1, \ldots, g_n)$ is the desired extension of Φ. ∎

Note. The true Kirszbraun theorem is a deeper result, which deals with Lipschitz maps with respect to the usual Euclidean metric in $\mathbf{R}^m$ and $\mathbf{R}^n$ (cf. [11, Theorem 2.10.43, p. 201]).

Proposition 10.3.4 *If $\Phi : \mathbf{R}^m \to \mathbf{R}^m$ is a Lipschitz map, then*

$$|\Phi(S)| \leq [\mathrm{Lip}(\Phi)]^m \cdot |S|$$

for each set $S \subset \mathbf{R}^m$; in particular, Φ is a Luzin map.

PROOF. The inequality, which is a direct consequence of Theorem 10.3.1, implies that $|\Phi(B)| < +\infty$ for each bounded set B and $|\Phi(N)| = 0$ for each negligible set N. ∎

Let $E \subset \mathbf{R}^m$. Recall that an injective continuous map $\Phi : E \to \mathbf{R}^m$ whose inverse $\Phi^{-1} : \Phi(E) \to \mathbf{R}^m$ is also continuous is called a *homeomorphism* (see [41, Chapter 7, Section 3, p. 131]). If both Φ and Φ^{-1} are Lipschitz maps, we say that Φ is a *lipeomorphism*. In the older terminology, the name *bi-Lipschitz map* is usually used in lieu of lipeomorphism.

Exercise 10.3.5 Let $E \subset \mathbf{R}^m$ and let $\Phi : E \to \mathbf{R}^m$ be a lipeomorphism. Show that there is a unique lipeomorphism $\Psi : E^- \to \mathbf{R}^m$ such that $\Psi(x) = \Phi(x)$ for each $x \in E$.

Exercise 10.3.6 Consider the *polar coordinates* given by the map

$$\Phi : (r, \theta) \mapsto (r \cos \theta, r \sin \theta)$$

for each $(r, \theta) \in \mathbf{R}^2$, and let $E = (0, 1] \times (0, 2\pi]$. Prove the following facts.

1. The map Φ on E is a Lipschitz homeomorphism.
2. For each $r \in (0, 1)$, the map Φ on $[r, 1] \times [r, 2\pi]$ is a lipeomorphism.
3. The map $\Phi^{-1} : \Phi(E) \to \mathbf{R}^m$ is Luzin but not Lipschitz.

Exercise 10.3.7 Let C and α be, respectively, the Cantor ternary set and Cantor ternary function (see Remark 5.3.12). Define a map $\Phi : [0, 1] \to \mathbf{R}$ by setting $\Phi(x) = x + \alpha(x)$ for each $x \in [0, 1]$, and show that the following statements are true.

1. The map Φ is a homeomorphism and $\Phi([0, 1]) = [0, 2]$.
2. The inverse map $\Phi^{-1} : [0, 2] \to \mathbf{R}$ is Lipschitz with $\mathrm{Lip}(\Phi^{-1}) = 1$.
3. The map Φ is not Luzin. *Hint.* The set $\alpha([0, 1] - C)$ is countable.

An important special case of Lipschitz maps are linear maps. The Lipschitz constant of a linear map Φ is equal to its norm:

$$\mathrm{Lip}(\Phi) = \sup\{|\Phi(x)| : x \in U[\mathbf{0}, 1]\}.$$

If Φ is a linear map of $\mathbf{R}^m$, we let $J_\Phi = |\Phi([0, 1]^m)|$ where $[0, 1]^m$ is the m-fold Cartesian product of the unit interval $[0, 1]$.

Proposition 10.3.8 *If Φ is a bijective linear map of $\mathbf{R}^m$, then*

$$|\Phi(E)| = J_\Phi |E|$$

for each set $E \subset \mathbf{R}^m$.

PROOF. By the linearity of Φ and Exercise 8.4.7, 1, we obtain that

$$|\Phi(E + z)| = |\Phi(E) + \Phi(z)| = |\Phi(E)|$$

for each $E \subset \mathbf{R}^m$ and each $z \in \mathbf{R}^m$. If K is a dyadic cube of diameter 2^{-k}, then $[0, 1]^m$ is the union of 2^{km} nonoverlapping translates of K. As Φ is a

Luzin map, we see that $J_\Phi = 2^{km}|\Phi(K)|$ or alternatively, $|\Phi(K)| = J_\Phi|K|$. Since each open set U is the union of a sequence $\{K_i\}$ of nonoverlapping dyadic cubes, we have

$$|\Phi(U)| = \sum_{i=1}^{\infty} |\Phi(K_i)| = J_\Phi \sum_{i=1}^{\infty} |K_i| = J_\Phi|U|.$$

An application of Proposition 3.2.8 completes the argument. $\blacksquare$

Corollary 10.3.9 *If Φ and Ψ are bijective linear maps of $\mathbf{R}^m$, then*

$$J_{\Phi \circ \Psi} = J_\Phi \cdot J_\Psi. \quad \blacksquare$$

An orthogonal map of $\mathbf{R}^m$ is called a *rotation*. A map $\Delta : \mathbf{R}^m \to \mathbf{R}^m$ is called a *diagonal map* determined by real numbers $d_1, \ldots, d_m$ whenever

$$\Delta(\xi_1, \ldots, \xi_m) = (\xi_1 d_1, \ldots, \xi_m d_m)$$

for each point $(\xi_1, \ldots, \xi_m)$ in $\mathbf{R}^m$. Clearly, each diagonal map is linear.

Lemma 10.3.10 *If Φ is a bijective linear map of $\mathbf{R}^m$, then $\Phi = \Theta \circ \Delta \circ \Psi$ where Θ and Ψ are rotations and Δ is a diagonal map determined by positive real numbers.*

PROOF. Since $B(x, y) = \Phi(x) \cdot \Phi(y)$ is a positively definite symmetric bilinear form on $\mathbf{R}^m$, there are orthonormal vectors $v_1, \ldots, v_m$ in $\mathbf{R}^m$ and positive real numbers $d_1, \ldots, d_m$ such that

$$B(v_i, v_j) = \begin{cases} d_i & \text{if } i = j, \\ 0 & \text{if } i \neq j, \end{cases}$$

for $i, j = 1, \ldots, m$ ([1, Chapter 7, Theorem (5.8), p. 255]). If $e_1, \ldots, e_m$ is the standard orthonormal base in $\mathbf{R}^m$, define a rotation Ψ by $\Psi(v_i) = e_i$ for $i = 1, \ldots, m$, and let

$$\Delta(\xi_1, \ldots, \xi_m) = (\xi_1 \sqrt{d_1}, \ldots, \xi_m \sqrt{d_m})$$

for each point $(\xi_1, \ldots, \xi_m)$ in $\mathbf{R}^m$. Now given $x = \sum_{i=1}^{m} \xi_i v_i$, we have

$$\|\Delta \circ \Psi(x)\|^2 = \sum_{i=1}^{m} d_i \xi_i^2 = B(x, x) = \|\Phi(x)\|^2.$$

Consequently, the map $\Theta : \Delta \circ \Psi(x) \mapsto \Phi(x)$ is a rotation and the lemma is proved. ∎

To each linear map Φ of $\mathbf{R}^m$ and each base of $\mathbf{R}^m$ corresponds an $m \times m$ matrix whose *determinant*, denoted by $\det \Phi$, depends only on Φ and not on the choice of a base (see [1, Chapter 1, Theorem (3.16), p. 23 and Chapter 4, Proposition (3.5), p. 116]).

Proposition 10.3.11 *If Φ is a bijective linear map of $\mathbf{R}^m$, then*

$$J_\Phi = |\det \Phi| .$$

PROOF. By Lemma 10.3.10, the map Φ is a composition of rotations Θ and Ψ and a diagonal map Δ determined by positive numbers $d_1, \ldots, d_m$. Since the unit ball $\{x \in \mathbf{R}^m : \|x\| < 1\}$ is invariant with respect to rotations and has a positive measure, Proposition 10.3.8 and [1, Chapter 1, Section 5, p. 124] imply that all the numbers J_Θ, J_Ψ, $|\det \Theta|$, and $|\det \Psi|$ are equal to one. Furthermore,

$$J_\Delta = \prod_{i=1}^{m} d_i = \det \Delta = |\det \Delta|$$

because $\Delta([0,1]^m) = \prod_{i=1}^{m}[0, d_i]$. An application of Corollary 10.3.9 and [1, Chapter 1, Theorem (3.16), p. 23] completes the proof. ∎

Exercise 10.3.12 Let Φ be a rotation of $\mathbf{R}^m$. Then $f \in \overline{\mathcal{R}}(\mathbf{R}^m, \lambda_m)$ if and only if $f \circ \Phi \in \overline{\mathcal{R}}(\mathbf{R}^m, \lambda_m)$, in which case $\int_{\mathbf{R}^m} f \, d\lambda_m = \int_{\mathbf{R}^m} f \circ \Phi \, d\lambda_m$.

Exercise 10.3.13 Show that Propositions 10.3.8 and 10.3.11 hold for any linear map Φ of $\mathbf{R}^m$ whether it is bijective or not. *Hint.* If V is an $(m-1)$-dimensional subspace of $\mathbf{R}^m$, find a rotation Θ so that $\Theta(V)$ is parallel to the coordinate axes, and use Exercise 8.4.8.

10.4 The Rademacher theorem

If U is an open subset of $\mathbf{R}^m$, we denote by $C_0^\infty(U)$ the family of all functions g on $\mathbf{R}^m$ that have continuous partial derivatives in $\mathbf{R}^m$ and vanish (i.e., are equal to zero) outside a compact subset of U.

Lemma 10.4.1 *Let U be an open set and let $K \subset U$ be a compact set. There is a function $g \in C_0^\infty(U)$ such that $0 \le g \le 1$ and $g(x) = 1$ for each $x \in K$.*

PROOF. Since K is compact, it is easy to find points $x_1, \ldots, x_k$ in K and an $\eta > 0$ so that

$$K \subset \bigcup_{j=1}^{k} U(x_j, \eta) \subset \bigcup_{j=1}^{k} U[x_j, \eta] \subset U.$$

For each $t \in \mathbf{R}$, let

$$\gamma(t) = \begin{cases} \exp\left(-\dfrac{1}{\eta^2 - t^2}\right) & \text{if } |t| < \eta, \\ 0 & \text{if } |t| \geq \eta, \end{cases}$$

and observe that γ is an infinitely differentiable function on $\mathbf{R}$. The nonnegative function

$$\theta : (\xi_1, \ldots, \xi_m) \mapsto \prod_{i=1}^{m} \gamma(\xi_i)$$

belongs to $C_0^\infty(\mathbf{R}^m)$. Since $\theta(x) > 0$ if and only if $x \in U(\mathbf{0}, \eta)$, the function

$$\psi : x \mapsto \sum_{j=1}^{k} \theta(x - x_j)$$

belongs to $C_0^\infty(U)$ and $\psi(x) > 0$ for each $x \in K$. In particular, the number $a = \inf_{x \in K} \psi(x)$ is positive. Now let

$$b = \int_{-\eta}^{\eta} \gamma \, d\lambda \quad \text{and} \quad \eta(t) = \frac{\eta}{a}(2t - a),$$

and define an increasing infinitely differentiable function α on $\mathbf{R}$ by the formula

$$\alpha(t) = \begin{cases} b^{-1} \int_{-\eta}^{\eta(t)} \gamma \, d\lambda & \text{if } t > 0, \\ 0 & \text{if } t \leq 0. \end{cases}$$

As $\alpha(t) = 1$ for each $t \geq a$, we see that $g = \alpha \circ \psi$ is the desired function. ∎

Lemma 10.4.2 *Let $f \in \overline{\mathcal{R}}(\mathbf{R}^m, \lambda_m)$ be such that $\int_{\mathbf{R}^m} fg \, d\lambda_m = 0$ for each $g \in C_0^\infty(\mathbf{R}^m)$. Then $f = 0$ almost everywhere in $\mathbf{R}^m$.*

PROOF. Proceeding towards a contradiction, we may assume that the set $\{x \in \mathbf{R}^m : f(x) > 0\}$ has a positive measure. By Exercise 3.3.9, there is a compact set K such that $|K| > 0$ and $f(x) > 0$ for each $x \in K$. Exercise 10.1.12 implies that there is an $\eta > 0$ such that $\int_B |f| < \int_K f$ for each measurable set B with $|B| < \eta$. Using Proposition 3.2.8, find an open set U so that $K \subset U$ and $|U - K| < \eta$. Now let $g \in C_0^\infty(U)$ be a function associated with U and K according to Lemma 10.4.1. Extending g by zero to $\mathbf{R}^m$ produces a function in $C_0^\infty(\mathbf{R}^m)$, still denoted by g. Since

$$\int_{\mathbf{R}^m} fg = \int_K f + \int_{U-K} fg \geq \int_K f - \int_{U-K} |f| > 0,$$

we have a contradiction. ∎

For the remainder of this section, we select a fixed Lipschitz function f on $\mathbf{R}^m$, and denote by $e_1, \ldots, e_m$ the standard orthonormal base in $\mathbf{R}^m$. We also choose a $v \in \mathbf{R}^m$ with $\|v\| = 1$ and let $V = \{x \in \mathbf{R}^m : v \cdot x = 0\}$. If $x \in \mathbf{R}^m$ and a finite limit

$$d_v f(x) = \lim_{t \to 0} \frac{f(x + tv) - f(x)}{t}$$

exists, we call it the *directional derivative* of f at x in the direction of v. For $i = 1, \ldots, m$, the equality $d_{e_i} f(x) = (\partial f / \partial \xi_i)(x)$ holds whenever either side is defined. If f is differentiable at x, then $d_v f(x)$ exists and is equal to $v \cdot \operatorname{grad} f(x)$.

Lemma 10.4.3 *The extended real-valued function*

$$\overline{d}_v f : x \mapsto \limsup_{t \to 0} \frac{f(x + tv) - f(x)}{t}$$

on $\mathbf{R}^m$ is measurable, and so is the set of all $x \in \mathbf{R}^m$ at which the directional derivative $d_v f(x)$ does not exist.

PROOF. Since f is continuous, it is easy to verify that

$$f_n : x \mapsto \sup\left\{ \frac{f(x + tv) - f(x)}{t} : 0 < |t| < \frac{1}{n} \right\}$$

is a lower semicontinuous extended real-valued function on $\mathbf{R}^m$ for $n = 1, 2, \ldots$. As $\overline{d}_v f = \lim f_n$, the measurability of $\overline{d}_v f$ follows from Exercise 4.2.2, 1 and Proposition 3.6.6. Since $|\overline{d}_v f| \leq \operatorname{Lip}(f)$, we see that

$$\{x \in \mathbf{R}^m : \overline{d}_v f(x) + \overline{d}_v(-f)(x) > 0\}$$

is the set of all $x \in \mathbf{R}^m$ at which the directional derivative $d_v f(x)$ does not exist. ∎

Lemma 10.4.4 *Let F be a measurable extended real-valued function defined almost everywhere in $\mathbf{R}^m$, and suppose that for each $x \in V$, the function $t \mapsto F(x + tv)$ belongs to $\overline{\mathcal{R}}(\mathbf{R}, \lambda)$ and*

$$\int_{\mathbf{R}} F(x + tv)\, d\lambda(t) = 0\,.$$

If $F \geq 0$ then $F = 0$ almost everywhere, and if $F \in \overline{\mathcal{R}}(\mathbf{R}^m, \lambda_m)$ then $\int_{\mathbf{R}^m} F\, d\lambda_m = 0$.

PROOF. If Φ is a rotation of $\mathbf{R}^m$ such that $\Phi(e_m) = v$, then $\Phi(\xi, 0) \in V$ and $\Phi(\xi, t) = \Phi(\xi, 0) + tv$ for each $\xi \in \mathbf{R}^{m-1}$ and $t \in \mathbf{R}$. By the assumptions, the function $t \mapsto F \circ \Phi(\xi, t)$ belongs to $\overline{\mathcal{R}}(\mathbf{R})$ and

$$\int_{\mathbf{R}} F \circ \Phi(\xi, t)\, d\lambda(t) = 0$$

for each $\xi \in \mathbf{R}^{m-1}$. If $F \geq 0$, then Exercise 10.1.14 implies that $F \circ \Phi$ belongs to $\overline{\mathcal{R}}(\mathbf{R}^m)$, and we have $\int_{\mathbf{R}^m} F \circ \Phi = 0$ by Theorem 10.1.13. In view of Exercise 10.3.12, the same conclusion is reached when $F \in \overline{\mathcal{R}}(\mathbf{R}^m)$. Applying this exercise again, we conclude that $\int_{\mathbf{R}^m} F = 0$. The lemma follows from Exercise 10.1.12, 3. ∎

Lemma 10.4.5 *The directional derivative $d_v f$ exists almost everywhere in $\mathbf{R}^m$; in particular, all partial derivatives $\partial f / \partial \xi_i$ exist almost everywhere in $\mathbf{R}^m$.*

PROOF. Choose an $x \in V$ and observe that $\varphi : t \mapsto f(x + tv)$ is an AC function on each one-cell. Thus, by Corollary 5.3.9, the directional derivative $d_v f(x + tv) = \varphi'(t)$ exists for λ-almost all $t \in \mathbf{R}$. Now if N is the set of all $z \in \mathbf{R}^m$ at which $d_v f(z)$ does not exist, it suffices to apply Lemma 10.4.4 to the characteristic function $F = \chi_N$ of N. ∎

Lemma 10.4.6 *For almost all $x \in \mathbf{R}^m$, we have $d_v f(x) = v \cdot \operatorname{grad} f(x)$.*

PROOF. Choose a $g \in C_0^\infty(\mathbf{R}^m)$, and observe that $F = g\, d_v f + f\, d_v g$ is a bounded measurable function defined almost everywhere in $\mathbf{R}^m$ that vanishes outside a compact set. It follows from Corollary 4.4.8 that F belongs to $\overline{\mathcal{R}}(\mathbf{R}^m)$. Fix an $x \in V$, and note that

$$\varphi : t \mapsto f(x + tv) \qquad \text{and} \qquad \gamma : t \mapsto g(x + tv)\,,$$

and hence also $\varphi\gamma$, are Lipschitz functions on $\mathbf{R}$. In particular, these functions are AC in each one-cell, and by Corollary 5.3.9, we have

$$F(x + tv) = \varphi'(t)\gamma(t) + \varphi(t)\gamma'(t) = (\varphi\gamma)'(t)$$

for λ-almost all $t \in \mathbf{R}$. There is a cell $[a, b]$ such that $\gamma(t) = 0$ for all $t \in \mathbf{R} - (a, b)$, and Theorem 5.3.15 implies that

$$\int_{\mathbf{R}} F(x + tv) \, d\lambda(t) = \int_a^b (\varphi\gamma)' \, d\lambda = \varphi(b)\gamma(b) - \varphi(a)\gamma(a) = 0.$$

Applying Lemma 10.4.4, we obtain

$$\int_{\mathbf{R}^m} (g \, d_v f + f \, v \cdot \operatorname{grad} g) = \int_{\mathbf{R}^m} F = 0.$$

Letting $v = e_i$, the previous equality yields

$$\int_{\mathbf{R}^m} \left(g \frac{\partial f}{\partial \xi_i} + f \frac{\partial g}{\partial \xi_i} \right) = 0$$

for $i = 1, \ldots, m$, and consequently

$$\int_{\mathbf{R}^m} (g \, v \cdot \operatorname{grad} f + f \, v \cdot \operatorname{grad} g) = 0.$$

We conclude that $\int_{\mathbf{R}^m} (d_v f - v \cdot \operatorname{grad} f) g = 0$, and the theorem follows from Lemma 10.4.2. ∎

Theorem 10.4.7 *The function f is differentiable almost everywhere in $\mathbf{R}^m$.*

PROOF. The unit sphere $S = \{v \in \mathbf{R}^m : \|v\| = 1\}$ has a countable dense subset D. By Lemma 10.4.6 and Proposition 3.5.3, 3, there is a negligible set N such that $d_v f(x) = v \cdot \operatorname{grad} f(x)$ for each $x \in \mathbf{R}^m - N$ and each $v \in D$. We fix a $z \in \mathbf{R}^m - N$ and prove the theorem by showing that f is differentiable at z. With no loss of generality, we may assume that $c = \operatorname{Lip}(f)$ is a positive number.

Choose an $\varepsilon > 0$ and let $L(x) = x \cdot \operatorname{grad} f(z)$ for every $x \in \mathbf{R}^m$. Observe that $\|x\| \leq \sqrt{m} \, |x|$ and $|L(x)| \leq mc|x|$ for all $x \in \mathbf{R}^m$. If $x \in \mathbf{R}^m$ and $x \neq z$, we have

$$\frac{f(x) - f(z) - L(x - z)}{\|x - z\|} = \frac{f(z + tv) - f(z)}{t} - L(v)$$

where $t = \|x - z\|$ is a positive number and $v = (x - z)/\|x - z\|$ belongs to S. Thus, letting

$$h(t, v) = \frac{f(z + tv) - f(z)}{t} - L(v),$$

it suffices to find a $\delta > 0$ such that $|h(t, v)| < \varepsilon$ for all positive $t < \delta$ and all $v \in S$. To this end, find a finite set $C \subset D$ so that for every $v \in S$ there is a $u_v \in C$ with

$$|v - u_v| < \frac{\varepsilon}{2c(m + 1)}.$$

Since $d_u f(z) = L(u)$ for each $u \in C$ and since C is finite, there is a $\delta > 0$ such that $|h(t, u)| < \varepsilon/2$ for every positive $t < \delta$ and every $u \in C$. Now if $0 < t < \delta$ and $v \in S$, we have

$$\begin{aligned}
|h(t, v)| &\leq |h(t, u_v)| + |h(t, v) - h(t, u_v)| \\
&< \frac{\varepsilon}{2} + \frac{1}{t}|f(z + tv) - f(z + tu_v)| + |L(v - u_v)| \\
&\leq \frac{\varepsilon}{2} + c|v - u_v| + mc|v - u_v| < \varepsilon. \quad \blacksquare
\end{aligned}$$

From Theorem 10.4.7 we obtain the following corollary called the *Rademacher theorem.*

Corollary 10.4.8 *Each Lipschitz map* $\Phi : \mathbf{R}^m \to \mathbf{R}^n$ *is differentiable almost everywhere in* $\mathbf{R}^m$. $\quad \blacksquare$

10.5 The main formula

Lemma 10.5.1 *Let* $\Gamma : U[0, 1] \to \mathbf{R}^m$ *be a continuous map such that*

$$|\Gamma(x) - x| \leq \varepsilon$$

for all $x \in \partial U[0, 1]$ *and a positive* $\varepsilon < 1$. *Then* $U[0, 1 - \varepsilon] \subset \Gamma(U[0, 1])$.

PROOF. Let $C = U[0, 1]$. Choose a $y \in C$ such that $\Gamma(x) \neq y$ for every $x \in C$, and define a map $\Phi : C \to C$ by setting

$$\Phi(x) = \frac{y - \Gamma(x)}{|y - \Gamma(x)|}$$

for each $x \in C$. By the *Brouwer fixed point theorem* (see [17] for a short and elementary proof), there is a $z \in C$ with $\Phi(z) = z$. Since $|z| = 1$ and $y - \Gamma(z) = z|y - \Gamma(z)|$, we have

$$|y| = \left| \Gamma(z) + z|y - \Gamma(z)| \right| = \left| z[1 + |y - \Gamma(z)|] - [z - \Gamma(z)] \right|$$

$$\geq 1 + |y - \Gamma(z)| - |z - \Gamma(z)| > 1 - \varepsilon. \quad \blacksquare$$

If $\Phi : \mathbf{R}^m \to \mathbf{R}^m$ is a map that is differentiable at $z \in \mathbf{R}^m$, we let

$$J_\Phi(z) = |\det D\Phi(z)|.$$

Thus, if Φ is a linear map, it follows from Proposition 10.3.11 and Exercise 10.3.13 that $J_\Phi(x) = J_\Phi$ for each $x \in \mathbf{R}^m$.

Proposition 10.5.2 *If $\Phi : \mathbf{R}^m \to \mathbf{R}^m$ is a continuous map that is differentiable at $z \in \mathbf{R}^m$, then*

$$\lim_{r \to 0+} \frac{|\Phi(U[z, r])|}{|U[z, r]|} = J_\Phi(z).$$

PROOF. If $\Lambda(x) = \Phi(x + z) - \Phi(z)$ for each $x \in \mathbf{R}^m$, then $\Lambda(0) = 0$, the map $\Lambda : \mathbf{R}^m \to \mathbf{R}^m$ is differentiable at $\mathbf{0}$ with $D\Lambda(\mathbf{0}) = D\Phi(z)$, and

$$|\Lambda(U[\mathbf{0}, r])| = |\Phi(U[\mathbf{0}, r] + z) - \Phi(z)| = |\Phi(U[z, r])|$$

for each $r > 0$. Thus, with no loss of generality, we may assume that $z = \mathbf{0}$ and $\Phi(\mathbf{0}) = \mathbf{0}$. We set $L = D\Phi(\mathbf{0})$, choose an $\varepsilon > 0$, and let $U_r = U[\mathbf{0}, r]$ for all $r > 0$.

If $J_\Phi(\mathbf{0}) = 0$, then the compact set $K = L(U_1)$ is negligible by Exercise 10.3.13. Using Proposition 3.2.8, we can find an $\eta > 0$ so that the set

$$W = \{x \in \mathbf{R}^m : \operatorname{dist}(x, K) < \eta\}$$

has measure less than $\varepsilon 2^m$. There is a $\theta > 0$ such that $|\Phi(x) - L(x)| < \eta|x|$ whenever $|x| < \theta$. If $|x| \leq r < \theta$, then

$$\left| \frac{1}{r}\Phi(x) - L\left(\frac{x}{r}\right) \right| < \frac{\eta}{r}|x| \leq \eta,$$

which implies that $(1/r)\Phi(x)$ belongs to W. Thus $\Phi(U_r) \subset rW$, and so

$$|\Phi(U_r)| < r^m \varepsilon 2^m = \varepsilon|U_r|$$

for each positive $r < \theta$ (use Exercise 8.4.7, 1 or apply Propositions 10.3.8 and 10.3.11). This proves that

$$\lim_{r \to 0+} \frac{|\Phi(U_r)|}{|U_r|} = 0.$$

If $J_\Phi(0) > 0$, observe that $\Psi = L^{-1} \circ \Phi$ is differentiable at 0, $\Psi(0) = 0$, and $D\Psi(0)$ is the identity map of $\mathbf{R}^m$. By Proposition 10.3.8, we have

$$|\Psi(U_r)| = |L^{-1}(\Phi[U_r])| = J_{L^{-1}}|\Phi(U_r)| = [J_\Phi(0)]^{-1}|\Phi(U_r)|$$

for each $r > 0$, and hence it suffices to show that

$$\lim_{r \to 0+} \frac{|\Psi(U_r)|}{|U_r|} = 1.$$

To this end, find a $\delta > 0$ so that $|\Psi(x) - x| \leq \varepsilon|x|$ for each $x \in U_\delta$. Select a positive $r \leq \delta$, and let $\Gamma(x) = \Psi(rx)/r$ for every $x \in \mathbf{R}^m$. If $|x| = 1$ then $rx \in U_\delta$, and we have

$$|\Gamma(x) - x| = \frac{1}{r}|\Psi(rx) - rx| \leq \frac{\varepsilon}{r}|rx| = \varepsilon.$$

If $y \in U_{r(1-\varepsilon)}$ then $y/r \in U_{1-\varepsilon}$, and so by Lemma 10.5.1, there is an $x \in U_1$ such that $\Gamma(x) = y/r$ or, alternatively, $\Psi(rx) = y$. It follows that $U_{r(1-\varepsilon)} \subset \Psi(U_r)$. On the other hand, when $|x| \leq r$ then

$$|\Psi(x)| \leq |x| + |\Psi(x) - x| \leq (1 + \varepsilon)|x| \leq (1 + \varepsilon)r,$$

and hence $\Psi(U_r) \subset U_{r(1+\varepsilon)}$. We conclude that

$$(1 - \varepsilon)^m \leq \frac{|\Psi(U_r)|}{|U_r|} \leq (1 + \varepsilon)^m$$

for each positive $r < \delta$, and the proposition is proved. $\blacksquare$

In complete analogy to the one-dimensional case (cf. Section 6.5), given $E \subset \mathbf{R}^m$ and $x \in \mathbf{R}^m$, we call the numbers

$$\overline{\Theta}(E, x) = \limsup_{r \to 0+} \frac{|E \cap U[x, r]|}{|U[x, r]|} \quad \text{and} \quad \underline{\Theta}(E, x) = \liminf_{r \to 0+} \frac{|E \cap U[x, r]|}{|U[x, r]|}$$

the *upper* and *lower density* of E at x, respectively. Clearly,

$$0 \leq \underline{\Theta}(E, x) \leq \overline{\Theta}(E, x) \leq 1,$$

and when $\underline{\Theta}(E, x) = \overline{\Theta}(E, x)$, this common value is called the *density* of E at x, denoted by $\Theta(E, x)$. If $\Theta(E, x) = 1$ or $\Theta(E, x) = 0$, the point x is called, respectively, a *density* or a *dispersion* point of E.

Proposition 10.5.3 *Let E be a measurable subset of $\mathbf{R}^m$. For almost all $x \in \mathbf{R}^m$, the set E has a density at x and*

$$\Theta(E, x) = \chi_E(x) = \lim \frac{|E \cap K_n|}{|K_n|}$$

for any sequence $\{K_n\}$ of cells containing x such that $\lim \lambda(K_n) = 0$ and $\inf s(K_n) > 0$.

PROOF. Let $\varphi(K) = |E \cap K|$ for each cell K. Applying Corollary 10.2.4 to the identity map $\Phi : x \mapsto x$ of E, we obtain that $\Theta(E, x) = \varphi'(x)$ exists for almost all $x \in \mathbf{R}^m$ and that $|E \cap K| = \int_K \varphi' d\lambda_m$ for each cell K. On the other hand, Theorem 4.1.1 implies that

$$|E \cap K| = \int_K \chi_E \, d\lambda_m$$

for each cell K. Now it follows from Theorem 4.1.8 that $\varphi' = \chi_E$ almost everywhere in any cell, and hence almost everywhere in $\mathbf{R}^m$. An application of Exercise 9.3.1 completes the proof. ∎

Remark 10.5.4 Replacing λ by λ_m, Lemma 6.5.4 and its proof remain valid in $\mathbf{R}^m$. This fact will be used on several occasions.

Lemma 10.5.5 *Let E be a measurable subset of an open set $U \subset \mathbf{R}^m$, and let g and h be measurable functions defined on U that have partial derivatives at each $x \in E$. If $g(x) = h(x)$ for all $x \in E$, then $(\partial g/\partial \xi_i)(x) = (\partial h/\partial \xi_i)(x)$ for $i = 1, \ldots, m$ and almost all $x \in E$.*

PROOF. Since the partial derivatives of measurable functions are measurable (Lemma 10.4.3), the set $C = \{x \in E : (\partial g/\partial \xi_m)(x) \neq (\partial h/\partial \xi_m)(x)\}$ is measurable. Proceeding towards a contradiction, assume that $|C| > 0$. By Exercise 10.1.15, there is a $\xi \in \mathbf{R}^{m-1}$ such that $\lambda(C_\xi) > 0$; in particular, C_ξ is uncountable. According to [42, Chapter 2, Exercise 27, p. 45], we can find a $t \in C_\xi$ and a sequence $\{t_n\}$ in $C_\xi - \{t\}$ so that $\lim t_n = t$. Consequently

$$\frac{\partial g}{\partial \xi_m}(\xi, t) = \lim \frac{g(\xi, t_n) - g(\xi, t)}{t_n - t} = \lim \frac{h(\xi, t_n) - h(\xi, t)}{t_n - t} = \frac{\partial h}{\partial \xi_m}(\xi, t),$$

which is a contradiction. The lemma follows by symmetry. ∎

Let $E \subset \mathbf{R}^m$ be a measurable set and let $\Phi : E \to \mathbf{R}^m$ be a Lipschitz map. By Kirszbraun's theorem, Φ can be extended to a Lipschitz map Ψ

defined on $\mathbf{R}^m$, which is differentiable almost everywhere in $\mathbf{R}^m$ by the Rademacher theorem. Almost everywhere in E, we define a function $\mathcal{J}_\Phi$ by setting $\mathcal{J}_\Phi(x) = J_\Psi(x)$ for each $x \in E$ at which Ψ is differentiable. It follows from Lemma 10.5.5 that almost everywhere in E the function $\mathcal{J}_\Phi$ is determined uniquely by the map Φ and does not depend on the extension Ψ.

Theorem 10.5.6   *Let E be a measurable set and let $\Phi : E \to \mathbf{R}^m$ be an injective Lipschitz map. If $f \in \overline{\mathcal{R}}(\Phi(E), \lambda_m)$, then $(f \circ \Phi)\mathcal{J}_\Phi$ belongs to $\overline{\mathcal{R}}(E, \lambda_m)$ and*

$$\int_E (f \circ \Phi)\mathcal{J}_\Phi \, d\lambda_m = \int_{\Phi(E)} f \, d\lambda_m \,.$$

PROOF. Let $\varphi(A) = |\Phi(E \cap A)|$ for each cell A. In view of Proposition 10.3.4, Corollary 10.2.4, and Theorem 10.2.6, it suffices to show that

$$\lim_{r \to 0+} \frac{\varphi(U[x,r])}{|U[x,r]|} = \mathcal{J}_{\Phi(x)}$$

for almost all $x \in E$. According to the definition of $\mathcal{J}_\Phi$, we may assume that Φ is a Lipschitz map of $\mathbf{R}^m$ and $\mathcal{J}_\Phi(x) = J_\Phi(x)$ for each $x \in E$ at which Φ is differentiable. It follows from Proposition 10.5.3 and Rademacher's theorem that almost all points of E are density points of E at which the map Φ is differentiable. Select such a point $x \in E$ and let $U_r = U[x, r]$ for each positive number r. Since

$$\limsup_{r \to 0+} \frac{\varphi(U_r)}{|U_r|} \le \lim_{r \to 0+} \frac{|\Phi(U_r)|}{|U_r|} = J_\Phi(x)$$

by Proposition 10.5.2, the desired equality holds when $J_\Phi(x) = 0$. Thus suppose that $J_\Phi(x) > a > 0$, let $b = \operatorname{Lip}(\Phi)$, and choose an $\varepsilon > 0$. Applying Proposition 10.5.2 again and using the fact that x is a density point of E, we can find a $\theta > 0$ so that

$$|\Phi(U_r)| > a|U_r| \qquad \text{and} \qquad |E \cap U_r| > (1 - \varepsilon)|U_r|$$

for each positive $r < \theta$. Fix such an r and observe that

$$|U_r - E| = |U_r| - |E \cap U_r| < \varepsilon|U_r| < \varepsilon a^{-1}|\Phi(U_r)| \,.$$

As $\Phi(U_r) - \Phi(E \cap U_r) \subset \Phi(U_r - E)$, Proposition 10.3.4 yields

$$|\Phi(U_r)| - |\Phi(E \cap U_r)| \le |\Phi(U_r - E)| \le b^m|U_r - E| < \varepsilon a^{-1}b^m|\Phi(U_r)| \,.$$

From the last inequality we obtain

$$1 - \varepsilon a^{-1} b^m < \frac{|\Phi(E \cap U_r)|}{|\Phi(U_r)|} \leq 1 \,,$$

which shows that

$$\lim_{r \to 0+} \frac{|\Phi(E \cap U_r)|}{|\Phi(U_r)|} = 1 \,.$$

Another application of Proposition 10.5.2 completes the argument:

$$\lim_{r \to 0+} \frac{\varphi(U_r)}{|U_r|} = \lim_{r \to 0+} \left(\frac{|\Phi(E \cap U_r)|}{|\Phi(U_r)|} \cdot \frac{|\Phi(U_r)|}{|U_r|} \right) = J_\Phi(x) \,. \quad \blacksquare$$

10.6 Almost differentiable maps

Let $E \subset \mathbf{R}^m$ be a measurable set and let $n \geq 1$ be an integer. A map Φ from E to $\mathbf{R}^n$ is called σ-*Lipschitz* if there is a countable family $\mathcal{C}$ of measurable sets such that $\bigcup \mathcal{C} = E$ and for each $C \in \mathcal{C}$, the map Φ restricted to C is Lipschitz. If the family $\mathcal{C}$ is also disjoint, we call it a *concassage* of Φ.

Exercise 10.6.1 Show that each σ-Lipschitz map has a concassage.

Let $E \subset \mathbf{R}^m$ be a measurable set and let $\Phi : E \to \mathbf{R}^m$ be a σ-Lipschitz map with a concassage $\mathcal{C}$. For each $C \in \mathcal{C}$, denote by Φ_C the restriction of Φ to C. Since Φ_C is a Lipschitz map, it can be extended to a Lipschitz map Ψ_C defined on $\mathbf{R}^m$ (Kirszbraun's theorem). Given a $C \in \mathcal{C}$, we set

$$\mathcal{J}_\Phi(x) = J_{\Psi_C}(x)$$

for each $x \in C$ at which Ψ_C is differentiable. This determines a function $\mathcal{J}_\Phi$ defined almost everywhere in E (Rademacher's theorem). The function $\mathcal{J}_\Phi$ depends on Φ, and may also depend on the choice of a concassage $\mathcal{C}$ and maps Ψ_C.

Proposition 10.6.2 *Let E be a measurable set and let $\Phi : E \to \mathbf{R}^m$ be a σ-Lipschitz map. Up to its values on a negligible set, the function $\mathcal{J}_\Phi$ is determined uniquely by the map Φ.*

PROOF. Let $\mathcal{C}$ and $\mathcal{D}$ be two concassages of Φ. For every $B \in \mathcal{C} \cup \mathcal{D}$ denote by Φ_B the restriction of Φ to B, and by Ψ_B a Lipschitz map defined on $\mathbf{R}^m$ that extends Φ_B. Now let $\mathcal{J}_\Phi^{\mathcal{C}}$ and $\mathcal{J}_\Phi^{\mathcal{D}}$ be defined by means of $\mathcal{C}$ and $\mathcal{D}$, respectively, and select a $C \in \mathcal{C}$ and $D \in \mathcal{D}$. Since $\Psi_C(x) = \Psi_D(x)$ for each $x \in C \cap D$, it follows from Rademacher's theorem and Lemma 10.5.5 that

$$\mathcal{J}_\Phi^C(x) = J_{\Psi_C}(x) = J_{\Psi_D}(x) = \mathcal{J}_\Phi^D(x)$$

for almost all $x \in C \cap D$. As $E = \bigcup_{(C,D) \in \mathcal{C} \times \mathcal{D}} (C \cap D)$, an application of Proposition 3.5.3, 3 completes the proof. ∎

Remark 10.6.3 Each Lipschitz map Φ defined on a measurable set E is also σ-Lipschitz. Since the singleton $\{E\}$ is a concassage of Φ, it follows from Proposition 10.6.2 that the functions $\mathcal{J}_\Phi$ defined in this section and Section 10.5 coincide almost everywhere in E.

Theorem 10.6.4 *Let E be a measurable set, and let $\Phi : E \to \mathbf{R}^m$ be an injective σ-Lipschitz map. If $f \in \overline{\mathcal{R}}(\Phi(E), \lambda_m)$, then $(f \circ \Phi)\mathcal{J}_\Phi$ belongs to $\overline{\mathcal{R}}(E, \lambda_m)$ and*

$$\int_E (f \circ \Phi)\mathcal{J}_\Phi \, d\lambda_m = \int_{\Phi(E)} f \, d\lambda_m \,.$$

PROOF. In view of Proposition 10.1.9, we may assume that f is nonnegative. Let $\mathcal{C}$ be a concassage of Φ and for each $C \in \mathcal{C}$, let Φ_C be the restriction of Φ to C. By Exercise 10.1.8 and Theorem 10.5.6, the function $(f \circ \Phi_C)\mathcal{J}_{\Phi_C}$ belongs to $\overline{\mathcal{R}}(C)$ and

$$\int_C (f \circ \Phi)\mathcal{J}_\Phi = \int_C (f \circ \Phi_C)\mathcal{J}_{\Phi_C} = \int_{\Phi_C(C)} f = \int_{\Phi(C)} f$$

for every $C \in \mathcal{C}$. This and Exercise 10.1.8 yield

$$\sum_{C \in \mathcal{C}} \int_C (f \circ \Phi)\mathcal{J}_\Phi = \sum_{C \in \mathcal{C}} \int_{\Phi(C)} f = \int_{\Phi(E)} f < +\infty \,.$$

Another application of Exercise 10.1.8 proves the theorem. ∎

Exercise 10.6.5 While σ-Lipschitz maps are generally not Luzin (they may even be discontinuous), show that they still satisfy the Luzin condition (see Remark 10.2.1).

If f is a function on a set $S \subset \mathbf{R}^m$ and $z \in S$, we let

$$(S) \limsup_{x \to z} f(x) = \lim_{r \to 0+} \sup\{f(x) : x \in (S - \{z\}) \cap U(z, r)\} \,.$$

Since $(S) \limsup_{x \to z} f(x) = \limsup_{x \to z} f(x)$ whenever z is an interior point of S, the following definition generalizes Definition 6.6.1.

Definition 10.6.6 Let $S \subset \mathbf{R}^m$ and let $n \geq 1$ be an integer. We say that a map $\Phi : S \to \mathbf{R}^n$ is *almost differentiable* at $z \in S$ if

$$(S) \limsup_{x \to z} \frac{|\Phi(x) - \Phi(z)|}{|x - z|} < +\infty.$$

In view of the Kirszbraun and Rademacher theorems and Remark 10.5.4, repeating verbatim the proofs of Lemma 6.6.7 and Theorem 6.6.8 yields the multidimensional version of *Stepanoff's theorem*.

Theorem 10.6.7 *Let $U \subset \mathbf{R}^m$ be an open set, and let $\Phi : U \to \mathbf{R}^n$ be a map that is almost differentiable at each point of a set $E \subset U$. Then Φ is differentiable at almost all $x \in E$.* ∎

The Stepanoff theorem is not needed in the present section. We shall use it later in Sections 11.7 and 12.2.

Proposition 10.6.8 *Let $E \subset \mathbf{R}^m$ be a measurable set and let $\Phi : E \to \mathbf{R}^n$ be a map that is almost differentiable at every $x \in E$. Then Φ is σ-Lipschitz.*

PROOF. The idea of the proof is similar to that employed in the proof of Theorem 6.6.8; however, we have the additional task of establishing measurability. For $i = 1, 2, \ldots$, let C_i be the set of all $z \in E$ such that

$$|\Phi(x) - \Phi(z)| \leq i|x - z|$$

for all $x \in E \cap U(z, 1/i)$. By the assumption, $E = \bigcup_{i=1}^{\infty} C_i$. If $\mathcal{K}_i$ is the family of all dyadic cubes of the same diameter 2^{-i}, then Φ is Lipschitz on each set of the countable family

$$\mathcal{D} = \{C_i \cap K : K \in \mathcal{K}_i, \ i = 1, 2, \ldots\}.$$

As $E = \bigcup \mathcal{D}$, it remains to show that each $D \in \mathcal{D}$ is measurable. We prove this by showing that C_i is measurable for $i = 1, 2, \ldots$. If $z \in E$, let

$$A_{i,z} = \{x \in E : |\Phi(x) - \Phi(z)| > i|x - z|\}$$

and set $B_i = \{z \in E : \operatorname{dist}(z, A_{i,z}) < 1/i\}$. Fix a $z \in B_i$, and find an $x \in A_{i,z}$ so that $|x - z| < 1/i$. Since Φ is continuous, there is a $\delta > 0$ such that

$$|x - y| < \frac{1}{i} \qquad \text{and} \qquad |\Phi(x) - \Phi(y)| > i|x - y|$$

for each $y \in E \cap U(z, \delta)$. The last inequality says that $x \in A_{i,y}$, and so $\operatorname{dist}(y, A_{i,y}) < 1/i$ for each $y \in E \cap U(z, \delta)$. This implies that the set B_i is a relatively open subset of E. In particular, B_i is measurable and so is

$$C_i = \left\{ z \in E : U\left(z, \frac{1}{i}\right) \cap A_{i,z} = \emptyset \right\} = E - B_i \,. \quad \blacksquare$$

The next theorem follows from Proposition 10.6.8 and Theorem 10.6.4.

Theorem 10.6.9   *Let E be a measurable set, and let $\Phi : E \to \mathbf{R}^m$ be an injective map that is almost differentiable at each $x \in E$. If $f \in \overline{\mathcal{R}}(\Phi(E), \lambda_m)$, then $(f \circ \Phi) \mathcal{J}_\Phi$ belongs to $\overline{\mathcal{R}}(E, \lambda_m)$ and*

$$\int_E (f \circ \Phi) \mathcal{J}_\Phi \, d\lambda_m = \int_{\Phi(E)} f \, d\lambda_m \,. \quad \blacksquare$$

Corollary 10.6.10   *Let E be a measurable set, and let $\Phi : E \to \mathbf{R}^m$ be an injective map that is almost differentiable at each $x \in E$. If $|\Phi(E)| < +\infty$, then $\mathcal{J}_\Phi$ belongs to $\overline{\mathcal{R}}(E, \lambda_m)$ and*

$$|\Phi(B)| = \int_B \mathcal{J}_\Phi \, d\lambda_m$$

for each measurable set $B \subset E$.

PROOF.   Applying Theorem 10.6.9 to the function $f = 1$ shows that $\mathcal{J}_\Phi$ belongs to $\overline{\mathcal{R}}(E)$. Choose a measurable set $B \subset E$. Then $\Phi(B)$ is a measurable set by Propositions 10.6.8, 10.3.4, and 10.2.2, and $\chi_{\Phi(B)} \circ \Phi(x) = \chi_B(x)$ for each $x \in E$. An application of Theorem 10.6.9 to the function $f = \chi_{\Phi(B)}$ yields

$$|\Phi(B)| = \int_{\Phi(E)} \chi_{\Phi(B)} = \int_E \chi_B \cdot \mathcal{J}_\Phi = \int_B \mathcal{J}_\Phi \,. \quad \blacksquare$$

Exercise 10.6.11   Let $U \subset \mathbf{R}^m$ be an open set, and let $\Phi : U \to \mathbf{R}^m$ be differentiable at each point of a measurable set $E \subset U$. Show that Φ is σ-Lipschitz on E and $\mathcal{J}_\Phi = J_\Phi$ almost everywhere in E.

Chapter 11

The gage integral

Guided by the ideas introduced in Section 6.7, we shall present a nontrivial m-dimensional extension of the Henstock–Kurzweil integral defined in Chapter 6. Throughout the rest of this book, we shall integrate only with respect to the Lebesgue volume.

11.1 A motivating example

Before generalizing the Henstock–Kurzweil integral to higher dimensions, we must make a basic decision: do we want to preserve Fubini's theorem or improve on the divergence theorem? Indeed, Example 11.1.2 below shows that the ability to integrate the divergence of an arbitrary differentiable vector field is *incompatible* with the Fubini theorem.

Replacing partitions by Perron partitions in Definition 8.1.1 produces the most obvious generalization of the McShane integral in $\mathbf{R}^m$, commonly referred to as the m-dimensional Henstock–Kurzweil integral. With little effort the reader can prove its basic properties, including the Fubini theorems analogous to Proposition 8.6.3 (see [27, Sections S6.3 and S6.4]) and Theorem 8.6.6. It follows that the unrestricted fundamental theorem of calculus (Theorem 6.1.2), the most prominent feature of the one-dimensional Henstock–Kurzweil integral, is lost in higher dimensions. Moreover, the utility of Fubini's theorem is severely limited, since there is no simple test for the existence of double integrals; the Tonelli theorem concerns only nonnegative functions and thus does not apply outside the realm of McShane integration. In view of this, we consider the m-dimensional Henstock–Kurzweil integral rather uninteresting, and shall not pursue its straightforward development; the reader who wishes to see more details is referred to [27].

Our goal is to obtain an integral that yields a divergence theorem for any differentiable vector field. The proof of Theorem 8.3.5 suggests that the integral defined by means of ε-shapely P-partitions may be the answer. While this is true in principle, a simple replacement of partitions by ε-shapely P-partitions in Definition 8.1.1 does not work well: the fact that the intersection of two ε-shapely cells need not be ε-shapely results in the loss of essential properties such as Proposition 8.1.4 (see [34, Example 7.2]). A key to the appropriate implementation is contained in Section 6.7. We shall see in the following sections that turning Theorem 6.7.5 into a definition leads to a well-behaved integral and very general divergence theorems (Theorems 11.7.5, 11.7.8, and 11.7.10 below).

The next lemma facilitates the construction of the example mentioned in the first paragraph.

Lemma 11.1.1   *Given a one-cell $B = [a, b]$, there is an infinitely differentiable increasing function $g : \mathbf{R} \to [0, 1]$ such that $g(t) = 0$ if $t \le a$ and $g(t) = 1$ if $t \ge b$.*

PROOF.   We modify the functions employed in the proof of Lemma 10.4.1. For each $t \in \mathbf{R}$, let

$$\gamma(t) = \begin{cases} \exp\left(-\frac{1}{1-t^2}\right) & \text{if } |t| < 1, \\ 0 & \text{if } |t| \ge 1, \end{cases}$$

and observe that γ is an infinitely differentiable function defined on $\mathbf{R}$. Setting $c = \int_{-1}^{1} \gamma \, d\lambda$ and $u(t) = (2t - b - a)/(b - a)$, the following formula

$$g(t) = \begin{cases} c^{-1} \int_{-1}^{u(t)} \gamma \, d\lambda & \text{if } t > a, \\ 0 & \text{if } t \le a \end{cases}$$

defines the desired function g on $\mathbf{R}$.   ∎

Example 11.1.2   For $n = 0, 1, \ldots$, let

$$A_n = [2^{-n-1}, 2^{-n}], \quad B_n = \left[\frac{4}{3}2^{-n-1}, \frac{5}{3}2^{-n-1}\right], \quad C_n = \left[\frac{5}{3}2^{-n-1}, \frac{4}{3}2^{-n}\right],$$

and let g_n be a function on $\mathbf{R}$ associated with B_n according to Lemma 11.1.1. Given (x, y) in $\mathbf{R}^2$, set

$$f(x, y) = \begin{cases} 0 & \text{if } y \leq 0, \\ y^2 \sin x & \text{if } y \geq 1, \end{cases}$$

and for $n = 0, 1, \ldots$, let

$$f(x, y) = g_n(y)y^2 \sin 8^n x + [1 - g_n(y)]y^2 \sin 8^{n+1} x$$

whenever $y \in A_n$. Note that f in $\mathbf{R} \times [0, 1]$ is obtained by a differentiable deformation of the function $y^2 \sin 8^n x$ on $\mathbf{R} \times C_n$ to the function $y^2 \sin 8^{n+1} x$ on $\mathbf{R} \times C_{n+1}$. A direct verification reveals that f is differentiable in $\mathbf{R}^2$. Thus $h = \partial f / \partial x$ is the divergence of a differentiable vector field $v = (f, 0)$.

Now assume that we have a multidimensional integral that satisfies the Fubini theorem and in dimension one coincides with the Henstock–Kurzweil integral of Chapter 6. According to [27, Section 6.1], the aforementioned m-dimensional Henstock–Kurzweil integral fulfills these conditions. We claim that with respect to such an integral the function h is not integrable over the cell $[0, 2\pi] \times [0, 1]$. To prove the claim, it suffices to consider the function $h_x : y \mapsto h(x, y)$ and show that the set

$$E = \{x \in [0, 2\pi] : h_x \notin \mathcal{R}_*([0, 1], \lambda)\}$$

has positive Lebesgue measure.

If $h_x \in \mathcal{R}_*([0, 1], \lambda)$, then writing dy instead of $d\lambda(y)$, it follows from Proposition 6.1.5 that

$$\int_0^1 h_x \, d\lambda = \lim \int_{2^{-n-1}}^1 \frac{\partial f}{\partial x}(x, y) \, dy = \sum_{n=0}^{\infty} \int_{A_n} \frac{\partial f}{\partial x}(x, y) \, dy = \sum_{n=0}^{\infty} c_n(x)$$

where

$$c_n(x) = 8^n \cos 8^n x \int_{A_n} y^2 g_n(y) \, dy + 8^{n+1} \cos 8^{n+1} x \int_{A_n} y^2 [1 - g_n(y)] \, dy.$$

We establish our assertion by showing that the series $\sum_{n=0}^{\infty} c_n(x)$ diverges for λ-almost all x in $[0, 2\pi]$. To this end, we present without proof the following fact, which is a consequence of a deeper result due to H. Weyl (see [22, Chapter 1, Theorem 4.1, p. 42]):

There is a λ-negligible set $N \subset [0, 2\pi]$ such that for each $x \in [0, 2\pi] - N$ and each $\varepsilon > 0$ we can find sequences $\{k_i\}$ and $\{n_i\}$ of positive integers so that $n_i < n_{i+1}$ and $|8^{n_i}x - 2k_i\pi| < \varepsilon$ for $i = 1, 2, \ldots$.

Choosing x in $[0, 2\pi] - N$ and letting $\varepsilon = \pi/24$, we obtain that

$$\max\{|8^{n_i}x - 2k_i\pi|, |8^{n_i+1}x - 16k_i\pi|\} < \frac{\pi}{3}$$

and consequently

$$c_{n_i}(x) > \frac{8^{n_i}}{2}\int_{A_{n_i}} y^2 g_{n_i}(y)\,dy + \frac{8^{n_i+1}}{2}\int_{A_{n_i}} y^2[1 - g_{n_i}(y)]\,dy$$

$$\geq \frac{8^{n_i}}{2}\int_{A_{n_i}} y^2\,dy > \frac{8^{n_i}}{2}(2^{-n_i-1})^2 2^{-n_i-1} = 2^{-4}$$

for $i = 1, 2, \ldots$. Thus $\{c_n(x)\}$ does not converge to zero, which implies that the series $\sum_{n=0}^{\infty} c_n(x)$ diverges ([42, Theorem 3.23, p. 60]).

11.2 Continuous additive functions

In order to present a unified treatment for all dimensions $m \geq 1$, it is convenient to let $\mathbf{R}^0 = \{0\}$ and set

$$\lambda_0(E) = \begin{cases} 1 & \text{if } E = \mathbf{R}^0, \\ 0 & \text{if } E = \emptyset. \end{cases}$$

Recall that a *hyperplane* of $\mathbf{R}^m$, or simply a *hyperplane*, is a translate of a subspace of $\mathbf{R}^m$ spanned by $m - 1$ coordinate axes (see Remark 8.4.4 and the paragraph preceding it). Making the obvious identification between a hyperplane H and $\mathbf{R}^{m-1}$, we extend the $(m - 1)$-dimensional Lebesgue measure λ_{m-1} to all subsets of H. In particular, a hyperplane of $\mathbf{R} = \mathbf{R}^1$ is a singleton $\{x\} \subset \mathbf{R}$ and $\lambda_0(\{x\}) = 1$.

As in dimension one, a *figure* is a finite, possibly empty, union of cells (cf. Section 6.7). A collection of figures is called *nonoverlapping* if their interiors are disjoint. If A and B are figures, then so are the sets $A \cup B$,

$$A \odot B = (A^\circ \cap B^\circ)^- \quad \text{and} \quad A \ominus B = (A - B)^-.$$

The *perimeter* of a figure A is the number

$$\|A\| = \sum_{H} \lambda_{m-1}(H \cap \partial A)$$

where the sum is taken over all hyperplanes H of $\mathbf{R}^m$. Since all but finitely many hyperplanes intersect ∂A in a λ_{m-1}-negligible set, the perimeter of A is, indeed, a finite number. This definition extends the previous definitions of perimeters introduced in Sections 6.7 and 8.3. If A and B are figures, then the boundaries of $A \cup B$, $A \odot B$, and $A \ominus B$ are contained in $(\partial A) \cup (\partial B)$, hence

$$\max\{\|A \cup B\|, \|A \odot B\|, \|A \ominus B\|\} \le \|A\| + \|B\|.$$

Let $A = \prod_{i=1}^m [a^i_-, a^i_+]$ be a cell, and let $I \subset \{1, \ldots, m\}$ have k elements where $0 \le k \le m$. A k-*dimensional face*, or simply a k-*face*, of A is the interval $\prod_{i=1}^m A_i$ where $A_i = [a^i_-, a^i_+]$ if $i \in I$, and A_i is the singleton $\{a^i_-\}$ or $\{a^i_+\}$ if $i \notin I$. In this terminology, the m-face of A is A, and the $(m-1)$-faces are the faces defined in Section 8.3; zero-dimensional faces are called *vertexes*.

A *division* of a figure A is a finite collection $\mathcal{D}$ of nonoverlapping cells such that $\bigcup \mathcal{D} = A$. By Lemma 7.1.6, each figure has a division. We say that a division $\mathcal{D}$ of a figure A is *cellular* if each pair of intersecting cells from $\mathcal{D}$ meets in a common k-face.

Lemma 11.2.1 *Each figure has a cellular division.*

PROOF. Let A be a figure and let B be a cell containing A. It follows from Lemma 7.1.6 that there is a division $\mathcal{C}$ of B such that a subfamily of $\mathcal{C}$ is a division of A. Suppose that each $C \in \mathcal{C}$ is the product of one-cells $C_1, \ldots, C_m$. For $i = 1, \ldots, m$, find segmentations $\mathcal{D}_i$ of $\{C_i : C \in \mathcal{C}\}$, and denote by $\mathcal{D}$ the collection of all cells $D = \prod_{i=1}^m D_i$ where $D_i \in \mathcal{D}_i$. It is easy to verify that $\mathcal{D}$ is a segmentation of $\mathcal{C}$ as well as a cellular division of B. Consequently the family $\{D \in \mathcal{D} : D \subset A\}$ is a cellular division of A. ∎

Exercise 11.2.2 Let $A \subset \mathbf{R}^m$, and let i be an integer with $1 \le i \le m$. For each $\xi = (\xi_1, \ldots, \xi_{i-1}, \xi_{i+1}, \ldots, \xi_m)$ in $\mathbf{R}^{m-1}$, we call the set

$$A^\xi = \{t \in \mathbf{R} : (\xi_1, \ldots, \xi_{i-1}, t, \xi_{i+1}, \ldots, \xi_m) \in A\}$$

the *section* of A determined by ξ. The set $A_{(i)} = \{\xi \in \mathbf{R}^{m-1} : A^\xi \ne \emptyset\}$ is called the *projection* of A into $\mathbf{R}^{m-1}$ in the direction of the ith coordinate axis (cf. Section 8.3). Assuming that A is a figure, prove the following claims.

1. The projection $A_{(i)}$ is a figure in $\mathbf{R}^{m-1}$.
2. Each section A^ξ is a figure in $\mathbf{R}$, and the nonnegative integer-valued function $\xi \mapsto \|A^\xi\|$ belongs to $\mathcal{R}(A_{(i)}, \lambda_{m-1})$.
3. If $\|A\|_i = \int_{A_{(i)}} \|A^\xi\| \, d\lambda_{m-1}(\xi)$, then $\|A\| = \sum_{i=1}^m \|A\|_i$.

Exercise 11.2.3 Let A be a figure and let B be a *cell*. Use Exercise 11.2.2 to show that $\|A \odot B\| \leq \|A\|$.

Let F be an additive function in a set $E \subset \mathbf{R}^m$. If a figure $B \subset E$ is the union of nonoverlapping cells $B_1, \ldots, B_k$, we let

$$F(B) = \sum_{i=1}^{k} F(B_i).$$

By Remark 7.2.6, the number $F(B)$ is independent of the choice of cells $B_1, \ldots, B_k$. Thus F determines uniquely a function defined on all subfigures of E, still denoted by F, that is *additive* in the following sense:

$$F(B \cup C) = F(B) + F(C)$$

for all nonoverlapping subfigures B and C of E. Denoting both functions by the same letter is legitimate because they coincide on the intersection of their domains, i.e., on the collection consisting of the empty set and all subcells of E.

Convention. Throughout the remainder of this book, we tacitly assume that each additive function in a set $E \subset \mathbf{R}^m$ has already been defined on all subfigures of E so that it is additive in the above sense.

Let v be a continuous vector field on a set $E \subset \mathbf{R}^m$ and let $A \subset E$ be a figure. By Proposition 8.3.2, the flux of v in E is an additive function in E. Extending the notation of Section 8.3, we denote by $\int_{\partial A} v \cdot n \, d\lambda_{m-1}$ the flux of v from the figure A. If $\mathcal{D}$ is a cellular division of A, then $\partial A = \bigcup_{k=1}^{n} D_k$ where each D_k is an $(m-1)$-face of a cell $D \in \mathcal{D}$. Moreover, each $(m-1)$-face of a cell $D \in \mathcal{D}$ that is not contained in ∂A is a common face of two adjacent cells from $\mathcal{D}$. Thus, if n_k denotes the exterior normal of D_k, we obtain

$$\int_{\partial A} v \cdot n \, d\lambda_{m-1} = \sum_{k=1}^{n} \int_{D_k} v \cdot n_k \, d\lambda_{m-1},$$

where the right side is the sum of genuine integrals. In particular, as in the case of a cell, the flux of v from A depends only on v restricted to ∂A. An important consequence of this is that Proposition 8.3.1 remains valid for figures. In view of Definition 10.1.5 and Proposition 8.3.2, it is clear that the divergence theorem (Theorem 8.3.5) is also true for figures.

Exercise 11.2.4 Let $v = (f_1, \ldots, f_m)$ be a continuous vector field defined on a figure A, and let i be an integer with $1 \leq i \leq m$. Recall from Exercise 11.2.2 that the projection $A_{(i)}$ and the sections A^ξ are figures in $\mathbf{R}^{m-1}$ and $\mathbf{R}$, respectively. If $\xi = (\xi_1, \ldots, \xi_{i-1}, \xi_{i+1}, \ldots, \xi_m)$ is a point of $\mathbf{R}^{m-1}$, let

$$f_i^\xi(t) = (f_i)^\xi(t) = f_i(\xi_1, \ldots, \xi_{i-1}, t, \xi_{i+1}, \ldots, \xi_m)$$

for each $t \in A^\xi$. Since A^ξ is a figure in $\mathbf{R}$, there is an integer $k(\xi) \geq 0$ and real numbers $a_1^\xi < b_1^\xi < \cdots < a_{k(\xi)}^\xi < b_{k(\xi)}^\xi$ such that $A^\xi = \bigcup_{j=1}^{k(\xi)} [a_j^\xi, b_j^\xi]$. In accordance with our notation (cf. Section 1.1), we have

$$f_i^\xi(A^\xi) = \sum_{j=1}^{k(\xi)} \left[f_i^\xi\left(b_j^\xi\right) - f_i^\xi\left(a_j^\xi\right) \right].$$

1. Show that the function $\xi \mapsto f_i^\xi(A^\xi)$ belongs to $\mathcal{R}(A_{(i)}, \lambda_{m-1})$, and that

$$\int_A v \cdot n \, d\lambda_{m-1} = \sum_{i=1}^{m} \int_{A_{(i)}} f_i^\xi(A^\xi) \, d\lambda_{m-1}(\xi).$$

2. Deduce from 1 that $\int_A v \cdot n \, d\lambda_{m-1}$ depends only on v restricted to ∂A.
3. Use 1 to show that the flux of v in A is an additive function in A.
4. Employing 1 and Exercise 11.2.2, prove Proposition 8.3.1 for figures.
5. Show that the divergence theorem for figures is a consequence of 1, the fundamental theorem of calculus, and Fubini's theorem.

Definition 11.2.5 An additive function F in a set $E \subset \mathbf{R}^m$ is called *continuous* if, given $\varepsilon > 0$, there is an $\eta > 0$ such that $|F(B)| < \varepsilon$ for each figure $B \subset E$ with $\|B\| < 1/\varepsilon$ and $|B| < \eta$.

Exercise 11.2.6 Let $m = 1$, and let F be a function defined on a *bounded* set $E \subset \mathbf{R}$. Use Proposition 6.7.3 to show that F is continuous if and only if the associated interval function F (see Section 1.1) is continuous in the sense of Definition 11.2.5.

Exercise 11.2.7 Show that each AC function in a set $E \subset \mathbf{R}^m$ is continuous, but not vice versa.

Proposition 11.2.8 *If v is a continuous vector field defined on a compact set $K \subset \mathbf{R}^m$, then the flux of v in K is an additive continuous function in K.*

PROOF. As the flux F of v in K is an additive function in K by Proposition 8.3.2, we only need to show that F is continuous. To this end, choose an $\varepsilon > 0$ and use the Stone–Weierstrass approximation theorem ([41, Chapter 9, Corollary 29, p. 174]) to find a continuously differentiable vector field w on $\mathbf{R}^m$ so that $|v(x) - w(x)| < \varepsilon^2/2$ for each $x \in K$. Select a $c > |\operatorname{div} w(x)|$ for all $x \in K$, and let $\eta = \varepsilon/(2c)$. Now if $B \subset K$ is a figure with $\|B\| < 1/\varepsilon$ and $|B| < \eta$, then Proposition 8.3.1 and the divergence theorem yield

$$|F(B)| \le \left| \int_{\partial B} (v - w) \cdot n \, d\lambda_{m-1} \right| + \left| \int_B \operatorname{div} w \, d\lambda_{m-1} \right|$$

$$\le \frac{\varepsilon^2}{2} \|B\| + c|B| < \varepsilon. \quad \blacksquare$$

11.3 Gages and calibers

A set $T \subset \mathbf{R}^m$ is called *thin* if $T = \bigcup_{j=1}^{\infty} T_j$ where each set T_j satisfies the following condition: given $\eta > 0$, there is a finite or infinite sequence $\{C_k\}$ of dyadic cubes of diameters less than η such that

$$T_j \subset \left(\bigcup_k C_k \right)^{\circ} \quad \text{and} \quad \sum_k \|C_k\| \le 4.$$

Remark 11.3.1 Clearly, each singleton $\{x\} \subset \mathbf{R}$ is a thin subset of $\mathbf{R}$. If $x = 0$, we must employ at least two dyadic cubes to show that $\{x\}$ is a thin set. Thus the bound 4 is actually attained.

Proposition 11.3.2 *The following statements are true.*

1. *A subset of a thin set is thin.*
2. *A countable union of thin sets is thin.*
3. *Each hyperplane is thin; in particular, a boundary of any figure is thin.*
4. *Each thin set is negligible, but not vice versa.*
5. *If $m = 1$, then the thin and countable subsets of $\mathbf{R}$ coincide.*

PROOF. The first statement is obvious, and the second follows from the fact that the union of countably many countable sets is countable ([42, Theorem 2.12, p. 29]).

If $m = 1$, then in view of Remark 11.3.1, the third statement follows from the second. Let $m > 1$ and let H be a hyperplane in $\mathbf{R}^m$. Choose a dyadic cube K of diameter $1/4$ and an integer $n \ge 3$. If $\mathcal{C}$ is the family of all dyadic cubes of the same diameters 2^{-n} that meet $H \cap K$, then $H \cap K \subset (\bigcup \mathcal{C})^\circ$. Since

$$\sum_{C \in \mathcal{C}} \|C\| \leq 2(2^{n-2} + 2)^{m-1} \cdot 2m2^{-n(m-1)}$$

$$\leq 2(2^{n-2} + 2^{n-2})^{m-1} \cdot 2m2^{-n(m-1)} = m2^{3-m} \leq 4,$$

we see that H is a thin set.

That every countable subset of $\mathbf{R}$ is a thin subset of $\mathbf{R}$ follows from Remark 11.3.1 and the second statement. Conversely, if an uncountable set $T \subset \mathbf{R}$ is the union of sets $T_1, T_2, \ldots$, then at least one T_j contains three distinct points x, y, z. Letting

$$\eta = \frac{1}{2} \min\{|x - y|, |x - z|, |y - z|\},$$

we see that the set T_j cannot be covered by fewer than three dyadic cubes of diameters less than η. This implies that T is not a thin subset of $\mathbf{R}$, and the fifth statement is established.

If $\{C_k\}$ is a sequence of dyadic cubes of diameters less than η, then

$$\left| \bigcup_k C_k \right| \leq \eta \sum_k [d(C_k)]^{m-1} \leq \frac{\eta}{2m} \sum_k \|C_k\|.$$

Using this and Proposition 3.5.3, 3, we conclude that each thin set is negligible. On the other hand, the Cantor set C of Example 5.3.11 is an uncountable λ-negligible subset of $\mathbf{R}$ that is not a thin subset of $\mathbf{R}$ by the fifth statement. It follows from Corollary 8.6.4 that $D = C \times [0, 1]^{m-1}$ is a negligible subset of $\mathbf{R}^m$. Elaborating on the idea employed in the previous paragraph, it is easy to show that D is not a thin set. $\blacksquare$

Note. The reader familiar with the *Hausdorff measures* will note that a set $T \subset \mathbf{R}^m$ is thin if and only if it is the union of countably many sets whose $(m - 1)$-dimensional Hausdorff measures are finite (cf. Section 12.5, in particular Proposition 12.5.9).

Definition 11.3.3 A nonnegative function δ on a set $E \subset \mathbf{R}^m$ is called a *gage* on E if the *null set* $Z_\delta = \{x \in E : \delta(x) = 0\}$ of δ is thin.

It follows from Proposition 11.3.2, 5 that in dimension one the previous definition of a gage agrees with that of Section 6.4 (the paragraph following the proof of Proposition 6.4.5). If δ is a gage on a set $E \subset \mathbf{R}^m$, then a partition $P = \{(A_1, x_1), \ldots, (A_p, x_p)\}$ anchored in E is called δ-*fine* whenever $A_i \subset U(x_i, \delta(x_i))$ for $i = 1, \ldots, p$; clearly, such a partition P must be anchored in the set $E - Z_\delta$.

Recall that a *caliber* is a sequence of positive real numbers. As in Section 6.7, given an $\varepsilon > 0$ and a caliber $\eta = \{\eta_j\}$, we say that a figure B is (ε, η)-*small* if B is the union of nonoverlapping, possibly empty, figures $B_1, \ldots, B_k$ such that $\|B_j\| < 1/\varepsilon$ and $|B_j| < \eta_j$ for $j = 1, \ldots, k$.

Lemma 11.3.4 *Let F be a continuous additive function in a set $E \subset \mathbf{R}^m$. Given $\varepsilon > 0$, there is a caliber η such that $|F(B)| < \varepsilon$ for each (ε, η)-small figure $B \subset E$.*

PROOF. Given an integer $j \geq 1$, find an $\eta_j > 0$ so that $|F(B)| < \varepsilon 2^{-j}$ for each figure $B \subset E$ with $\|B\| < 2^j/\varepsilon$ and $|B| < \eta_j$. Let $\eta = \{\eta_j\}$ and let B be an (ε, η)-small subfigure of E. By definition, B is the union of nonoverlapping figures $B_1, \ldots, B_k$ such that $\|B_j\| < 1/\varepsilon < 2^j/\varepsilon$ and $|B_j| < \eta_j$ for $j = 1, \ldots, k$. Consequently

$$|F(B)| \leq \sum_{j=1}^{k} |F(B_j)| < \sum_{j=1}^{k} \varepsilon 2^{-j} < \varepsilon. \quad \blacksquare$$

Definition 11.3.5 Let A be a figure, let $\varepsilon > 0$, and let η be a caliber. A partition P in A is called a *partition of A mod (ε, η)* whenever the figure $A \ominus \bigcup P$ is (ε, η)-small.

The next lemma, due to E.J. Howard ([18, Lemma 5]), improves on the earlier ideas of Besicovitch ([2, Theorem 1]).

Lemma 11.3.6 *Let Δ be a gage on a cell $K = \prod_{j=1}^{m}[r_j, s_j]$ where $r_j < s_j$ are integers for $j = 1, \ldots, m$. Then for each positive $\varepsilon < 1/4$ and each caliber η there is a Δ-fine Perron partition $\{(K_1, x_1), \ldots, (K_p, x_p)\}$ of K mod (ε, η) such that $K_1, \ldots, K_p$ are dyadic cubes.*

PROOF. Let $\eta = \{\eta_j\}$. By the assumptions, the null set Z_Δ of Δ is the union of a sequence $\{T_j\}$ of sets such that for each $j = 1, 2, \ldots$, we can find a countable family $\mathcal{C}_j$ of dyadic cubes of diameters less than $\eta_j/2$ for which

$$T_j \subset \left(\bigcup \mathcal{C}_j\right)^{\circ} \quad \text{and} \quad \sum_{C \in \mathcal{C}_j} \|C\| \leq 4.$$

Thus, if $\mathcal{E} \subset \mathcal{C}_j$ is a finite family and $E = \bigcup \mathcal{E}$, then

$$\|E\| \leq \sum_{C \in \mathcal{E}} \|C\| \leq 4 < \frac{1}{\varepsilon},$$

$$|E| < \frac{\eta_j}{2} \sum_{C \in \mathcal{E}} [d(C)]^{m-1} \leq \frac{\eta_j}{4m} \sum_{C \in \mathcal{E}} \|C\| \leq \eta_j.$$

By the fundamental property of dyadic cubes, there is a nonoverlapping family $\mathcal{C} \subset \bigcup_{j=1}^{\infty} \mathcal{C}_j$ such that $\bigcup \mathcal{C} = \bigcup_{j=1}^{\infty} (\bigcup \mathcal{C}_j)$, and we have

$$Z_\Delta = \bigcup_{j=1}^{\infty} T_j \subset \bigcup_{j=1}^{\infty} \left(\bigcup \mathcal{C}_j \right)^{\circ} \subset \left(\bigcup \mathcal{C} \right)^{\circ}.$$

Now define a positive function Δ_+ on K as follows:

$$\Delta_+(x) = \begin{cases} \Delta(x) & \text{if } x \in K - Z_\Delta, \\ \min\{d(C) : C \in \mathcal{C}, \, x \in C\} & \text{if } x \in Z_\Delta. \end{cases}$$

By Lemma 7.3.2, there is a Δ_+-fine P-partition $\{(K_1, x_1), \ldots, (K_q, x_q)\}$ of K where $K_1, \ldots, K_q$ are dyadic cubes. Denote by $\mathcal{D}$ the family of all $C \in \mathcal{C}$ such that $C \subset K$ and $K_i \subset C$ for an $i = 1, \ldots, q$. Since $\mathcal{C}$ is a nonoverlapping family, $\mathcal{D}$ is finite (it contains at most q elements). If K_i overlaps with a $D \in \mathcal{D}$, then either $K_i \subset D$ or D is a proper subset of K_i; for both K_i and D are dyadic cubes. The latter case is, however, impossible since D contains a K_j that does not overlap with K_i. We conclude that for $i = 1, \ldots, q$, either $K_i \subset \bigcup \mathcal{D}$ or K_i overlaps with no $D \in \mathcal{D}$. Thus, after a suitable reordering, $\bigcup \mathcal{D} = K \ominus \bigcup_{i=1}^{p} K_i$ for an integer p with $0 \leq p \leq q$. As $Z_\Delta \subset (\bigcup \mathcal{C})^{\circ}$, the definition of Δ_+ implies that $K_i \subset \bigcup \mathcal{D}$ whenever $x_i \in Z_\Delta$. It follows that $\{x_1, \ldots, x_p\}$ is a subset of $K - Z_\Delta$, and so $P = \{(K_1, x_1), \ldots, (K_p, x_p)\}$ is a Δ-fine P-partition in K. We complete the proof by showing that the set $K \ominus \bigcup P$ is (ε, η)-small. To this end, define inductively the families $\mathcal{D}_j \subset \mathcal{D}$ by setting

$$\mathcal{D}_j = \mathcal{D} \cap \mathcal{C}_j - \bigcup_{i=1}^{j-1} \mathcal{D}_i$$

for $j = 1, 2, \ldots$. The family $\mathcal{D}$, being a finite subfamily of $\mathcal{C}$, is the union of a finite, possibly empty, disjoint collection $\{\mathcal{D}_1, \ldots, \mathcal{D}_k\}$. If $D_j = \bigcup \mathcal{D}_j$, then $D_1, \ldots, D_k$ are nonoverlapping figures and $\bigcup \mathcal{D} = \bigcup_{j=1}^{k} D_j$. Since each family $\mathcal{D}_j$ is a finite subfamily of $\mathcal{C}_j$, we have $\|D_j\| < 1/\varepsilon$ and $|D_j| < \eta_j$ for $j = 1, \ldots, k$, which establishes our assertion. $\blacksquare$

Proposition 11.3.7 *Let δ be a gage on a figure A. Then for each positive $\varepsilon < 1/4$ and each caliber η there is a δ-fine Perron partition $\{(K_1, x_1), \ldots, (K_p, x_p)\}$ of A mod (ε, η) such that $K_1, \ldots, K_p$ are dyadic cubes.*

PROOF. Assume first that A is a cell. For $j = 1, \ldots, m$, find integers $r_j < s_j$ so that A is contained in the cell $K = \prod_{j=1}^{m}[r_j, s_j]$. Define a gage Δ on K by setting

$$\Delta(x) = \begin{cases} \min\{\delta(x), \text{dist}(x, \partial A)\} & \text{if } x \in A, \\ \text{dist}(x, \partial A) & \text{if } x \in K - A. \end{cases}$$

By Lemma 11.3.6, there is a Δ-fine P-partition $Q = \{(K_1, x_1), \ldots, (K_q, x_q)\}$ of K mod (ε, η) such that $K_1, \ldots, K_q$ are dyadic cubes. Thus $P = \{(K_i, x_i) : x_i \in A\}$ is a δ-fine P-partition in A, and since

$$A \ominus \bigcup P = A \odot \left(K \ominus \bigcup Q \right),$$

Exercise 11.2.3 implies that P is a P-partition of A mod (ε, η).

Now let A be a figure. Then A is the union of nonoverlapping cells, say $A_1, \ldots, A_n$. If $\eta = \{\eta_j\}_{j=1}^{\infty}$, let $\eta^s = \{\eta_{nj-s+1}\}_{j=1}^{\infty}$ for $s = 1, \ldots, n$. Applying the first part of the proof to A_s and η^s, find a δ-fine P-partition $P_s = \{(K_1^s, x_1^s), \ldots, (K_{p_s}^s, x_{p_s}^s)\}$ of A_s mod (ε, η^s) such that $K_1^s, \ldots, K_{p_s}^s$ are dyadic cubes. It is clear that $P = \bigcup_{s=1}^{n} P_s$ is the desired P-partition. ∎

Corollary 11.3.8 *Let δ be a gage on a figure A, let $0 < \varepsilon < 1/4$, and let η be a caliber. Each δ-fine ε-shapely Perron partition P in A is a subset of a δ-fine ε-shapely Perron partition Q of A mod (ε, η).*

PROOF. By Proposition 11.3.7, there is a δ-fine ε-shapely P-partition S of $A \ominus \bigcup P$ mod (ε, η), and it suffices to let $Q = P \cup S$. ∎

11.4 The g-integral

If $P = \{(A_1, x_1), \ldots, (A_p, x_p)\}$ is a partition and f is a function defined on $\{x_1, \ldots, x_p\}$, we denote by $\sigma(f, P)$ the Stieltjes sum $\sigma(f, P; \lambda_m)$, i.e.,

$$\sigma(f, P) = \sum_{i=1}^{p} f(x_i)|A_i|.$$

In general, when no confusion is possible, we shall suppress λ_m in our notation.

Definition 11.4.1 A function f on a figure A is called *gage integrable* (abbreviated as *g-integrable*) over A if there is a real number I having the following property: given $\varepsilon > 0$, we can find a gage δ on A and a caliber η so that

$$|\sigma(f, P) - I| < \varepsilon$$

for each δ-fine ε-shapely Perron partition P of A mod (ε, η).

It follows from Proposition 11.3.7 that the number I of Definition 11.4.1 is determined uniquely by the function f. We call it the *g-integral* of f over A, temporarily denoted by $\int_A^* f$. The family of all g-integrable functions over A is denoted by $\mathcal{R}_*(A)$ or $\mathcal{R}_*(A, \lambda_m)$. If $m = 1$, then Theorem 6.7.5 shows that over one-cells the g-integral coincides with the P-integral of Henstock and Kurzweil with respect to λ (cf. Definition 6.1.1). Thus the notation $\mathcal{R}_*(A, \lambda)$ is consistent with that of Section 6.1.

We shall not attempt the tedious task of establishing all standard properties of the g-integral. We prove only a few examples to give the reader a feeling of how to manipulate gages and calibers.

Proposition 2.1.3 and its proof translate verbatim. We state the Cauchy test for integrability, but leave its straightforward proof to the reader.

Proposition 11.4.2 *A function f on a figure A belongs to $\mathcal{R}_*(A)$ if and only if for each $\varepsilon > 0$ there is a gage δ on A and a caliber η such that*

$$|\sigma(f, P) - \sigma(f, Q)| < \varepsilon$$

for all δ-fine ε-shapely P-partitions P and Q of A mod (ε, η). ∎

Proposition 11.4.3 *If A is a figure and $f \in \mathcal{R}_*(A)$, then $f \in \mathcal{R}_*(B)$ for each figure $B \subset A$.*

PROOF. The proof is similar to that of Proposition 2.1.9. Choose a positive $\varepsilon < 1/4$, and find a gage δ in A and a caliber $\eta = \{\eta_j\}_{j=1}^{\infty}$ so that

$$|\sigma(f, P) - \sigma(f, Q)| < \varepsilon$$

for any δ-fine ε-shapely P-partitions P and Q of A mod (ε, η). If B is a subfigure of A and $C = A \ominus B$, let $\eta_B = \{\eta_{2j}\}_{j=1}^{\infty}$ and $\eta_C = \{\eta_{2j-1}\}_{j=1}^{\infty}$. By Proposition 11.3.7, there is a δ-fine ε-shapely P-partition P_C of C mod (ε, η_C). Now if P_B and Q_B are δ-fine ε-shapely P-partitions of B mod (ε, η_B), then

$P = P_B \cup P_C$ and $Q = Q_B \cup P_C$ are δ-fine ε-shapely P-partitions of A mod (ε, η). Thus

$$\varepsilon > |\sigma(f, P) - \sigma(f, Q)| = |\sigma(f, P_B) - \sigma(f, Q_B)|$$

and $f \in \mathcal{R}_*(B)$ by Cauchy's test. $\blacksquare$

Proposition 11.4.4 *Let f be a function on a figure A which is the union of nonoverlapping figures $A_1, \ldots, A_n$. If $f \in \mathcal{R}_*(A_i)$ for $i = 1, \ldots, n$, then $f \in \mathcal{R}_*(A)$ and*

$$\int_A^* f = \sum_{i=1}^n \int_{A_i}^* f .$$

PROOF. Choose an $\varepsilon > 0$, and let $\varepsilon' = \varepsilon/[n(1 + c\varepsilon)]$ where

$$c = \max\{\|A_1\|, \ldots, \|A_n\|\} .$$

Given a positive integer $i \leq n$, find a gage δ_i on A_i and a caliber $\eta^i = \{\eta_j^i\}_{j=1}^\infty$ so that

$$\left| \sigma(f, P) - \int_{A_i}^* f \right| < \varepsilon'$$

for each δ_i-fine ε'-shapely P-partition P of A_i mod (ε', η^i). With no loss of generality, we may assume that $\delta_i(x) \leq \text{dist}(x, \partial A_i)$ for each $x \in A_i$, and define a gage δ on A by setting $\delta(x) = \delta_i(x)$ whenever $x \in A_i$ for $i = 1, \ldots, n$. If $\eta_j = \min\{\eta_j^1, \ldots, \eta_j^n\}$ for $j = 1, 2, \ldots$, then $\eta = \{\eta_j\}$ is a caliber. Let P be a δ-fine ε-shapely P-partition of A mod (ε, η). By the choice of δ, we see that

$$P_i = \{(B, x) \in P : x \in A_i\}$$

is a δ_i-fine ε'-shapely P-partition in A_i, and that $P = \bigcup_{i=1}^n P_i$. By definition, $A \ominus \bigcup P = \bigcup_{j=1}^k B_j$ where $B_1, \ldots, B_k$ are figures with $\|B_j\| < 1/\varepsilon$ and $|B_j| < \eta_j$ for $j = 1, \ldots, k$. If $i \leq n$ is a positive integer, then

$$A_i \ominus \bigcup P_i = \bigcup_{j=1}^k (A_i \odot B_j) ,$$

$$\|A_i \odot B_j\| \leq \|A_i\| + \|B_j\| < c + \frac{1}{\varepsilon} < \frac{1}{\varepsilon'} ,$$

and $|A_i \odot B_j| < \eta_j \leq \eta_j^i$ for $j = 1, \ldots, k$. Thus P_i is a partition of A_i mod (ε', η^i), and the proposition follows from the inequality

$$\left| \sigma(f, P) - \sum_{i=1}^n \int_{A_i}^* f \right| \leq \sum_{i=1}^n \left| \sigma(f, P_i) - \int_{A_i}^* f \right| < n\varepsilon' < \varepsilon . \quad \blacksquare$$

If $A \subset \mathbf{R}$ is a one-dimensional figure, it follows from Theorem 6.7.5, Proposition 11.4.4, and Definition 10.1.5 that $\mathcal{R}(A, \lambda) \subset \mathcal{R}_*(A, \lambda)$ and that $\int_A^* f = \int_A f \, d\lambda$ for each $f \in \mathcal{R}(A, \lambda)$. The next theorem shows that the same is true for figures in all dimensions.

Theorem 11.4.5 *If A is a figure, then $\mathcal{R}(A) \subset \mathcal{R}_*(A)$ and*

$$\int_A^* f = \int_A f$$

for each $f \in \mathcal{R}(A)$.

PROOF. With no loss of generality, we may assume that A is a cell. Choose an $f \in \mathcal{R}(A)$ and $\varepsilon > 0$. Using the Vitali–Carathéodory theorem, we can find $g < +\infty$ and $h > -\infty$ in $\overline{\mathcal{R}}(A)$ so that g is upper semicontinuous, h is lower semicontinuous, $g \leq f \leq h$, and $\int_A (h - g) < \varepsilon$. There is a positive function δ on A such that

$$g(y) < f(x) + \varepsilon \qquad \text{and} \qquad h(y) > f(x) - \varepsilon$$

for each $x, y \in A$ with $|x - y| < \delta(x)$. Let $P = \{(A_1, x_1), \ldots, (A_p, x_p)\}$ be a δ-fine partition in A. Integrating the above inequalities yields

$$\int_{A_i} g \leq \int_{A_i} f \leq \int_{A_i} h \, ,$$

$$\int_{A_i} g - \varepsilon |A_i| \leq f(x_i)|A_i| \leq \int_{A_i} h + \varepsilon |A_i|$$

for $i = 1, \ldots, p$. According to Corollary 5.1.10 and Exercise 11.2.7, the map $B \mapsto \int_B f$ is a continuous additive function in A. Thus, by Lemma 11.3.4, there is a caliber η such that $|\int_B f| < \varepsilon$ for each (ε, η)-small figure $B \subset A$. Consequently, if P is a partition of $A \bmod (\varepsilon, \eta)$ and $B = A \ominus \bigcup P$, then

$$\left| \sigma(f, P) - \int_A f \right| \leq \left| \int_B f \right| + \sum_{i=1}^{p} \left| f(x_i)|A_i| - \int_{A_i} f \right|$$

$$\leq \varepsilon + \sum_{i=1}^{p} \left[\varepsilon |A_i| + \int_{A_i} (h - g) \right] < \varepsilon(|A| + 2)$$

and the theorem is proved. ∎

Exercise 11.4.6   Modify Definition 11.4.1 by omitting the word *Perron*, and show that the modified definition is equivalent to Definition 8.1.1.

The proof of the following corollary is left to the reader; it is identical to that of Proposition 6.3.1.

Corollary 11.4.7   *Let f and g be functions defined on a figure A such that $f = g$ almost everywhere in A. Then f belongs to $\mathcal{R}_*(A)$ if and only if g does, in which case*

$$\int_A^* f = \int_A^* g \, . \quad \blacksquare$$

In view of Theorem 11.4.5, we may denote the g-integral of a g-integrable function f over a figure A by $\int_A f$ or $\int_A f \, d\lambda_m$. The symbol $\int_A^* f$ is no longer needed and shall not be used. In the same vein, we say *integral* instead of g-integral, and reserve the term g-integral for emphasis only. Corollary 11.4.7 allows us to introduce, in the obvious way, the family $\overline{\mathcal{R}}_*(A) = \overline{\mathcal{R}}_*(A, \lambda_m)$ of all g-integrable extended real-valued functions defined almost everywhere in a figure A. If A is a figure and $f \in \overline{\mathcal{R}}_*(A)$, then it follows from Proposition 11.4.4 that the map $B \mapsto \int_B f$ is an additive function in A, called the *indefinite integral* of f.

Proposition 11.4.8   *If A is a figure and $f \in \overline{\mathcal{R}}_*(A)$, then the indefinite integral of f is a continuous additive function in A.*

PROOF.   It suffices to prove the continuity of the indefinite integral F of an $f \in \mathcal{R}_*(A)$. To this end, choose a positive $\varepsilon < 1/4$ and find a gage δ on A and a caliber $\eta = \{\eta_1, \eta_2, \ldots\}$ so that $|\sigma(f, P) - F(A)| < \varepsilon/2$ for each δ-fine ε-shapely P-partition P of A mod (ε, η). Select a figure $B \subset A$ with $\|B\| < 1/\varepsilon$ and $|B| < \eta_1$, and let $C = A \ominus B$ and $\theta = \{\eta_2, \eta_3, \ldots\}$. By Propositions 11.3.7 and 11.4.3, there is a δ-fine ε-shapely P-partition Q of C mod (ε, θ) such that $|\sigma(f, Q) - F(C)| < \varepsilon/2$. Since Q is also a δ-fine ε-shapely P-partition of A mod (ε, η), we obtain

$$|F(B)| = |F(A) - F(C)| \leq |F(A) - \sigma(f, Q)| + |\sigma(f, Q) - F(C)| < \varepsilon \, ,$$

which establishes the continuity of F. $\blacksquare$

Theorem 11.4.9   *A function f on a figure A is g-integrable over A if and only if there is an additive continuous function F in A having the following* *property: given $\varepsilon > 0$, we can find a gage δ on A so that*

$$\sum_{i=1}^{p}\left|f(x_i)|A_i| - F(A_i)\right| < \varepsilon$$

for each δ-fine ε-shapely P-partition $\{(A_1, x_1), \ldots, (A_p, x_p)\}$ in A. In this case, F is the indefinite integral of f.

PROOF. Suppose that a function F satisfying the conditions of the theorem exists, choose an $\varepsilon > 0$, and find a gage δ on A corresponding to $\varepsilon/2$. By Lemma 11.3.4, there is a caliber η such that $|F(B)| < \varepsilon/2$ for each (ε, η)-small figure $B \subset A$. Thus, if $P = \{(A_1, x_1), \ldots, (A_p, x_p)\}$ is a δ-fine ε-shapely P-partition of A mod (ε, η), we have

$$|\sigma(f, P) - F(A)| \le \sum_{i=1}^{p}\left|f(x_i)|A_i| - F(A_i)\right| + \left|F\left(A \ominus \bigcup P\right)\right| < \varepsilon.$$

Consequently, $f \in \mathcal{R}_*(A)$ and $\int_A f = F(A)$. Since F satisfies the condition of the theorem in every subfigure of A, we see that F is the indefinite integral of f.

The proof of the converse is similar to that of Henstock's lemma. Assuming that $f \in \mathcal{R}_*(A)$, denote by F its indefinite integral, which is a continuous and additive function by Proposition 11.4.8. Choose a positive $\varepsilon < 1/4$, and find a gage δ on A and a caliber $\eta = \{\eta_j\}$ so that $|\sigma(f, P) - F(A)| < \varepsilon/3$ for each δ-fine ε-shapely P-partition of A mod (ε, η). Let $\eta^0 = \{\eta_{2j-1}\}$ and select a δ-fine ε-shapely P-partition $\{(A_1, x_1), \ldots, (A_p, x_p)\}$ of A mod (ε, η^0). In view of Corollary 11.3.8, it suffices to show that

$$\sum_{i=1}^{p}\left|f(x_i)|A_i| - F(A_i)\right| < \varepsilon.$$

To this end, let $\eta^i = \{\eta_{2(pj-i+1)}\}_{j=1}^{\infty}$ for $i = 1, \ldots, p$, and find δ-fine ε-shapely P-partitions P_i of A_i mod (ε, η^i) so that $|\sigma(f, P_i) - F(A_i)| < \varepsilon/(3p)$. The existence of the partitions P_i is guaranteed by Propositions 11.3.7 and 11.4.3. After a suitable reordering, we may assume that there is an integer k with $0 \le k \le p$ such that $f(x_i)|A_i| - F(A_i)$ is nonnegative for $i = 1, \ldots, k$ and negative for $i = k+1, \ldots, p$. Since

$$Q_+ = \{(A_1, x_1), \ldots, (A_k, x_k)\} \cup \bigcup_{i=k+1}^{p} P_i,$$

$$Q_- = \{(A_{k+1}, x_{k+1}), \ldots, (A_p, x_p)\} \cup \bigcup_{i=1}^{k} P_i$$

are δ-fine ε-shapely P-partitions of A mod (ε, η), we obtain

$$\frac{\varepsilon}{3} > |\sigma(f, P) - F(A)|$$

$$\geq \sum_{i=1}^{k} [f(x_i)|A_i| - F(A_i)] - \left| \sum_{i=k+1}^{p} [\sigma(f, P_i) - F(A_i)] \right|$$

$$\geq \left| \sum_{i=1}^{k} \left| f(x_i)|A_i| - F(A_i) \right| - (p - k)\frac{\varepsilon}{3p} \right.,$$

and similarly

$$\frac{\varepsilon}{3} > \sum_{i=k+1}^{p} \left| f(x_i)|A_i| - F(A_i) \right| - k\frac{\varepsilon}{3p}.$$

Adding the last two inequalities completes the proof. $\blacksquare$

The following *divergence theorem* is already a significant generalization of Theorem 8.3.5. More prominent generalizations will be given later in Theorems 11.7.5, 11.7.8, and 11.7.10.

Theorem 11.4.10 *Let T be a thin set, and let v be a continuous vector field on a figure A that is differentiable in $A° - T$. Then $\operatorname{div} v$ belongs to $\overline{\mathcal{R}}_*(A, \lambda_m)$ and*

$$\int_A \operatorname{div} v \, d\lambda_m = \int_{\partial A} v \cdot n \, d\lambda_{m-1}.$$

PROOF. By Proposition 11.2.8, the flux F of v is an additive continuous function in A, and we may assume that $A \neq \emptyset$. Let $f = \operatorname{div} v$ and choose an $\varepsilon > 0$. Given $x \in A° - T$, use Lemma 8.3.4 to find a $\delta_x > 0$ so that

$$|f(x)|B| - F(B)| < \frac{\varepsilon^2}{2m|A|} d(B)\|B\|$$

for each cell $B \subset A \cap U(x, \delta_x)$ with $x \in B$. Define a gage δ on A by setting

$$\delta(x) = \begin{cases} \delta_x & \text{if } x \in A° - T, \\ 0 & \text{if } x \in T \cup \partial A, \end{cases}$$

and choose a δ-fine ε-shapely P-partition $P = \{(A_1, x_1), \dots, (A_p, x_p)\}$ in A. Observing that

$$d(A_i)\|A_i\| \leq 2m[d(A_i)]^m = 2m|A_i|\frac{1}{s(A_i)} \leq \frac{2m}{\varepsilon}|A_i|$$

for $i = 1, \ldots, m$, we obtain

$$\sum_{i=1}^{p} \left| f(x_i)|A_i| - F(A_i) \right| < \frac{\varepsilon}{|A|} \sum_{i=1}^{p} |A_i| \leq \varepsilon .$$

An application of Theorem 11.4.9 completes the argument. ∎

Remark 11.4.11 Let A be a cell. It follows from Example 11.1.2 and Theorem 11.4.10 that Fubini's Theorem 8.6.6 is false for certain functions in $\mathcal{R}_*(A, \lambda_m)$. Nonetheless, a weaker version of Theorem 8.6.6 may still hold for each function $f \in \mathcal{R}_*(A, \lambda_m)$. Finding a suitable Fubini-type theorem for $\mathcal{R}_*(A, \lambda_m)$ is an open problem. This problem is also open (and more important) for the families $\mathcal{FR}_*(A, \lambda_m)$ and $\mathcal{BVR}_*(A, \lambda_m)$, which we shall encounter in Chapters 12 and 13, respectively.

11.5 Improper integrals

We show that the family of all g-integrable functions on a figure A is closed with respect to the formation of improper integrals. This result is analogous to Theorem 6.1.6; however, its proof is more involved.

We say that a sequence $\{A_n\}$ of figures *converges* to a set $X \subset \mathbf{R}^m$, and write $A_n \to X$, if the following condition is satisfied:

$$\sup \|A_n\| < +\infty \qquad \text{and} \qquad \lim |(X - A_n) \cup (A_n - X)| = 0 .$$

It is easy to verify that if $\{A_n\}$ and $\{B_n\}$ are sequences of figures converging, respectively, to subsets X and Y of $\mathbf{R}^m$, then

$$A_n \cup B_n \to X \cup Y \qquad \text{and} \qquad A_n \odot B_n \to X \cap Y .$$

Lemma 11.5.1 *Let F be an additive function in a figure A.*

1. *F is continuous whenever $F(A) = \lim F(A_n)$ for each sequence $\{A_n\}$ of subfigures of A converging to A.*
2. *If F is continuous, then $F(B) = \lim F(B_n)$ for each sequence $\{B_n\}$ of subfigures of A converging to a figure $B \subset A$.*

PROOF. If F is not continuous, then there is an $\varepsilon > 0$ and a sequence $\{B_n\}$ of subfigures of A such that $\|B_n\| < 1/\varepsilon$, $|B_n| < 1/n$, and $|F(B_n)| \geq \varepsilon$ for $n = 1, 2, \ldots$. It follows that $A \ominus B_n \to A$ while for each n, we have

$$|F(A) - F(A \ominus B_n)| = |F(B_n)| \geq \varepsilon .$$

Let F be continuous and let $\{B_n\}$ be a sequence of subfigures of A converging to a figure $B \subset A$. Choose an $\varepsilon > 0$ so that $\|B\| + \|B_n\| < 1/\varepsilon$ for $n = 1, 2, \ldots$. There is an $\eta > 0$ such that $|F(C)| < \varepsilon$ for each figure $C \subset A$ with $\|C\| < 1/\varepsilon$ and $|C| < \eta$. Since $\max\{\|B \ominus B_n\|, \|B_n \ominus B\|\} \leq 1/\varepsilon$ and $\lim |(B \ominus B_n) \cup (B_n \ominus B)| = 0$, we obtain

$$|F(B) - F(B_n)| = |F(B \ominus B_n) - F(B_n \ominus B)|$$
$$\leq |F(B \ominus B_n)| + |F(B_n \ominus B)| < 2\varepsilon$$

for all sufficiently large n. ∎

Theorem 11.5.2 *Let f be a function on a figure A such that $f \in \mathcal{R}_*(B)$ for each figure $B \subset A^\circ$, and suppose that a finite $\lim \int_{A_n} f$ exists for each sequence $\{A_n\}$ of subfigures of A° converging to A. Then all these limits have the same value I, the function f belongs to $\mathcal{R}_*(A)$, and $\int_A f = I$.*

PROOF. For each figure $B \subset A^\circ$, we let $F(B) = \int_B f$. If $\{B_n\}$ and $\{C_n\}$ are sequences of subfigures of A° converging to A, then so is the sequence $\{E_n\}$ where $E_{2n-1} = B_n$ and $E_{2n} = C_n$ for $n = 1, 2, \ldots$. It follows that $\lim F(B_n) = \lim F(C_n)$, and we denote this common value by I.

Let B be a subfigure of A, let $C = A \ominus B$, and let $\{E_n\}$ be a sequence of subfigures of A° converging to A. If $\limsup F(B \odot E_n) = +\infty$, then for $k = 1, 2, \ldots$, there is an integer $n_k \geq 1$ such that $F(B \odot E_{n_k}) \geq k - F(C \odot E_k)$. The sets $D_k = (B \odot E_{n_k}) \cup (C \odot E_k)$ are subfigures of A° converging to A and $\lim F(D_k) = +\infty$. This contradiction and symmetry imply that the sequence $\{F(B \odot E_n)\}$ is bounded.

Now choose subsequences $\{B_{n\pm}\}$ of $\{B \odot E_n\}$ and $\{C_{n\pm}\}$ of $\{C \odot E_n\}$ so that the $\lim F(B_{n\pm}) = b_\pm$ and $\lim F(C_{n\pm}) = c_\pm$ exist. Without loss of generality, we may assume that $b_- \leq b_+$ and $c_- \leq c_+$. Since $B_{n\pm} \cup C_{n\pm}$ are subfigures of A° converging to A, we obtain

$$I = b_- + c_- \leq b_+ + c_+ = I$$

and hence $b_- = b_+$ and $c_- = c_+$. It follows that a finite $\lim F(B \odot E_n) = b$ exists, and an argument analogous to the first part of the proof shows that b does not depend on the choice of $\{E_n\}$. Moreover, by Proposition 11.4.8 and Lemma 11.5.1, we see that $b = F(B)$ whenever $B \subset A^\circ$.

As A is a finite union of cells, it is easy to construct a sequence $\{A_n\}$ of subfigures of A° converging to A. According to the previous paragraph, setting

$$G(B) = \begin{cases} \lim F(B \odot A_n) & \text{if } B \subset A \text{ is a figure,} \\ 0 & \text{if } B \subset A \text{ is a degenerate interval} \end{cases}$$

defines an additive function G in A that extends F. To show that G is continuous, select a sequence $\{B_k\}$ of subfigures of A converging to A. There is a subsequence $\{A_{n_k}\}$ of $\{A_n\}$ such that $|G(B_k) - F(B_k \odot A_{n_k})| < 1/n$ for $n = 1, 2, \ldots$. Now $B_k \odot A_{n_k}$ are subfigures of A° converging to A, so

$$G(A) = \lim F(B_k \odot A_{n_k}) = \lim G(B_k)$$

and the continuity of G is a consequence of Lemma 11.5.1.

Using the fundamental property of dyadic cubes, construct a countable family $\{K_1, K_2, \ldots\}$ of nonoverlapping cells so that $A^\circ = \bigcup_{n=1}^{\infty} K_n$. Choose an $\varepsilon > 0$ and use Theorem 11.4.9 to find gages δ_n on K_n so that

$$\sum_{i=1}^{p} \left| f(x_i)|B_i| - F(B_i) \right| < \frac{\varepsilon}{2^n}$$

for each δ_n-fine ε-shapely P-partition $\{(B_1, x_1), \ldots, (B_p, x_p)\}$ in K_n. With no loss of generality, we may assume that $\delta_n(x) \leq \text{dist}(x, \partial K_n)$ for each $x \in K_n$ and $n = 1, 2, \ldots$. Define a gage δ on A by the formula

$$\delta(x) = \begin{cases} \delta_n(x) & \text{if } x \in K_n, \\ 0 & \text{if } x \in \partial A. \end{cases}$$

If $P = \{(C_1, x_1), \ldots, (C_q, x_q)\}$ is a δ-fine ε-shapely P-partition in A, the choice of δ implies that $P_n = \{(C_i, x_i) : x_i \in K_n\}$ is a δ_n-fine ε-shapely P-partition in K_n. As $P = \bigcup_{n=1}^{\infty} P_n$, we have

$$\sum_{i=1}^{q} \left| f(x_i)|C_i| - G(C_i) \right| = \sum_{n=1}^{\infty} \sum_{x_i \in K_n} \left| f(x_i)|C_i| - F(C_i) \right|$$

$$< \sum_{n=1}^{\infty} \varepsilon 2^{-n} = \varepsilon,$$

and since $G(A) = I$, another application of Theorem 11.4.9 completes the argument. ∎

11.6 Connections with the McShane integral

It is an easy exercise to verify that Theorem 9.3.4 and its proof hold for
g-integrable functions. Thus each g-integrable function is measurable by
Proposition 9.3.3.

Theorem 11.6.1 *Let f be a function on a figure A. Then $f \in \mathcal{R}(A)$ if and
only if both f and $|f|$ belong to $\mathcal{R}_*(A)$.* ∎

Theorem 11.6.1 is the same as Theorem 6.3.4, and so is its proof. Immediate corollaries are results analogous to Corollary 6.3.5 and Propositions 6.3.7
and 6.3.8.

Let F be an additive function in a figure A. We say that F is of
bounded variation if there is a division $\mathcal{D}$ of A such that $V(F, D) < +\infty$
for each $D \in \mathcal{D}$ (cf. Definition 9.1.1). When F is of bounded variation, then
$V(F, B) < +\infty$ for each cell $B \subset A$, and it follows from Proposition 9.1.3
that the functions $VF : B \mapsto V(F, B)$ and $VF - F$ are volumes in A (cf.
Corollary 9.1.4).

Theorem 11.6.2 *Let A be a figure, and let F be the indefinite integral of
$f \in \mathcal{R}_*(A)$. Then $f \in \mathcal{R}(A)$ if and only if F is of bounded variation, in which
case VF is the indefinite integral of $|f|$.*

PROOF. In view of Propositions 11.4.3 and 11.4.4, we may assume that A
is a cell. If $f \in \mathcal{R}(A)$ then $|f| \in \mathcal{R}(A)$, and we have

$$\sum_{D \in \mathcal{D}} |F(D)| \leq \sum_{D \in \mathcal{D}} \int_D |f| = \int_A |f|$$

for each division $\mathcal{D}$ of A. It follows that $V(F, A) \leq \int_A |f| < +\infty$.

Conversely, let $V(F, A) < +\infty$ and choose a positive $\varepsilon < 1/4$. Find a
division $\{D_1, \ldots, D_n\}$ of A so that

$$V(F, A) < \sum_{k=1}^{n} |F(D_k)| + \frac{\varepsilon}{3}.$$

By Theorem 11.4.9, there is a gage δ on A such that

$$\sum_{i=1}^{p} \left| f(x_i)|A_i| - F(A_i) \right| < \frac{\varepsilon}{3}$$

for each δ-fine ε-shapely P-partition $\{(A_1, x_1), \ldots, (A_p, x_p)\}$ in A. We may
assume that $\delta(x) \leq \text{dist}(x, \partial D_k)$ for each $x \in D_k$ and $k = 1, \ldots, n$. Use

Proposition 11.4.8 and Lemma 11.3.4 to find a caliber η so that $|F(B)| <$ $\varepsilon/(3n)$ for each (ε, η)-small figure $B \subset A$. According to Proposition 11.3.7, there is a δ-fine ε-shapely P-partition $P = \{(A_1, x_1), \ldots, (A_p, x_p)\}$ of A mod (ε, η). For $k = 1, \ldots, n$, let $C_k = \bigcup_{x_i \in D_k} A_i$ and observe that by the choice of δ and Exercise 11.2.3, the figures $B_k = D_k \ominus C_k$ are (ε, η)-small. Thus

$$|\sigma(|f|, P) - V(F, A)|$$

$$\leq \sum_{i=1}^{p} \left| |f(x_i)| \cdot |A_i| - |F(A_i)| \right| + V(F, A) - \sum_{k=1}^{n} \sum_{x_i \in D_k} |F(A_i)|$$

$$\leq \sum_{i=1}^{p} \left| f(x_i)|A_i| - F(A_i) \right| + V(F, A) - \sum_{k=1}^{n} |F(C_k)|$$

$$< \frac{\varepsilon}{3} + \left[V(F, A) - \sum_{k=1}^{n} |F(D_k)| \right] + \sum_{k=1}^{n} |F(B_k)|$$

$$< \frac{2\varepsilon}{3} + n\frac{\varepsilon}{3n} = \varepsilon,$$

and we see that $|f| \in \mathcal{R}_*(A)$ and $\int_A |f| = V(F, A)$. Now $f \in \mathcal{R}(A)$ by Theorem 11.6.1, and repeating the previous argument for a cell $B \subset A$ completes the proof. $\blacksquare$

Definition 11.6.3 An additive continuous function F in a figure A is called AC_* if, given a negligible set $E \subset A$ and $\varepsilon > 0$, there is a gage δ on E such that $\sum_{i=1}^{p} |F(A_i)| < \varepsilon$ for each ε-shapely Perron partition $\{(A_1, x_1), \ldots, (A_p, x_p)\}$ in A anchored in E that is δ-fine.

It follows from Proposition 6.4.6 that Definitions 11.6.3 and 6.4.1 coincide if $A \subset \mathbf{R}$ is a one-cell. By Proposition 5.1.9, 3 and Exercise 11.2.7, each AC function is AC_*, but not vice versa. As in dimension one, AC_* functions provide a *partial descriptive definition* of the g-integral. To my knowledge, a *full descriptive definition* of the g-integral in dimensions larger than one is not known.

Theorem 11.6.4 *Let F be an additive function in a figure A that is derivable almost everywhere in A. Then F is AC_* if and only if $F' \in \overline{\mathcal{R}}_*(A)$ and F is the indefinite integral of F'.*

We omit the proof, which is identical to that of Theorem 6.4.4.

Exercise 11.6.5   Let A be a figure. Combine Theorems 9.3.7 and 11.6.4 to show that $\overline{\mathcal{R}}(A) \subset \overline{\mathcal{R}}_*(A)$ without referring to Theorem 11.4.5.

Exercise 11.6.6   Let F be an additive function of bounded variation in a figure A. Following the proof of Proposition 6.4.5, show that F is AC_* whenever it is AC.

11.7 Almost derivable functions

Let F be a function defined on all subcells of a figure A, and let $x \in A$. Given positive numbers δ and $\eta < 1$, denote by $A_{\delta,\eta}$ the collection of all cells $B \subset A \cap U(x, \delta)$ with $x \in B$ and $s(B) > \eta$. Since $A_{\delta,\eta} \neq \emptyset$, the extended real number

$$|D\eta|F(x) = \inf_{\delta>0} \left[\sup_{B\in A_{\delta,\eta}} \frac{|F(B)|}{|B|} \right]$$

is nonnegative. If $|D\eta|F(x) < +\infty$ for each $\eta \in (0,1)$, we say that F is *almost derivable* at x. Almost derivability resembles almost differentiability defined in Sections 6.6 and 10.6. The two concepts, which coincide in dimension one, are different but closely related in higher dimensions.

Exercise 11.7.1   Let F be a function defined on all subcells of a figure A, and let $x \in A$. Show that $\sup_{\eta\in(0,1)} |D\eta|F(x) < +\infty$ if and only if the extended real numbers $D_{\lambda_m}F(x)$ and $D^{\lambda_m}F(x)$ are finite.

Proposition 11.7.2   *Let T be a thin set, and let F be an additive continuous function in a figure A. If F is almost derivable at each $x \in A - T$, then F is AC_*.*

PROOF.   The proof is similar to that of Proposition 6.6.3. Choose a negligible set $E \subset A$ and a positive $\varepsilon < 1$. For $n = 1, 2, \ldots$, set

$$E_n = \{x \in E - T : n - 1 \leq |D\varepsilon|F(x) < n\}$$

and find open sets U_n so that $E_n \subset U_n$ and $|U_n| < \varepsilon 2^{-n}/n$. Given $x \in E_n$, there is a $\delta_n(x) > 0$ such that $U(x, \delta_n(x)) \subset U_n$ and $|F(B)| < n|B|$ for each cell $B \subset A \cap U(x, \delta_n(x))$ with $x \in B$ and $s(B) > \varepsilon$. Since $E - T$ is the disjoint union of the sets E_n, the formula

$$\delta(x) = \begin{cases} \delta_n(x) & \text{if } x \in E_n \text{ and } n = 1, 2, \ldots, \\ 0 & \text{if } x \in E \cap T, \end{cases}$$

defines a gage on E. For an ε-shapely P-partition $\{(A_1, x_1), \ldots, (A_p, x_p)\}$ in A anchored in E that is δ-fine, we obtain

$$\sum_{i=1}^{p} |F(A_i)| = \sum_{n=1}^{\infty} \sum_{x_i \in E_n} |F(A_i)| < \sum_{n=1}^{\infty} \sum_{x_i \in E_n} n|A_i|$$

$$\leq \sum_{n=1}^{\infty} n|U_n| < \sum_{n=1}^{\infty} \varepsilon 2^{-n} = \varepsilon,$$

which proves the proposition. ∎

The assumption of continuity cannot be omitted in Proposition 11.7.2. This is clear in dimension one; however, the following two-dimensional example is instructive.

Example 11.7.3 Let $m = 2$ and $A = [0,1]^2$. For $n = 1, 2, \ldots$, let

$$A_+^n = [3 \cdot 2^{-n-1}, 2^{-n+1}] \times [2^{-n-1}, 2^{-n}],$$

$$A_-^n = [2^{-n-1}, 2^{-n}] \times [3 \cdot 2^{-n-1}, 2^{-n+1}],$$

and define a function G on all intervals $B \subset A - \{\mathbf{0}\}$ by setting

$$G(B) = \sum_{n=1}^{\infty} \frac{1}{n|A_+^n|} |B \cap A_+^n| - \sum_{n=1}^{\infty} \frac{1}{n|A_-^n|} |B \cap A_-^n|.$$

Since all but finitely many terms in the above sums are equal to zero, G is a well-defined additive function in $A - \{\mathbf{0}\}$, which we extend to an additive function F in A as follows. We let $F(B) = 0$ for each degenerate interval $B \subset A$, and if B is a subcell of A, we set

$$F(B) = \begin{cases} G(B) & \text{if } \mathbf{0} \notin B, \\ G(A \ominus B) & \text{if } \mathbf{0} \in B. \end{cases}$$

A simple check reveals that F is almost derivable at each $x \in A - \{\mathbf{0}\}$, but not at $\mathbf{0}$. As the cells $A_\pm^n$ are positioned symmetrically with respect to the diagonal of A, it is easy to verify that $\lim F(B_n) = 0$ for each sequence $\{B_n\}$ of *subcells* of A with $\lim |B_n| = 0$ (draw a picture!). On the other hand, observing that

$$\sum_{n=1}^{\infty} \|A_+^n\| = 4 \sum_{n=1}^{\infty} 2^{-n-1} = 2 \quad \text{and} \quad \sum_{n=1}^{\infty} F(A_+^n) = \sum_{n=1}^{\infty} \frac{1}{n} = +\infty,$$

we conclude without difficulty that F is not continuous.

Whether an additive function F in a figure A that is almost derivable, or even derivable, at *every* $x \in A$ is continuous appears to be an open problem if $m \geq 2$. For $m = 1$, an affirmative answer follows from Exercise 11.2.6.

Lemma 11.7.4 *Let v be a continuous vector field defined on a figure A, and let F be the flux of v in A.*

1. *If v is almost differentiable at $z \in A$, then F is almost derivable at z.*
2. *If v is differentiable at $z \in A^\circ$, then F is derivable at z and $F'(z) = \operatorname{div} v(z)$.*

PROOF. If v is almost differentiable at $z \in A$, then there are positive numbers c and δ such that $|v(x) - v(z)| \leq c|x - z|$ for each $x \in A \cap U(z, \delta)$ (cf. Definition 10.6.6). Select a positive $\eta < 1$ and a cell $B \subset A \cap U(z, \delta)$ with $z \in B$ and $s(B) > \eta$, and observe that $|v(x) - v(z)| \leq cd(B)$ for each $x \in B$. By the divergence theorem, the flux of a constant vector field from any cell is zero. Thus

$$|F(B)| = \left| \int_{\partial B} [v(x) - v(z)] \cdot n \, d\lambda_{m-1}(x) \right|$$

$$\leq cd(B)\|B\| \leq 2mc[d(B)]^m \leq \frac{2mc}{\eta}|B|$$

by Proposition 8.3.1, and consequently $|D\eta|F(z) < +\infty$.

Suppose that v is differentiable at z and choose positive numbers η and ε. Let $\{B_n\}$ be a sequence of subcells of A containing z for which $\lim d(B_n) = 0$ and $\inf s(B_n) > \eta$. By Lemma 8.3.4, there is a $\delta > 0$ such that

$$\left| \operatorname{div} v(z)|B| - F(B) \right| < \frac{\varepsilon\eta}{2m} d(B)\|B\|$$

for each cell $B \subset A \cap U(z, \delta)$ containing z. Hence

$$\left| \operatorname{div} v(z)|B_n| - F(B_n) \right| < \frac{\varepsilon\eta}{2m} d(B_n)\|B_n\| \leq \varepsilon\eta[d(B_n)]^m < \varepsilon|B_n|$$

for all sufficiently large n. We conclude that $\lim[F(B_n)/|B_n|] = \operatorname{div} v(z)$, and the lemma follows from Exercise 9.3.1. ∎

Theorem 11.7.5 *Let T be a thin set, and let v be a continuous vector field on a figure A that is almost differentiable at each $x \in A^\circ - T$. Then $\operatorname{div} v$ belongs to $\overline{\mathcal{R}}_*(A)$ and*

$$\int_A \operatorname{div} v \, d\lambda_m = \int_{\partial A} v \cdot n \, d\lambda_{m-1}.$$

PROOF. By Proposition 11.2.8, the flux F of v in A is an additive continuous function in A. It follows from Lemma 11.7.4 and Stepanoff's Theorem 10.6.7 that F is almost derivable in $A - T$, and $F' = \operatorname{div} v$ almost everywhere in A. Since F is AC_* by Proposition 11.7.2, an application of Theorem 11.6.4 completes the proof. ∎

The following theorem, which resembles Stepanoff's theorem, justifies the term "almost derivable." We shall not prove it in this book, but note that it follows directly from a more general result of A.J. Ward (see [44, Chapter IV, Section 11; particularly the small print on p. 139]).

Theorem 11.7.6 *Let F be an additive function in a figure A, and let E be a subset of A. If F is almost derivable at all $x \in E$, then it is derivable at almost all $x \in E$.* ∎

Combining Theorems 11.7.6 and 11.6.4 with Proposition 11.7.2 yields the next corollary.

Corollary 11.7.7 *Let T be a thin set, and let F be an additive continuous function in a figure A that is almost derivable at each $x \in A - T$. Then $F' \in \overline{\mathcal{R}}_*(A)$ and F is the indefinite integral of F'.* ∎

Let v be a continuous vector field defined on a figure A, let F be the flux of v in A, and let $x \in A$. If F is derivable at x, we call the derivate $F'(x)$ the *mean divergence* of v at x, denoted by $\operatorname{DIV} v(x)$. By Lemma 11.7.4, 2, the mean divergence $\operatorname{DIV} v(x)$ exists and equals $\operatorname{div} v(x)$ whenever v is differentiable at x. On the other hand, Example 11.7.9 below shows that $\operatorname{DIV} v(x)$ may well exist also when v is not differentiable at x.

Note. The mean divergence of vector fields was studied by V.L. Shapiro ([46]), but only in the context of McShane integration. His general approach is different from ours.

A mere rewording of Corollary 11.7.7 provides a result called the *mean divergence theorem*.

Theorem 11.7.8 *Let T be a thin set, and let v be a continuous vector field on a figure A whose flux is almost derivable at each $x \in A^\circ - T$. Then $\operatorname{DIV} v$ belongs to $\overline{\mathcal{R}}_*(A, \lambda_m)$ and*

$$\int_A \operatorname{DIV} v \, d\lambda_m = \int_{\partial A} v \cdot n \, d\lambda_{m-1} . \quad ∎$$

The following example, constructed by Z. Buczolich, shows that the mean divergence theorem is more general than Theorem 11.7.5, already for vector fields whose mean divergence is integrable in the sense of McShane.

Example 11.7.9 Let $m = 2$ and $A = [0,1]^2$. For $\delta > 0$ and $t \in \mathbf{R}$, set

$$
\varphi_\delta(t) = \begin{cases} \exp\left(-\dfrac{t^2}{\delta^2 - t^2}\right) & \text{if } |t| < \delta, \\ 0 & \text{if } |t| \geq \delta. \end{cases}
$$

Since φ_δ is continuously differentiable in $\mathbf{R}$, the map $u_\delta : x \mapsto (\varphi_\delta(\|x\|), 0)$ is a continuously differentiable vector field in $\mathbf{R}^2$ that vanishes outside the set $U_\delta = \{x \in \mathbf{R}^2 : \|x\| < \delta\}$. If $x = (\xi_1, \xi_2)$ is a point of U_δ, then

$$
\operatorname{div} u_\delta(x) = -2\delta^2 \xi_1 \frac{\varphi_\delta(\|x\|)}{(\delta^2 - \|x\|^2)^2}.
$$

By symmetry, $\int_{U_\delta} \operatorname{div} u_\delta \, d\lambda_2 = 0$; to be precise, this follows from Theorem 10.5.6 applied to the map $(\xi_1, \xi_2) \mapsto (-\xi_1, \xi_2)$. Observing that in the interval $(-\delta, \delta)$ the function $\varphi_\delta(t)(\delta^2 - t^2)^{-2}$ attains its maximum at $t = 0$, we obtain

$$
\int_{U_\delta} |\operatorname{div} u_\delta| \, d\lambda_2 \leq \frac{2}{\delta} |U_\delta| = 2\pi\delta < 7\delta.
$$

Enumerating a dense countable subset of A°, it is easy to construct inductively sequences $\{z_k\}$ of points in A° and $\{\varepsilon_k\}$ of positive numbers so that the following conditions are satisfied:

1. $\varepsilon_1 \leq 1/2$ and $\varepsilon_{n+1} \leq \varepsilon_k/2$ for $n = 1, 2, \ldots$;
2. $U[z_1, \varepsilon_1], U[z_2, \varepsilon_2], \ldots$ are disjoint subcells of A°;
3. $E = \bigcup_{k=1}^\infty U(z_k, \varepsilon_k)$ is a dense subset of A.

Claim 1. Each neighborhood of $x \in A - E$ contains some z_k.

Proof. Suppose that there is an $x \in A - E$ and an $\varepsilon > 0$ such that $U(x, \varepsilon)$ contains no z_k. Then $U(x, \varepsilon/2)$ meets only finitely many sets $U(z_k, \varepsilon_k)$, say $U(z_1, \varepsilon_1), \ldots, U(z_r, \varepsilon_r)$. If $x \in \partial A$, we already have a contradiction. If $x \in A^\circ$, we may assume that $U(x, \varepsilon) \subset A$. Since $x \notin E$ and $\bigcup_{k=1}^r U(z_k, \varepsilon_k) = (\bigcup_{k=1}^r U[z_k, \varepsilon_k])^\circ$, we see that $U(x, \varepsilon/2) \not\subset \bigcup_{k=1}^r U[z_k, \varepsilon_k]$. Thus there is a $y \in U(x, \varepsilon/2)$ and an $\eta > 0$ such that

$$
U(y, \eta) \subset U(x, \varepsilon/2) - \bigcup_{k=1}^r U(z_k, \varepsilon_k) = U(x, \varepsilon/2) - E \subset A - E,
$$

a contradiction again.

Claim 2. For each $\varepsilon > 0$ and each nonnegative integer p there is an integer $q > p$ such that $\text{dist}(x, \{z_{p+1}, \ldots, z_q\}) < \varepsilon$ for each $x \in A - E$.

Proof. Since $A - E$ is compact, given $\varepsilon > 0$, there are points $x_1, \ldots, x_r$ in $A - E$ such that $A - E \subset \bigcup_{i=1}^{r} U(x_i, \varepsilon/2)$. By Claim 1, each $U(x_i, \varepsilon/2)$ contains a z_{k_i} with $k_i > p$. Hence $q = \max\{k_1, \ldots, k_r\}$ is the desired integer.

Using Claim 2, construct inductively a strictly increasing sequence $\{p_i\}$ of integers such that $p_1 = 0$ and $\text{dist}(x, \{z_{p_i+1}, \ldots, z_{p_{i+1}}\}) < 2^{-i}$ for $i = 1, 2, \ldots$ and each $x \in A - E$. Given an integer $k \geq 1$, let $\delta_k = \varepsilon_k^3$, $v_k = u_{\delta_k}$, $V_k = U_{\delta_k}$, and set $c_k = 1/i$ whenever $p_i < k \leq p_{i+1}$. Further let

$$v(x) = \sum_{k=1}^{\infty} c_k v_k(x - z_k)$$

for each $x \in A$, and set $D = \bigcup_{k=1}^{\infty} U(z_k, \delta_k)$. Clearly, $v(x) = \mathbf{0}$ for every point $x \in A - E$.

Claim 3. The vector field v is continuous in A and continuously differentiable in E. The set $A - E$ has positive measure and v is almost differentiable at no point of $A - E$.

Proof. Clearly, v is continuously differentiable in E. Let $x \in A - E$. Given $\varepsilon > 0$, there is an integer $i > 1/\varepsilon$ and an $\eta > 0$ such that $U(x, \eta) \cap V_k = \emptyset$ for $k = 1, \ldots, p_i$. Thus, for each $y \in U(x, \eta)$, we can find an integer $k(y) > p_i$ with

$$|v(y) - v(x)| = c_{k(y)}|v_{k(y)}(y - z_{k(y)})| \leq \frac{1}{i} < \varepsilon.$$

It follows that v is continuous at x. On the other hand, given an integer $i \geq 1$, we can find an integer k_i so that $p_i < k_i \leq p_{i+1}$ and $|z_{k_i} - x| < 2^{-i}$. Hence

$$\frac{|v(z_{k_i}) - v(x)|}{|z_{k_i} - x|} \geq 2^i c_{k_i} = \frac{2^i}{i}$$

for $i = 1, 2, \ldots$, and we see that v is not almost differentiable at x. Finally, the inequality

$$|E| = 4 \sum_{k=1}^{\infty} \varepsilon_k^2 \leq 4\varepsilon_1^2 \sum_{k=1}^{\infty} 4^{-k} \leq \frac{4}{3}\varepsilon_1^2 \leq \frac{1}{3} < |A|$$

establishes the claim.

We show next that the mean divergence DIV v exists at each $x \in A$. To this end, define a function g on A by setting

$$g(x) = \sum_{n=1}^{\infty} |\operatorname{div} v_k(x - z_k)|$$

for every $x \in A$. Theorem 10.5.6 yields

$$\sum_{k=1}^{\infty} \int_A |\operatorname{div} v_k(x - z_k)| \, d\lambda_2(x) = \sum_{k=1}^{\infty} \int_{V_k} |\operatorname{div} v_k| \, d\lambda_2 \leq 7 \sum_{k=1}^{\infty} \delta_k < +\infty,$$

and so $g \in \mathcal{R}(A, \lambda_2)$ by Theorem 2.3.13.

Claim 4. If F is the flux of v in A, then $|F(B)| \leq \int_B g \, d\lambda_2$ for each subfigure B of A.

Proof. Let $B \subset A$ be a figure. Making the obvious identification between $\mathbf{R}^{m-1}$ and a hyperplane of $\mathbf{R}^m$, the flux $\int_{\partial B} v_k(x - z_k) \cdot n \, d\lambda_1(x)$ is the sum of $(m-1)$-dimensional integrals over the $(m-1)$-cells contained in ∂B. Since $|v(x)| = \sum_{k=1}^{\infty} c_k |v_k(x - z_k)|$ for all $x \in A$, Theorems 2.3.13 and 8.3.5 imply that

$$|F(B)| \leq \sum_{k=1}^{\infty} c_k \left| \int_{\partial B} v_k(x - z_k) \cdot n \, d\lambda_1 \right|$$

$$\leq \sum_{n=1}^{\infty} c_k \int_B |\operatorname{div} v_k(x - z_k)| \, d\lambda_2(x) \leq \int_B g \, d\lambda_2.$$

By Lemma 11.7.4, 2, the flux F is derivable at each point of E. Select an $x \in A - E$ and an $\eta > 0$, and choose a cell $B \subset A$ with $x \in B$ and $s(B) > \eta$. If $B \cap D = \emptyset$, then $F(B) = 0$. Otherwise, there is a least integer $p \geq 1$ such that B meets $U(z_p, \delta_p)$. As $x \notin U(z_p, \varepsilon_p)$, we see that $d(B) > \varepsilon_p - \delta_p$ and hence

$$|B| > \eta[d(B)]^2 > \eta(\varepsilon_p - \delta_p)^2 = \eta \varepsilon_p^2 (1 - \varepsilon_p^2)^2.$$

Since Claim 4 gives

$$|F(B)| \leq \int_B g \, d\lambda_2 \leq \sum_{k=p}^{\infty} \int_{V_k} |\operatorname{div} v_k| \, d\lambda_2 < 7 \sum_{n=p}^{\infty} \delta_k \leq 8\varepsilon_p^3,$$

we have

$$\left| \frac{F(B)}{|B|} \right| < \frac{8}{\eta} \cdot \frac{\varepsilon_p}{(1 - \varepsilon_p)^2}.$$

If the diameter of B is getting smaller and B keeps intersecting D, then p approaches infinity. Thus the last inequality implies that F is derivable at x and DIV $v(x) = F'(x) = 0$. From the mean divergence theorem we obtain

$$\int_{\partial C} v \cdot n \, d\lambda_1 = \int_C \text{DIV} \, v \, d\lambda_2 = \int_{C \cap E} \text{div} \, v \, d\lambda_2$$

for each figure $C \subset A$.

Let A be a figure and let v be a vector field defined on ∂A. To keep the exposition simple, we defined the flux $\int_{\partial A} v \cdot n \, d\lambda_{m-1}$ only for continuous v. It is clear, however, that the flux of v from A can be defined as long as the function $v \cdot n$ is integrable with respect to the volume λ_{m-1} on each $(m-1)$-cell contained in ∂A. For example, the dominated convergence theorem implies that we can define the flux of any bounded vector field that is the pointwise limit of continuous vector fields.

We say that a vector field v defined on a set $E \subset \mathbf{R}^m$ is *fluxing* if the flux $\int_{\partial B} v \cdot n \, d\lambda_{m-1}$ is defined for each cell, and hence for each figure, $B \subset E$. Arguing as in the proof of Proposition 8.3.2, it is easy to verify that the flux of a fluxing vector field is an additive function. Corollary 11.7.7 implies the following obvious variation of the mean divergence theorem.

Theorem 11.7.10 *Let T be a thin set, and let v be a fluxing vector field on a figure A whose flux is continuous. If the flux of v is almost derivable at each $x \in A^\circ - T$, then DIV v belongs to $\overline{\mathcal{R}}_*(A, \lambda_m)$ and*

$$\int_A \text{DIV} \, v \, d\lambda_m = \int_{\partial A} v \cdot n \, d\lambda_{m-1} . \quad \blacksquare$$

Example 11.7.11 We adhere to the notation of Example 11.7.9 and agree that by Claims 1–4, we mean the claims of that example. Let

$$w(x) = \sum_{k=1}^{\infty} v_k(x - z_k)$$

for every $x \in A$. It is clear that w is continuously differentiable in E, and it follows from Claim 1 that w is discontinuous at each $x \in A - E$; indeed, $w(x) = \mathbf{0}$ while $w(z_k) = (1, 0)$ for $k = 1, 2, \ldots$. Since w is the sum of continuous vector fields and $|w(x)| \leq 1$ for each $x \in A$, the vector field w is fluxing by Theorem 2.3.13. Denote by G the flux of w in A. A closer look at the proof of Claim 4 reveals that $|G(B)| \leq \int_B g \, d\lambda_2$ for each figure $B \subset A$. Thus G is continuous according to Exercise 11.2.7. Proceeding as in Example 11.7.9, we can show that G is derivable at each $x \in A$ and $\mathrm{DIV}\, w(x) = G'(x) = 0$ for every $x \in A - E$. It follows from Theorem 11.7.10 that

$$\int_{\partial C} w \cdot n \, d\lambda_1 = \int_C \mathrm{DIV}\, w \, d\lambda_2 = \int_{C \cap E} \mathrm{div}\, w \, d\lambda_2$$

for each figure $C \subset A$.

We shall not pursue the study of fluxing vector fields any further, and return back to considering continuous vector fields only. The reader should keep in mind, however, that often this is an assumption of convenience rather than necessity.

Chapter 12

The $\mathcal{F}$-integral

The results of Chapter 11 show that the g-integral has many desirable properties. Yet, there is a disturbing asymmetry in its definition: while only cells enter into partitions, figures are needed to define the approximation by Stieltjes sums. We remove the asymmetry by showing that an equally viable integral can be based on partitions that use figures instead of cells. This symmetrization leads to a change of variables theorem and motivates the new developments discussed in Chapter 13.

12.1 Shape and regularity

Let A be a figure in $\mathbf{R}^m$. If A is nonempty, let

$$s(A) = \frac{|A|}{[d(A)]^m} \qquad \text{and} \qquad r(A) = \frac{|A|}{d(A)\|A\|},$$

and set $s(A) = r(A) = 0$ if $A = \emptyset$. The numbers $s(A)$ and $r(A)$ are called, respectively, the *shape* and *regularity* of A. The shape of a figure is a direct generalization of the earlier defined shape of a cell (see the end of Section 7.2). On the other hand, regularity is a new concept, whose relationship to shape will be investigated in this section.

Lemma 12.1.1 *For $j = 1, \ldots, n$, let $a_j \geq 0$ and $p_j > 0$ be real numbers. If $p_1 + \cdots + p_n = 1$, then $a_1^{p_1} \cdots a_n^{p_n} \leq p_1 a_1 + \cdots + p_n a_n$. In particular,*

$$(a_1 \cdots a_n)^{\frac{1}{n}} \leq \frac{1}{n}(a_1 + \cdots + a_n).$$

PROOF. If $n = 2$, let $\varphi(t) = p_1 t + p_2 a_2 - t^{p_1} a_2^{p_2}$ for each $t \geq 0$, and observe that φ attains its minimum at $t = a_2$. Since $\varphi(a_2) = 0$, the desired inequality follows. Now assuming that the lemma holds for $n - 1$, we obtain

$$a_1^{p_1} \cdots a_n^{p_n} = a_1^{p_1} \left(a_2^{\frac{p_2}{1-p_1}} \cdots a_n^{\frac{p_n}{1-p_1}} \right)^{1-p_1}$$

$$\leq p_1 a_1 + (1 - p_1) a_2^{\frac{p_2}{1-p_1}} \cdots a_n^{\frac{p_n}{1-p_1}}$$

$$\leq p_1 a_1 + \cdots + p_n a_n . \quad \blacksquare$$

Lemma 12.1.2 *Let $E \subset \mathbf{R}^m$ be a bounded measurable set, and let $f_1, \ldots, f_n$ be nonnegative functions in $\mathcal{R}(E)$. Then $\prod_{j=1}^n f_j^{1/n}$ belongs to $\mathcal{R}(E)$ and*

$$\int_E \prod_{j=1}^n f_j^{\frac{1}{n}} \leq \prod_{j=1}^n \left(\int_E f_j \right)^{\frac{1}{n}} .$$

PROOF. With no loss of generality, we may assume that E is a cell. Suppose first that the functions $f_1, \ldots, f_n$ are bounded. Then $\prod_{j=1}^n f_j^{1/n}$ belongs to $\mathcal{R}(E)$ by Corollary 2.2.6, and we only need to prove the inequality. If there is a j with $\int_E f_j = 0$, then $f_j = 0$ almost everywhere in E (Theorem 4.1.7) and both sides of the inequality are equal to zero. Thus we may assume that $\int_E f_j > 0$ for all j, and let $g_j = f_j / \int_E f_j$. Since each $\int_E g_j$ equals one, the desired result follows by integrating the inequality

$$(g_1 \cdots g_n)^{\frac{1}{n}} \leq \frac{1}{n}(g_1 + \cdots + g_n)$$

established in Lemma 12.1.1.

If $f_1, \ldots, f_n$ are arbitrary functions in $\mathcal{R}(E)$, let $f_{j,k} = \min\{f_j, k\}$ for $j = 1, \ldots, n$ and $k = 1, 2, \ldots$. The first part of the proof yields

$$\int_E \prod_{j=1}^n (f_{j,k})^{\frac{1}{n}} \leq \prod_{j=1}^n \left(\int_E f_{j,k} \right)^{\frac{1}{n}} \leq \prod_{j=1}^n \left(\int_E f_j \right)^{\frac{1}{n}} < +\infty,$$

and an application of the monotone convergence theorem completes the argument. $\blacksquare$

Exercise 12.1.3 Show that Lemma 12.1.2 holds for any measurable set E, bounded or not.

Recall from Exercise 11.2.2, 1 that the projections $A_{(1)}, \ldots, A_{(m)}$ of a figure A in $\mathbf{R}^m$ are figures in $\mathbf{R}^{m-1}$.

Lemma 12.1.4 *If A is a figure in $\mathbf{R}^m$ and $m \geq 2$, then*

$$|A|^{m-1} \leq \prod_{i=1}^{m} \lambda_{m-1}(A_{(i)}).$$

PROOF. The lemma is true if $m = 2$, since $A \subset A_{(1)} \times A_{(2)}$. Proceeding inductively, assume that $m \geq 3$ and that the lemma holds for $m - 1$. Let K be the projection of A to the first coordinate axis. Thus K consists of all $t \in \mathbf{R}$ for which the set $A^t = \{\xi \in \mathbf{R}^{m-1} : (t, \xi) \in A\}$ is nonempty. Using a suitable division of A (cf. Lemma 11.2.1), it is easy to see that K and A^t are figures in $\mathbf{R}$ and $\mathbf{R}^{m-1}$, respectively. For $j = 2, \ldots, m$, the figure K is also the projection of $A_{(j)}$ to the first coordinate axis; i.e., the set $A^t_{(j)} = \{\xi \in \mathbf{R}^{m-2} : (t, \xi) \in A_{(j)}\}$ is nonempty if and only if $t \in K$. Moreover, $A^t_{(j)}$ is the projection of A^t to $\mathbf{R}^{m-2}$ in the direction of the jth coordinate axis. Thus

$$[\lambda_{m-1}(A^t)]^{m-2} \leq \prod_{j=2}^{m} \lambda_{m-2}(A^t_{(j)})$$

according to the induction hypothesis, and the inclusion $A^t \subset A_{(1)}$ yields

$$[\lambda_{m-1}(A^t)]^{m-1} \leq \lambda_{m-1}(A_{(1)}) \prod_{j=2}^{m} \lambda_{m-2}(A^t_{(j)}).$$

By Fubini's theorem (Theorem 10.1.13) and Lemma 12.1.2, we obtain

$$|A| = \int_K \lambda_{m-1}(A^t) \, d\lambda(t)$$

$$\leq [\lambda_{m-1}(A_{(1)})]^{\frac{1}{m-1}} \int_K \prod_{j=2}^{m} [\lambda_{m-2}(A^t_{(j)})]^{\frac{1}{m-1}} \, d\lambda(t)$$

$$\leq [\lambda_{m-1}(A_{(1)})]^{\frac{1}{m-1}} \prod_{j=2}^{m} \left[\int_K \lambda_{m-2}(A^t_{(j)}) \, d\lambda(t) \right]^{\frac{1}{m-1}}$$

$$= \left[\prod_{j=1}^{m} \lambda_{m-1}(A_{(j)}) \right]^{\frac{1}{m-1}}. \quad \blacksquare$$

Exercise 12.1.5 Show that Lemma 12.1.2 holds for any set $A \subset \mathbf{R}^m$. *Hint.* Assume first that A is an open set, and observe that it is the union of an increasing sequence of figures. If A is an arbitrary bounded set, choose an $\varepsilon > 0$ and use Proposition 3.2.8 to find open sets $U_i \subset \mathbf{R}^{m-1}$ so that $A_{(i)} \subset U_i$ and $\lambda_{m-1}(U_i) < \lambda_{m-1}(A_{(i)}) + \varepsilon$ for $i = 1, \ldots, m$.

Proposition 12.1.6 *If A is a figure, then*

$$|A|^{\frac{m-1}{m}} \le \frac{1}{2m}\|A\| \quad and \quad r(A) \le \frac{1}{2m}[s(A)]^{\frac{1}{m}}$$

and these estimates cannot be improved.

PROOF. A direct verification shows that the inequalities hold if $m = 1$. If $m > 1$, then Lemmas 12.1.1 and 12.1.4 together with Exercise 11.2.2, 3 give

$$|A|^{\frac{m-1}{m}} \le \left[\prod_{i=1}^{m} \lambda_{m-1}(A_{(i)})\right]^{\frac{1}{m}} \le \frac{1}{m}\sum_{i=1}^{m} \lambda_{m-1}(A_{(i)})$$

$$\le \frac{1}{2m}\sum_{i=1}^{m} \|A\|_i = \frac{1}{2m}\|A\|.$$

If $A \ne \emptyset$, it follows that

$$[r(A)]^m = \frac{|A|^{m-1}}{\|A\|^m} \cdot \frac{|A|}{[d(A)]^m} \le (2m)^{-m}s(A).$$

A straightforward calculation shows that the equalities occur whenever A is a cube. ∎

Remark 12.1.7 Let A be a cell. Then $\|A\| \le 2m[d(A)]^{m-1}$ and in view of Proposition 12.1.6, we obtain

$$\frac{1}{2m}s(A) \le r(A) \le \frac{1}{2m}[s(A)]^{\frac{1}{m}}.$$

Thus, for cells, regularity and shape provide essentially the same information. For figures, however, regularity is a *finer* indicator than shape. Indeed, if

$$B_n = \left(\bigcup_{k=1}^{n}\left[\frac{1}{2k}, \frac{1}{2k-1}\right]\right) \times [0,1]^{m-1},$$

then $s(B_n) \ge 1/2$ and $r(B_n) \le 1/(2n)$ for $n = 1, 2 \ldots$.

12.2 The F-integral

We denote by $\mathcal{F}$ the family of all figures in $\mathbf{R}^m$. An *F-partition* is a collection, possibly empty,

$$P = \{(A_1, x_1), \ldots, (A_p, x_p)\}$$

where $A_1, \ldots, A_p$ are nonoverlapping *figures*, and $x_i \in A_i$ for $i = 1, \ldots, p$. Clearly, every P-partition is an *F*-partition, and all concepts associated with

P-partitions translate, in the obvious way, to $\mathcal{F}$-partitions. Given $\varepsilon > 0$, we say that P is ε-*regular* whenever $r(A_i) > \varepsilon$ for $i = 1, \ldots, p$.

Definition 12.2.1 A function f on a figure A is called $\mathcal{F}$-*integrable* over A if there is a real number I having the following property: given $\varepsilon > 0$, we can find a gage δ on A and a caliber η so that

$$|\sigma(f, P) - I| < \varepsilon$$

for each δ-fine ε-regular $\mathcal{F}$-partition P of A mod (ε, η).

For a figure A, we denote by $\mathcal{FR}_*(A)$ or $\mathcal{FR}_*(A, \lambda_m)$ the family of all $\mathcal{F}$-integrable functions on A. It follows from Proposition 12.1.6 that $\mathcal{FR}_*(A) \subset \mathcal{R}_*(A)$ and that for each $f \in \mathcal{FR}_*(A)$, the integral $\int_A f$ is equal to the number I of Definition 12.2.1; for emphasis, I is sometimes called the $\mathcal{F}$-*integral* of f over A.

When ε-shapely P-partitions are replaced by ε-regular $\mathcal{F}$-partitions, it is straightforward to verify that, with the exception of Theorem 11.4.10, all results of Sections 11.4 and 11.5 together with their proofs hold for the $\mathcal{F}$-integral. As before, we shall quote these results directly, leaving it to the reader to make the appropriate translations applicable to the $\mathcal{F}$-integral.

By Theorem 11.4.5, the $\mathcal{F}$-integral extends the McShane integral over figures, and Corollary 11.4.7 allows us to introduce the family $\overline{\mathcal{FR}}_*(A) = \overline{\mathcal{FR}}_*(A, \lambda_m)$ of all $\mathcal{F}$-integrable extended real-valued functions defined almost everywhere in a figure A. Since $\overline{\mathcal{FR}}_*(A) \subset \overline{\mathcal{R}}_*(A)$, the *indefinite integral* of an $f \in \overline{\mathcal{FR}}_*(A)$, defined in the usual way, is an additive function in A. As Theorem 11.4.9 will be applied to the $\mathcal{F}$-integral on several occasions, we reformulate it without proof.

Theorem 12.2.2 *A function f on a figure A is $\mathcal{F}$-integrable over A if and only if there is an additive continuous function F in A having the following property: given $\varepsilon > 0$, we can find a gage δ on A so that*

$$\sum_{i=1}^{p} \left| f(x_i)|A_i| - F(A_i) \right| < \varepsilon$$

for each δ-fine ε-regular $\mathcal{F}$-partition $\{(A_1, x_1), \ldots, (A_p, x_p)\}$ in A. In this case, F is the indefinite integral of f. ∎

The divergence theorem is also true for the $\mathcal{F}$-integral once we realize that Lemma 8.3.4 is valid for figures (we omit the proof, which is identical to that for the cells). Employing a lemma related to Lemma 4.3.8, we establish directly a more general version of the divergence theorem corresponding to Theorem 11.7.5.

Lemma 12.2.3   *Let E be a negligible set and let $\varepsilon > 0$. On the family of all subsets of $\mathbf{R}^m$, there is a function β satisfying the following conditions.*

1. *$0 \leq \beta(B) < \varepsilon$ for every $B \subset \mathbf{R}^m$.*
2. *$\beta(B) + \beta(C) = \beta(B \cup C)$ for each pair B, C of measurable sets with $|B \cap C| = 0$; in particular, β is a volume in $\mathbf{R}^m$.*
3. *For each $x \in E$ and each integer $n \geq 1$, there is a $\theta > 0$ such that $\beta(B) \geq n|B|$ whenever $B \subset U(x, \theta)$.*

PROOF.   Use Proposition 3.2.8 to find open sets U_k so that $E \subset U_{k+1} \subset U_k$ and $|U_k| < \varepsilon 2^{-k}$ for $k = 1, 2, \ldots$. Define a function β by letting

$$\beta(B) = \sum_{k=1}^{\infty} |B \cap U_k|$$

for each $B \subset \mathbf{R}^m$, and observe that β satisfies conditions 1 and 2. If $x \in E$ and $n \geq 1$ is an integer, then there is a $\theta > 0$ such that $U(x, \theta) \subset U_n$. It follows that

$$\beta(B) = n|B| + \sum_{k=n+1}^{\infty} |B \cap U_k| \geq n|B|$$

whenever $B \subset U(x, \theta)$. ∎

Exercise 12.2.4   Prove the following facts about the function β constructed in the proof of Lemma 12.2.3.

1. β is a metric measure in $\mathbf{R}^m$ (see Remark 3.2.7).
2. β is a measure on the σ-algebra in $\mathbf{R}^m$ consisting of all measurable sets (see Remark 3.3.6).

Theorem 12.2.5   *Let T be a thin set, and let v be a continuous vector field on a figure A that is almost differentiable at each point of $A^\circ - T$. Then $\operatorname{div} v$ belongs to $\overline{\mathcal{FR}}_*(A, \lambda_m)$ and*

$$\int_A \operatorname{div} v \, d\lambda_m = \int_{\partial A} v \cdot n \, d\lambda_{m-1} .$$

PROOF. The proof refines that of Theorem 11.4.10. By Proposition 11.2.8, the flux F of v is an additive continuous function in A, and we may assume that $A \neq \emptyset$. According to Stepanoff's Theorem 10.6.7, there is a negligible set $E \subset A^\circ - T$ such that v is differentiable at every $x \in A^\circ - E$. We let

$$f(x) = \begin{cases} \operatorname{div} v(x) & \text{if } x \in A^\circ - (E \cup T), \\ 0 & \text{if } x \in (E \cup T \cup \partial A), \end{cases}$$

and prove the theorem by showing that F satisfies the conditions of Theorem 12.2.2 with respect to the function f.

To this end, choose an $\varepsilon > 0$ and a function β associated with E and $\varepsilon/2$ according to Lemma 12.2.3. Keep in mind that $d(B)\|B\| < |B|/\varepsilon$ for each figure B with $r(B) > \varepsilon$. If $x \in A^\circ - (E \cup T)$, use the version of Lemma 8.3.4 that applies to figures, and find an $\eta_x > 0$ so that

$$\left| f(x)|B| - F(B) \right| < \frac{\varepsilon^2}{2|A|} d(B)\|B\| < \frac{\varepsilon}{2|A|}|B|$$

whenever $B \subset A \cap U(x, \eta_x)$ is a figure with $x \in B$ and $r(B) > \varepsilon$. If $x \in E$, there are positive numbers θ_x and c_x such that

$$|v(y) - v(x)| \leq c_x|y - x|$$

for every $y \in A \cap U(x, \theta_x)$. Making θ_x smaller, we may assume that $\beta(B) \geq (c_x/\varepsilon)|B|$ for each $B \subset U(x, \theta_x)$. Thus, when $B \subset A \cap U(x, \theta_x)$ is a figure with $r(B) > \varepsilon$, Proposition 8.3.1 and Lemma 8.3.3 imply that

$$\left| f(x)|B| - F(B) \right| = \left| \int_{\partial B} v(y) \cdot n \, d\lambda_{m-1}(y) \right|$$

$$= \left| \int_{\partial B} [v(y) - v(x)] \cdot n \, d\lambda_{m-1}(y) \right|$$

$$\leq c_x d(B)\|B\| < \frac{c_x}{\varepsilon}|B| \leq \beta(B).$$

Now define a gage δ on A by setting

$$\delta(x) = \begin{cases} \eta_x & \text{if } x \in A^\circ - (E \cup T), \\ \theta_x & \text{if } x \in E, \\ 0 & \text{if } x \in (T \cup \partial A), \end{cases}$$

and choose a δ-fine ε-regular $\mathcal{F}$-partition $\{(A_1, x_1), \ldots, (A_p, x_p)\}$ in A. The previous inequalities yield

$$\sum_{i=1}^{p} \left| f(x_i)|A_i| - F(A_i) \right| < \frac{\varepsilon}{2|A|} \sum_{x_i \in E} |A_i| + \sum_{x_i \notin E} \beta(A_i) < \varepsilon,$$

and the theorem follows from Theorem 12.2.2. $\blacksquare$

Exercise 12.2.6   Prove the following version of Henstock's lemma for the McShane integral.

Let α be a volume in a cell A, and let $f \in \mathcal{R}(A, \alpha)$. Given $\varepsilon > 0$, there is a positive function δ on A such that

$$\sum_{i=1}^{p} \left| f(x_i)\alpha(A_i) - \int_{A_i} f \, d\alpha \right| < \varepsilon$$

for each δ-fine $\mathcal{F}$-partition $\{(A_1, x_1), \ldots, (A_p, x_p)\}$ in A.

12.3 Derivability relative to $\mathcal{F}$

The *shape* $s(B)$ can be defined for any bounded set $B \subset \mathbf{R}^m$ by setting

$$s(B) = \begin{cases} \dfrac{|B|}{[d(B)]^m} & \text{if } d(B) > 0, \\ 0 & \text{if } d(B) = 0. \end{cases}$$

Using this definition, we show that the Vitali covering theorem holds also for Vitali covers consisting of *arbitrary closed sets*. A family $\mathcal{C}$ of closed sets is called a *Vitali cover* of a set $E \subset \mathbf{R}^m$ if there is a positive function γ on E such that for each $x \in E$ and each $\eta > 0$ we can find a $C \in \mathcal{C}$ with $x \in C$, $C \subset U(x, \eta)$, and $s(C) > \gamma(x)$.

Proposition 12.3.1   *If a family $\mathcal{C}$ of closed sets is a Vitali cover of a set $E \subset \mathbf{R}^m$, then there is a countable disjoint family $\mathcal{D} \subset \mathcal{C}$ such that $E - \bigcup \mathcal{D}$ is negligible.*

PROOF.   Avoiding a triviality, suppose that $|E| > 0$. For each set $C \in \mathcal{C}$ with $s(C) > 0$, select a cube C^* of diameter $d(C)$ containing C and observe that $|C| = s(C)|C^*|$.

Assume first that $|E| < +\infty$, and that $s(C) > \varepsilon$ for each $C \in \mathcal{C}$ and a fixed positive $\varepsilon < 1$. Use Proposition 3.2.8 to find an open set U such that

$E \subset U$ and $|U| < (1 + \varepsilon)|E|$. As the family $\mathcal{C}^\star$ of all sets $C^\star$ contained in U is a Vitali cover of E consisting of cubes, we claim that there is a disjoint collection $\{C_1^\star, \ldots, C_n^\star\} \subset \mathcal{C}^\star$ such that $|E - \bigcup_{k=1}^n C_k^\star| < \varepsilon^3|E|$. Indeed, by Proposition 9.2.4, there is a countable disjoint family $\{C_1^\star, C_2^\star, \ldots\} \subset \mathcal{C}^\star$ with $|E - \bigcup_k C_k^\star| = 0$; since $\sum_k |C_k^\star| \le |U| < +\infty$, there is an integer $n \ge 1$ such that $\sum_{k>n} |C_k^\star| < \varepsilon^3|E|$ and our claim follows. Let $\mathcal{D}_1 = \{C_1, \ldots, C_n\}$ and observe that

$$
\left| E - \bigcup \mathcal{D}_1 \right| \le \left| E - \bigcup_{k=1}^n C_k^\star \right| + \sum_{k=1}^n |C_k^\star - C_k|
$$

$$
< \varepsilon^3|E| + (1-\varepsilon)\sum_{k=1}^n |C_k^\star|
$$

$$
< \varepsilon^3|E| + (1-\varepsilon)|U| < c|E|
$$

where $c = 1 - \varepsilon^2 + \varepsilon^3$. Since $\bigcup \mathcal{D}_1$ is a closed set, the family of all sets in $\mathcal{C}$ disjoint from $\bigcup \mathcal{D}_1$ is a Vitali cover of $E - \bigcup \mathcal{D}_1$. By the above argument, there is a finite collection $\mathcal{D}_2 \subset \mathcal{C}$ such that $\mathcal{D}_1 \cup \mathcal{D}_2$ is a disjoint family and

$$
\left| E - \left(\bigcup \mathcal{D}_1 \right) \cup \left(\bigcup \mathcal{D}_2 \right) \right| = \left| \left(E - \bigcup \mathcal{D}_1 \right) - \bigcup \mathcal{D}_2 \right|
$$

$$
< c \left| E - \bigcup \mathcal{D}_1 \right| < c^2|E| .
$$

Proceeding inductively, we construct finite families $\mathcal{D}_1, \mathcal{D}_2, \ldots$ so that $\mathcal{D} = \bigcup_{n=1}^\infty \mathcal{D}_n$ is a disjoint subfamily of $\mathcal{C}$ and

$$
\left| E - \bigcup \mathcal{D} \right| \le \left| E - \bigcup_{n=1}^p \left(\bigcup \mathcal{D}_n \right) \right| < c^p|E|
$$

for $p = 1, 2, \ldots$. As $0 < c < 1$, we have $|E - \bigcup \mathcal{D}| = 0$.

The proof of the general case, which is identical to that of Proposition 9.2.4, is left to the reader. ∎

Let F be a function defined on the family of all subfigures of a set E, and let $x \in E$. We set

$$
\mathcal{F}D_{\lambda_m} F(x) = \inf_{\eta>0} \sup_{\delta>0} \left[\inf \frac{F(C)}{|C|} \right]
$$

where the infimum inside the brackets is taken over all figures $C \subset E \cap U(x, \delta)$ such that $x \in C$ and $r(C) > \eta$. If $-\mathcal{F}D_{\lambda_m}(-F)(x) = \mathcal{F}D_{\lambda_m} F(x) \ne \pm\infty$, we denote this common value by $\mathcal{F}F'(x)$ and say that F is *derivable at* x *relative* to $\mathcal{F}$.

Theorem 12.3.2 *Let* A *be a figure and* $f \in \overline{\mathcal{F}\mathcal{R}}_*(A)$. *If* F *is the indefinite integral of* f, *then for almost all* $x \in A$ *the function* F *is derivable at* x *relative to* $\mathcal{F}$ *and* $\mathcal{F}F'(x) = f(x)$.

PROOF. While the proof is similar to the proofs of Theorems 5.3.5 and 9.3.4, technical differences warrant its presentation. Let E be the set of all $x \in A$ for which either F is not derivable at x relative to $\mathcal{F}$ or $\mathcal{F}F'(x) \neq f(x)$. Given $x \in E$, we can find a $\gamma(x) > 0$ so that for each $\beta > 0$ there is a figure $C \subset A \cap U(x, \beta)$ with $x \in C$, $r(C) > \gamma(x)$, and

$$\left| f(x)|C| - F(C) \right| \geq \gamma(x)|C|.$$

Fix an integer $n \geq 2m$, let $E_n = \{x \in E : \gamma(x) \geq 1/n\}$, and choose an $\varepsilon > 0$. By Theorem 12.2.2, there is a gage δ on A such that

$$\sum_{i=1}^{p} \left| f(x_i)|A_i| - F(A_i) \right| < \frac{\varepsilon}{n}$$

for each δ-fine $(1/n)$-regular $\mathcal{F}$-partition $\{(A_1, x_1), \dots, (A_p, x_p)\}$ in A. Let $\mathcal{C}$ be the collection of all figures $C \subset A$ such that $C \subset U(x_C, \delta(x_C))$ for a point $x_C \in C \cap E_n$, $r(C) > 1/n$, and

$$\left| f(x_C)|C| - F(C) \right| \geq \frac{1}{n}|C|.$$

In view of Proposition 12.1.6, it is easy to verify that $\mathcal{C}$ is a Vitali cover of $E_n - Z_\delta$ where Z_δ is the null set of δ. Since $|Z_\delta| = 0$, it follows from Proposition 12.3.1 that there is a countable disjoint subfamily $\mathcal{D}$ of $\mathcal{C}$ such that $|E_n - \bigcup \mathcal{D}| = 0$. For each finite family $\mathcal{T} \subset \mathcal{D}$, the collection $\{(D, x_D) : D \in \mathcal{T}\}$ is a δ-fine $(1/n)$-regular $\mathcal{F}$-partition in A. Hence

$$\sum_{D \in \mathcal{T}} |D| \leq n \sum_{D \in \mathcal{T}} \left| f(x_D)|D| - F(D) \right| < \varepsilon$$

and consequently

$$|E_n| = \left| E_n \cap \left(\bigcup \mathcal{D} \right) \right| \leq \sum_{D \in \mathcal{D}} |D| \leq \varepsilon.$$

The arbitrariness of ε implies that E_n is negligible, and according to Proposition 3.5.3, 3, so is $E = \bigcup_{n=2m}^{\infty} E_n$. ∎

Note. In the proof of Theorem 12.3.2 we set $n \geq 2m$ to avoid the logically correct but vacuous cases arising when $n = 1, \ldots, 2m-1$; for $r(C) \leq 1/(2m)$ for each figure C.

Definition 12.3.3 An additive continuous function F in a figure A is called $\mathcal{F}AC_*$ if, given a negligible set $E \subset A$ and $\varepsilon > 0$, there is a gage δ on E such that $\sum_{i=1}^{p} |F(A_i)| < \varepsilon$ for each ε-regular $\mathcal{F}$-partition $\{(A_1, x_1), \ldots, (A_p, x_p)\}$ in A anchored in E that is δ-fine.

The next theorem is the anticipated partial descriptive definition of the $\mathcal{F}$-integral. Its proof is identical to that of Theorem 6.4.4.

Theorem 12.3.4 *Let F be an additive function in a figure A that is derivable almost everywhere in A relative to $\mathcal{F}$. Then F is $\mathcal{F}AC_*$ if and only if $\mathcal{F}F'$ belongs to $\overline{\mathcal{F}\mathcal{R}}_*(A)$ and F is the indefinite integral of $\mathcal{F}F'$.* ∎

Each $\mathcal{F}AC_*$ function is AC_*, but the following example shows that the converse is false, already in dimension one. Moreover, we shall see that $\mathcal{F}\mathcal{R}_*(A)$ is a *proper subset* of $\mathcal{R}_*(A)$ for each one-cell $A \subset \mathbf{R}$.

Example 12.3.5 Assume that $m = 1$. If $D = [a, b]$ is a cell, let

$$D_+^n = [a + 2^{-2n}|D|, a + 2^{-2n+1}|D|],$$

$$D_-^n = [a + 2^{-2n+1}|D|, a + 2^{-2n+2}|D|]$$

for $n = 1, 2, \ldots$. Given $x \in \mathbf{R}$, set

$$f_D(x) = \begin{cases} \pm \frac{1}{n} \cdot \frac{|D|}{|D_\pm^n|} & \text{if } x \in (D_\pm^n)^\circ \text{ and } n = 1, 2, \ldots, \\ 0 & \text{otherwise,} \end{cases}$$

and deduce from Theorem 6.1.6 that $f_D \in \mathcal{R}_*(D, \lambda)$. The formula

$$F_D(x) = \begin{cases} \int_a^x f_D & \text{if } x \in D, \\ 0 & \text{if } x \in \mathbf{R} - D \end{cases}$$

defines a continuous function F_D such that $|F_D(x)| \leq |D|$ for each $x \in \mathbf{R}$.

Let C be the Cantor ternary set (see Remark 5.3.12), let $U_1, U_2, \ldots$ be an enumeration of all connected components of $[0, 1] - C$ (i.e., of the segments

$\Delta_{r,s}$ defined in Example 5.3.11), and let $D_k = U_k^-$ for $k = 1, 2, \ldots$. If

$$f = \sum_{k=1}^{\infty} f_{D_k} \qquad \text{and} \qquad F = \sum_{k=1}^{\infty} F_{D_k},$$

then $\mathcal{F}F'(x) = F'(x) = f(x)$ exists for all but countably many $x \in \mathbf{R} - C$. Since $\sum_{k=1}^{\infty} |D_k| = 1$ and $|F_{D_k}(x)| \leq |D_k|$ for all $x \in \mathbf{R}$ and $k = 1, 2, \ldots$, the function F is continuous in $\mathbf{R}$ ([42, Theorems 7.10 and 7.12, pp. 148 and 150]).

Claim 1. The function f belongs to $\mathcal{R}_*([0, 1])$, and F is the indefinite integral of f. In particular, F is AC_*.

Proof. Let $T = \bigcup_{k=1}^{\infty} \partial D_k$ and choose an $\varepsilon > 0$. For $k = 1, 2, \ldots$, use Henstock's lemma to find a positive function δ_k on D_k so that

$$\sum_{i=1}^{p} \left| f(x_i)|A_i| - F(A_i) \right| < \varepsilon 2^{-k}$$

for each δ_k-fine P-partition $\{(A_1, x_1), \ldots, (A_p, x_p)\}$ in D_k. With no loss of generality, we may assume that $\delta_k(x) \leq \mathrm{dist}(x, T)$ whenever $x \in U_k$. There is an integer $s \geq 1$ such that $\sum_{k=s+1}^{\infty} |D_k| < \varepsilon$. We let $A = \bigcup_{k=1}^{s} D_k$ and define a gage δ on $[0, 1]$ by setting

$$\delta(x) = \begin{cases} \delta_k(x) & \text{if } x \in U_k \text{ and } k = 1, 2, \ldots, \\ \mathrm{dist}(x, A) & \text{if } x \in C - T, \\ 0 & \text{if } x \in T. \end{cases}$$

When $\{(A_1, x_1), \ldots, (A_p, x_p)\}$ is a δ-fine P-partition in $[0, 1]$, then the set $\bigcup_{x_i \in C} \partial A_i$ is disjoint from A and it meets each D_k in at most two points. Thus

$$\sum_{i=1}^{p} \left| f(x_i)|A_i| - F(A_i) \right| \leq \sum_{k=1}^{\infty} \sum_{x_i \in D_k} \left| f(x_i)|A_i| - F(A_i) \right| + \sum_{x_i \in C} |F(A_i)|$$

$$< \sum_{k=1}^{\infty} \varepsilon 2^{-k} + 2 \sum_{k=s+1}^{\infty} |D_k| < 3\varepsilon$$

and the claim follows from Exercise 11.2.6 and Theorem 12.2.2.

Claim 2. The function F is not $\mathcal{F}AC_*$ in $[0, 1]$ and $f \notin \mathcal{F}\mathcal{R}_*([0, 1], \lambda)$.

PROOF. Choose a gage δ on $[0, 1]$, and use Exercise 3.3.12 to show that $K = C - Z_\delta$ is a G_δ set. Since $K = \bigcup_{j=1}^\infty \{x \in K : \delta(x) > 1/j\}$, the Baire category theorem ([41, Chapter 7, Corollary 16, p. 139, and Chapter 8, Problem 38, p. 155]) implies that there is an open segment U and an integer $r \geq 1$ such that $E = \{x \in K \cap U : \delta(x) > 1/r\}$ is a nonempty dense subset of $K \cap U$. As Z_δ is a countable set and C is a perfect set ([42, Section 2.44, p. 42]), the set E is also dense in $C \cap U$. Select a cell $D_k = [a, b]$ with $a \in C \cap U$, and construct sequences $\{x_n\}$ and $\{y_n\}$ in E so that

$$a - 2^{-2n}|D_k| < x_n < y_n < x_{n+1} < a$$

for $n = 1, 2, \ldots$. If $A_n = [x_n, y_n] \cup (D_k)_+^n$, then

$$d(A_n) < 3 \cdot 2^{-2n}|D_k|, \quad \lambda(A_n) > 2^{-2n}|D_k|, \quad \text{and} \quad \|A_n\| = 4.$$

It follows that $\{(A_p, x_p), \ldots, (A_{p+q}, x_{p+q})\}$ is a $(1/12)$-regular $\mathcal{F}$-partition in $[0, 1]$ anchored in C, which is δ-fine whenever p is sufficiently large. As

$$\sum_{n=p}^\infty |F(A_n)| = \sum_{n=p}^\infty |F[(D_k)_+^n]| = |D_k| \sum_{n=p}^\infty \frac{1}{n} = +\infty$$

for $p = 1, 2, \ldots$, we conclude that F is not $\mathcal{F}AC_*$ in $[0, 1]$. Now assume that $f \in \mathcal{F}R_*([0, 1])$. Since the continuous additive function F in $[0, 1]$ is the indefinite integral of f by Claim 1, a contradiction follows from Theorem 12.3.4. $\blacksquare$

Remark 12.3.6 At this point it is easy to define *almost derivability relative to $\mathcal{F}$* and prove results analogous to Propositions 11.7.2 and Lemma 11.7.4. Whether a useful analog of Theorem 11.7.6 is also true appears unknown.

12.4 Integration by parts

While the one-dimensional $\mathcal{F}$-integral is less general than the P-integral of Henstock and Kurzweil, it still shares the special properties of the P-integral established in Section 6.1. Indeed, the fundamental theorem of calculus is a special case of the divergence theorem, and modifying the proofs of Theorems 6.1.4 and 6.1.6 for the $\mathcal{F}$-integral is merely a challenging exercise; the same is true about Proposition 6.1.9. It turns out, however, that the integration by parts theorem for the $\mathcal{F}$-integral may require more stringent assumptions than Theorem 6.2.1.

Theorem 12.4.1   *Let $A = [a, b]$ be a cell and let $f \in \mathcal{FR}_*(A)$. For each $x \in A$, set $F(x) = \int_a^x f$. If g is a Lipschitz function on A, then fg belongs to $\mathcal{FR}_*(A)$, Fg' belongs to $\overline{\mathcal{R}}(A)$, and*

$$\int_a^b fg = F(b)g(b) - \int_a^b Fg'.$$

PROOF.   Let $c = \mathrm{Lip}(g)$. Since g is the difference of increasing Lipschitz functions $x \mapsto g(x) + cx$ and $x \mapsto cx$, we may assume that g is already increasing. As F is continuous and g is AC, Theorems 2.2.8, 5.3.15, and 2.3.12 imply that Fg' belongs to $\overline{\mathcal{R}}(A)$. In view of Remark 5.3.16, we have $F \in \mathcal{R}(A, g)$ and $\int_a^x F\, dg = \int_a^x Fg'$ for all $x \in A$. The formula

$$H(x) = F(x)g(x) - \int_a^x F\, dg$$

defines a continuous function H on A. We show that the continuous additive function in A associated with H according to Section 1.1, still denoted by H, satisfies the conditions of Theorem 12.2.2 with respect to the function fg.

To this end, choose an $\varepsilon > 0$ and find an $\eta > 0$ so that $|F(x) - F(y)| < \varepsilon^2$ for each $x, y \in A$ with $|x - y| < \eta$. By Theorem 12.2.2 and Exercise 12.2.6, there is a gage $\delta \leq \eta/2$ on A such that

$$\sum_{i=1}^p \left| f(x_i)|A_i| - F(A_i) \right| < \varepsilon \quad \text{and} \quad \sum_{i=1}^p \left| F(x_i)g(A_i) - \int_{A_i} F\, dg \right| < \varepsilon$$

for each δ-fine ε-regular $\mathcal{F}$-partition $P = \{(A_1, x_1), \ldots, (A_p, x_p)\}$ in A. For such a partition P, we obtain

$$\sum_{i=1}^p \left| f(x_i)g(x_i)|A_i| - H(A_i) \right| \leq \sum_{i=1}^p |g(x_i)| \cdot \left| f(x_i)|A_i| - F(A_i) \right|$$

$$+ \sum_{i=1}^p |g(x_i)F(A_i) - (Fg)(A_i) + F(x_i)g(A_i)|$$

$$+ \sum_{i=1}^p \left| F(x_i)g(A_i) - \int_{A_i} F\, dg \right| < \varepsilon[g(b) + 1] + \sum_{i=1}^p |c_i|$$

where $c_i = g(x_i)F(A_i) + F(x_i)g(A_i) - (Fg)(A_i)$ for $i = 1, \ldots, p$. Fix an integer i with $1 \leq i \leq p$, and assume that $[a_1, b_1], \ldots, [a_k, b_k]$ are the connected components of A_i. Since

$$c_i = \sum_{j=1}^{k} \Big(g(x_i)[F(b_j) - F(a_j)] + F(x_i)[g(b_j) - g(a_j)]$$

$$-[F(b_j)g(b_j) - F(a_j)g(a_j)] \Big)$$

$$= \sum_{j=1}^{k} \Big([g(x_i) - g(b_j)] \cdot [F(b_j) - F(a_j)]$$

$$+ [F(x_i) - F(a_j)] \cdot [g(b_j) - g(a_j)] \Big),$$

we have

$$|c_i| \le c\varepsilon^2 \sum_{j=1}^{k} \Big(|x_i - b_j| + |b_j - a_j| \Big) \le 2kc\varepsilon^2 d(A_i) = c\varepsilon^2 \|A_i\| d(A_i)| < c\varepsilon |A_i|.$$

Therefore

$$\sum_{i=1}^{p} \Big| f(x_i)g(x_i)|A_i| - H(A_i) \Big| < \varepsilon[1 + g(b) + c|A|]$$

and an application of Theorem 12.2.2 completes the proof. ∎

Remark 12.4.2 Our proof of Theorem 12.4.1 utilizes the hypothesis that the function g is Lipschitz. Nonetheless, it appears unknown whether the theorem is still true if $\int_a^b fg'$ is replaced by $\int_a^b f \, dg$ where g is an increasing function or an increasing AC function, which is not Lipschitz. It is also unclear if a theorem analogous to Theorem 6.2.3 holds for the $\mathcal{F}$-integral. A more detailed study of the one-dimensional $\mathcal{F}$-integral can be found in [4].

12.5 The quasi-Hausdorff measure

Let $E \subset \mathbf{R}^m$. For any $\eta > 0$, we set

$$h_\eta(E) = \inf \sum_i [d(C_i)]^{m-1}$$

where the infimum is taken over all finite or infinite sequences $\{C_i\}$ of nonempty sets $C_i \subset \mathbf{R}^m$ such that $E \subset \bigcup_i C_i$ and $d(C_i) < \eta$ for all i (from

Section 3.1 recall the convention that $x^0 = 1$ for each $x \geq 0$). The extended real number

$$h(E) = \sup_{\eta > 0} h_\eta(E)$$

is called the *quasi-Hausdorff measure* of E, or more precisely, the $(m-1)$-*dimensional quasi-Hausdorff measure* of E.

Exercise 12.5.1   Let $\eta > 0$. Prove the following statements.

1. $h_\eta(\emptyset) = h(\emptyset) = 0$. *Hint*. Note that an empty sequence is a special case of a finite sequence.
2. $h_\eta(A) \leq h_\eta(B)$ and $h(A) \leq h(B)$ for each $A \subset B \subset \mathbf{R}^m$.
3. If $\{E_n\}$ is a sequence of subsets of $\mathbf{R}^m$, then

$$h_\eta \left(\bigcup_{n=1}^{\infty} E_n \right) \leq \sum_{n=1}^{\infty} h_\eta(E_n) \quad \text{and} \quad h \left(\bigcup_{n=1}^{\infty} E_n \right) \leq \sum_{n=1}^{\infty} h(E_n).$$

4. $h(A \cup B) = h(A) + h(B)$ for each $A, B \subset \mathbf{R}^m$ with $\text{dist}(A, B) > 0$.

Exercise 12.5.1 shows that h is a *metric measure* in $\mathbf{R}^m$ in the sense of Remark 3.2.7. Consequently, it is possible to define the σ-algebra of h-measurable subsets of $\mathbf{R}^m$ (see Remark 3.3.6) by means of the Carathéodory test for measurability (Theorem 3.3.17). This is a standard procedure due to Carathéodory, which we shall not pursue because our use of the quasi-Hausdorff measure is only peripheral. The interested reader is referred to [10, Section 1.1].

Exercise 12.5.2   Let $m = 1$ and $E \subset \mathbf{R}$. Show that $h(E) < +\infty$ if and only if the set E is finite, in which case $h(E)$ equals the number of points in E.

Note. Using only *infinite* sequences $\{C_i\}$ would result in the infinite zero-dimensional quasi-Hausdorff measure for each subset of $\mathbf{R}$ (cf. Exercise 12.5.2). It is easy to see, however, that finite sequences are not needed when $m \geq 2$.

Remark 12.5.3   The definition of the quasi-Hausdorff measure justifies the term "Hausdorff-type evaluation" of the Lebesgue measure introduced prior to Theorem 10.3.1. If E is a subset of a hyperplane H in $\mathbf{R}^m$, then it follows from Theorem 10.3.1 and Exercise 12.5.2 that $h(E) = \lambda_{m-1}(E)$ where λ_{m-1}

is the $(m - 1)$-*dimensional Lebesgue measure in* H *(see the first paragraph of Section 11.2). In particular,*

$$\|A\| = h(\partial A)$$

whenever A *is a figure in* $\mathbf{R}^m$.

Remark 12.5.4 The commonly used $(m - 1)$-dimensional Hausdorff measure $\mathcal{H}(E)$ of a set $E \subset \mathbf{R}^m$ is defined differently. Denote by $d_e(C)$ the Euclidean diameter of a set $C \subset \mathbf{R}^m$ (see Remark 10.3.2), and set

$$\omega_{m-1} = \lambda_{m-1}\left(\left\{x \in \mathbf{R}^{m-1} : \|x\| < \frac{1}{2}\right\}\right).$$

For any $\eta > 0$, let

$$\mathcal{H}_\eta(E) = \omega_{m-1}\inf\sum_i[d_e(C_i)]^{m-1}$$

where the infimum is taken over all finite or infinite sequences $\{C_i\}$ of nonempty sets $C_i \subset \mathbf{R}^m$ such that $E \subset \bigcup_i C_i$ and $d_e(C_i) < \eta$ for all i. The extended real number

$$\mathcal{H}(E) = \sup_{\eta>0}\mathcal{H}_\eta(E)$$

is called the $(m - 1)$-*dimensional Hausdorff measure* of E, or simply the *Hausdorff measure* of E. Properties 1–4 listed in Exercise 12.5.1 are easily verified for the Hausdorff measure $\mathcal{H}$. Since $d(C) \le d_e(C) \le \sqrt{m}\,d(C)$ for every set $C \subset \mathbf{R}^m$, we see that

$$h(E) \le \frac{1}{\omega_{m-1}}\mathcal{H}(E) \le m^{\frac{m-1}{2}}h(E).$$

Moreover, it follows from Remark 10.3.2 that $h(E) = \mathcal{H}(E)$ whenever E is a subset of a hyperplane. Thus, for every figure A in $\mathbf{R}^m$, Remark 12.5.3 implies that $\|A\| = \mathcal{H}(\partial A)$.

Note. Unlike the quasi-Hausdorff measure h, the Hausdorff measure $\mathcal{H}$ is rotation invariant and reflects correctly our geometric intuition about the $(m - 1)$-dimensional volume in $\mathbf{R}^m$. For instance, if $m = 2$ and

$$E = \{(x, y) \in \mathbf{R}^2 : |x| + |y| = 1\},$$

then $\mathcal{H}(E) = 4\sqrt{2}$ as expected, while $h(E) = 4$. On the other hand, computations involving h are generally simpler than those involving $\mathcal{H}$. In most cases, we are not interested in the exact value of $\mathcal{H}(E)$; we only need to know whether $0 < \mathcal{H}(E) < +\infty$. Consequently, for our purposes it is more convenient to work with the quasi-Hausdorff measure h.

Exercise 12.5.5　Prove that each set of finite quasi-Hausdorff measure is negligible.

Lemma 12.5.6　Let $E \subset \mathbf{R}^m$ be such that $0 < h(E) < +\infty$. If $m \geq 2$, then $E = A \cup B$ where

$$\max\{h(A), h(B)\} \leq \frac{2}{3}h(E).$$

PROOF.　There is a positive number $\vartheta \leq 1$ such that $\eta < h_\eta(E)/6$ whenever $0 < \eta < \vartheta$. Choose a positive $\eta < \vartheta$ and let $\varepsilon = [h_\eta(E)/6] - \eta^{m-1}$. Since $\varepsilon > 0$, there is a sequence $\{C_i\}$ of nonempty subsets of $\mathbf{R}^m$ whose diameters are less than η and for which

$$E \subset \bigcup_{i=1}^{\infty} C_i \quad \text{and} \quad \sum_{i=1}^{\infty} [d(C_i)]^{m-1} < h_\eta(E) + \varepsilon.$$

As $h_\eta(E) \leq \sum_{i=1}^{\infty}[d(C_i)]^{m-1}$, the choice of η implies that

$$\sum_{i=1}^{p} [d(C_i)]^{m-1} < \frac{1}{2}h_\eta(E) \leq \sum_{i=1}^{p+1}[d(C_i)]^{m-1}$$

for an integer $p \geq 1$. Letting

$$A = E \cap \left(\bigcup_{i=1}^{p} C_i\right) \quad \text{and} \quad B = E \cap \left(\bigcup_{i=p+1}^{\infty} C_i\right),$$

we see immediately that $h_\eta(A) < h_\eta(E)/2 \leq h(E)/2$. Moreover,

$$h_\eta(B) \leq \sum_{i=p+1}^{\infty} [d(C_i)]^{m-1} < h_\eta(E) + \varepsilon - \sum_{i=1}^{p+1}[d(C_i)]^{m-1} + \eta^{m-1}$$

$$\leq \frac{1}{2}h_\eta(E) + \varepsilon + \eta^{m-1} = \frac{2}{3}h_\eta(E) \leq \frac{2}{3}h(E)$$

and the lemma follows from the arbitrariness of η.　∎

Corollary 12.5.7　Let $E \subset \mathbf{R}^m$ be such that $0 < h(E) < +\infty$, and let $\varepsilon > 0$. If $m \geq 2$, then $E = \bigcup_{k=1}^{n} E_k$ where $0 < h(E_k) < \varepsilon$ for $k = 1, \ldots, n$.

PROOF. Select sets A and B that are associated with E according to Lemma 12.5.6. Then $h(A) > 0$ and $h(B) > 0$ by Exercise 12.5.1, 3. Thus it suffices to apply Lemma 12.5.6 n times where $n \geq 1$ is an integer such that $(2/3)^n < \varepsilon/h(E)$. ∎

Lemma 12.5.8 *If $E \subset \mathbf{R}^m$ and $0 < \eta \leq 1$, then*

$$h_\eta(E) \leq \frac{1}{2m} \inf \sum_j \|K_j\| \leq 3^m h_\eta(E)$$

where the infimum is taken over all finite or infinite sequences $\{K_j\}$ of dyadic cubes such that $E \subset (\bigcup_j K_j)^\circ$ and $d(K_j) < \eta$ for all j.

PROOF. Since $\|K\| = 2m[d(K)]^{m-1}$ for each cube K, the first inequality is obvious. In proving the second inequality, we may assume that $h_\eta(E) < +\infty$. Given $\varepsilon > 0$, there is a finite or infinite sequence $\{C_i\}$ of nonempty subsets of $\mathbf{R}^m$ of diameters less than η such that

$$E \subset \bigcup_i C_i \qquad \text{and} \qquad \sum_i [d(C_i)]^{m-1} < h_\eta(E) + \varepsilon .$$

Suppose that $d(C_j) = 0$ for some integers $j \geq 1$. If $m = 1$, replacing each C_j by any set of positive diameter does not change the sum $\sum_i [d(C_i)]^{m-1}$. If $m > 1$, there is a $\theta > 0$ such that replacing each C_j by a set D_j with $0 < [d(D_j)]^{m-1} < 2^{-i}\theta$ does not alter the inequality

$$\sum_i [d(C_j)]^{m-1} < h_\eta(E) + \varepsilon .$$

Thus we may assume that $d(C_i) > 0$ for all i. Now for $i = 1, 2, \ldots$, find a nonnegative integer n_i so that $2^{-n_i-1} \leq d(C_i) < 2^{-n_i}$, and denote by $\mathcal{K}_i$ the collection of all dyadic cubes K with $d(K) = 2^{-n_i-1}$ and $K \cap C_i \neq \emptyset$. Observe that $C_i \subset (\bigcup \mathcal{K}_i)^\circ$ and that $d(K) \leq d(C_i) < \eta$ for each $K \in \mathcal{K}_i$. Since $d(C_i) < 2^{-n_i}$, the family $\mathcal{K}_i$ contains at most 3^m elements. Enumerating the countable family $\bigcup_i \mathcal{K}_i$ as $\{K_1, K_2, \ldots\}$, we obtain

$$E \subset \bigcup_i C_i \subset \bigcup_i \left(\bigcup \mathcal{K}_i\right)^\circ \subset \left(\bigcup_j K_j\right)^\circ$$

and

$$\frac{1}{2m} \sum_j \|K_j\| = \sum_i \sum_{K \in \mathcal{K}_i} [d(K)]^{m-1}$$

$$\leq 3^m \sum_i [d(C_i)]^{m-1} < 3^m [h_\eta(E) + \varepsilon].$$

The lemma follows from the arbitrariness of ε. $\blacksquare$

Proposition 12.5.9   *A set $T \subset \mathbf{R}^m$ is thin if and only if $T = \bigcup_{n=1}^\infty T_n$ where $h(T_n) < +\infty$ for $n = 1, 2, \ldots$.*

PROOF.   That a thin set T satisfies the condition of the proposition follows directly from Lemma 12.5.8. If $m = 1$, the converse is a consequence of Exercise 12.5.2 and Proposition 11.3.2, 5. If $m \geq 2$, then using Corollary 12.5.7, we may assume that $h(T_n) \leq 2 \cdot 3^{-m}/m$. Another application of Lemma 12.5.8 shows that T is a thin set. $\blacksquare$

12.6 Solids

A *solid* is a bounded set $S \subset \mathbf{R}^m$ for which $h(\partial S) < +\infty$. By Remark 12.5.3, each figure is a solid, and we show that every solid can be approximated by figures. Specifically, we prove that for every solid S there is a sequence of figures $\{A_n\}$ converging to S in the sense of Section 11.5.

Exercise 12.6.1   Prove the following facts.

1. Each solid is a measurable set.
2. If B and C are solids, then so are B^-, B°, $B \cup C$, $B \cap C$, and $B - C$.
3. If a solid S is negligible, then $h(S) < +\infty$. *Hint.* Observe that every negligible set has an empty interior.

As for figures, we say that solids C and D *overlap* whenever $C^\circ \cap D^\circ \neq \emptyset$. A function F defined on the family of all subsolids of a set $E \subset \mathbf{R}^m$ is called

1. *additive* if $F(B \cup C) = F(B) + F(C)$ for all solids $B, C \subset E$ that do not overlap;
2. *continuous* if, given $\varepsilon > 0$, there is an $\eta > 0$ such that $|F(B)| < \varepsilon$ for each solid $B \subset E$ with $h(\partial B) < 1/\varepsilon$ and $|B| < \eta$.

If F is an additive continuous function on the family of all subsolids of a set $E \subset \mathbf{R}^m$, then its restriction to the family of all subfigures of E is an additive continuous function in E in the sense of Definition 11.2.5.

Exercise 12.6.2 Let F be an additive continuous function defined on the family of all subsolids of a set $E \subset \mathbf{R}^m$. Show that the following statements are true.

1. If $S \subset E$ is a negligible solid, then $F(S) = 0$.
2. $F(A \cup B) = F(A) + F(B)$ for any solids $A, B \subset E$ with $|A \cap B| = 0$.
3. If $A, B \subset E$ are solids and $|(A - B) \cup (B - A)| = 0$, then $F(A) = F(B)$.

Lemma 12.6.3 *Let F be an additive continuous function defined on the family of all subsolids of a set $E \subset \mathbf{R}^m$. If $\{C_n\}$ is a sequence of subsolids of a solid $C \subset E$ such that $\sup h(\partial C_n) < +\infty$ and $\lim |C - C_n| = 0$, then $\lim F(C_n) = F(C)$.*

PROOF. The proof is similar to that of Lemma 11.5.1. Choose an $\varepsilon > 0$ so that $h(\partial C) + h(\partial C_n) < 1/\varepsilon$ for $n = 1, 2, \ldots$. There is an $\eta > 0$ such that $|F(B)| < \varepsilon$ for each solid $B \subset E$ with $h(\partial B) < 1/\varepsilon$ and $|B| < \eta$. By Exercise 12.5.1, 2 and 3, we have

$$h[\partial(C - C_n)] \le h[(\partial C) \cup (\partial C_n)] \le h(\partial C) + h(\partial C_n) < \frac{1}{\varepsilon},$$

and it follows from the assumptions that

$$|F(C) - F(C_n)| = |F(C - C_n)| < \varepsilon$$

for all sufficiently large n. ∎

Lemma 12.6.4 *Let C be a solid with $C^\circ \ne \emptyset$, and let $c > h(\partial C)$. There is a sequence $\{A_n\}$ of subfigures of C° such that*

$$\lim \rho_n = 0, \quad \lim |C - A_n| = 0, \quad \text{and} \quad \sup \|A_n\| \le \kappa c$$

where $\rho_n = \sup_{x \in \partial A_n} \mathrm{dist}(x, \partial C)$ and $\kappa \ge 1$ is a constant depending only on the dimension m. In particular, $A_n \to C$.

PROOF. Fix an integer $n \ge 1$. By Lemma 12.5.8, we can find dyadic cubes $K_1, K_2, \ldots$ of diameters less than $1/n$ so that

$$\partial C \subset \left(\bigcup_i K_i\right)^\circ \quad \text{and} \quad \sum_i \|K_i\| < 2m \cdot 3^m c.$$

With no loss of generality, we may assume that each K_i meets ∂C. For $i = 1, 2, \ldots$, let $\mathcal{K}_i$ be the collection of all dyadic cubes of the same diameter

$d(K_i)$ that meet K_i, and let $L_i = \bigcup \mathcal{K}_i$. Since the interiors of the cubes L_i cover the compact set ∂C, there is an integer $p \geq 1$ such that ∂C is contained in the interior of the figure $B = \bigcup_{i=1}^{p} L_i$. We have

$$\|B\| \leq \sum_{i=1}^{p} \sum_{K \in \mathcal{K}_i} \|K\| = 3^m \sum_{i=1}^{p} \|K_i\| < 2m \cdot 3^{2m} c,$$

$$|B| \leq \sum_{i=1}^{p} \sum_{K \in \mathcal{K}_i} |K| = 3^m \sum_{i=1}^{p} |K_i| \leq \frac{1}{n} \cdot \frac{3^m}{2m} \sum_{i=1}^{p} \|K_i\| < \frac{1}{n} \cdot 3^{2m} c,$$

and $\operatorname{dist}(x, \partial C) \leq 2/n$ for each $x \in B$. Let $d = \min\{d(K_1), \ldots, d(K_p)\}$, and let $\mathcal{K}$ be the family of all dyadic cubes of the same diameter d that meet C but not B°. As C is bounded, the family $\mathcal{K}$ is finite and $A_n = \bigcup \mathcal{K}$ is a figure. Each $K \in \mathcal{K}$ is contained in C°; for K is connected ([42, Chapter 2, Exercise 21(c), p. 45]), $K \cap C \neq \emptyset$, and $K \cap \partial C = \emptyset$. It follows that $A_n \subset C^\circ$, $C - A_n \subset B$, and $\partial A_n \subset \partial B$. Thus letting $\kappa = 2m \cdot 3^{2m}$ completes the proof. ∎

Proposition 12.6.5 *Each additive continuous function F in a set $E \subset \mathbf{R}^m$ has a unique extension to an additive continuous function on the family of all subsolids of E.*

PROOF. The uniqueness of the extension is a direct consequence of Lemmas 12.6.3 and 12.6.4. Let $\{A_n\}$ be a sequence of subfigures of a set $X \subset E$ converging to X. Then $\sup \|A_n\| < +\infty$ and

$$\lim_{m,n \to \infty} |(A_m \ominus A_n) \cup (A_n \ominus A_m)| \leq \lim_{n \to \infty} |X - A_n| + \lim_{m \to \infty} |X - A_m| = 0.$$

Since

$$|F(A_m) - F(A_n)| = |F(A_m \ominus A_n) - F(A_n \ominus A_m)|$$

$$\leq |F(A_m \ominus A_n)| + |F(A_n \ominus A_m)|$$

for $m, n = 1, 2, \ldots$, the continuity of F implies that the sequence $\{F(A_n)\}$ is Cauchy and hence convergent ([42, Theorem 3.11(c), p. 53]). If $\{B_n\}$ is another sequence of figures converging to X, then so is the sequence $\{C_n\}$ where $C_{2n-1} = A_n$ and $C_{2n} = B_n$ for $n = 1, 2, \ldots$. Consequently, $\lim F(B_n) = \lim F(A_n)$.

Using Lemma 12.6.4, associate to every solid $C \subset E$ a sequence $\{C_n\}$ of subfigures of C converging to C. In view of the previous paragraph, the value of $\lim F(C_n)$ depends only on C and not on the choice of $\{C_n\}$. It follows that the map $G : C \mapsto \lim F(C_n)$ is an additive extension of F to the family of all subsolids of E. We complete the proof by showing that G is continuous.

Let κ be the constant from Lemma 12.6.4. Given $\varepsilon > 0$, find an $\eta > 0$ so that $|F(A)| < \varepsilon$ for each figure $A \subset E$ with $\|A\| < \kappa/\varepsilon$ and $|A| < \eta$. Now let $C \subset E$ be a solid with $h(\partial C) < 1/\varepsilon$ and $|C| < \eta$. By Lemma 12.6.4, there is a sequence $\{A_n\}$ of subfigures of C such that $A_n \to C$ and $\sup \|A_n\| < \kappa/\varepsilon$. Since $G(C) = \lim F(A_n)$ and $|F(A_n)| < \varepsilon$ for $n = 1, 2, \ldots$, we see that $|G(C)| \leq \varepsilon$. This establishes the continuity of G. $\blacksquare$

12.7 Change of variables

In this section we show that a theorem similar to Theorem 10.5.6 holds for the $\mathcal{F}$-integral. This is a recent result obtained jointly by A. Novikov and the author (see [32]). We begin with a proposition analogous to Proposition 10.3.4.

Proposition 12.7.1 *If* $\Phi : \mathbf{R}^m \to \mathbf{R}^m$ *is a Lipschitz map, then*

$$h[\Phi(E)] \leq [\mathrm{Lip}(\Phi)]^{m-1} h(E)$$

for each set $E \subset \mathbf{R}^m$.

PROOF. If $\mathrm{Lip}(\Phi) = 0$ then $\Phi(E)$ is a singleton, and the desired inequality is readily verifiable. If $\mathrm{Lip}(\Phi) > 0$, choose an $\eta > 0$ and set $\vartheta = \eta/\mathrm{Lip}(\Phi)$. Let $\{C_i\}$ be a finite or infinite sequence of nonempty subsets of $\mathbf{R}^m$ such that $E \subset \bigcup_i C_i$ and $d(C_i) < \vartheta$ for all i, and let $D_i = \Phi(E \cap C_i)$. Since $\Phi(E) \subset \bigcup_i D_i$ and $d(D_i) \leq \mathrm{Lip}(\Phi)d(C_i) < \eta$, we see that

$$h_\eta[\Phi(E)] \leq \sum_{D_i \neq \emptyset} [d(D_i)]^{m-1} \leq [\mathrm{Lip}(\Phi)]^{m-1} \sum_i [d(C_i)]^{m-1} .$$

Consequently

$$h_\eta[\Phi(E)] \leq [\mathrm{Lip}(\Phi)]^{m-1} h_\vartheta(E) \leq [\mathrm{Lip}(\Phi)]^{m-1} h(E) ,$$

and the proposition follows from the arbitrariness of η. $\blacksquare$

Exercise 12.7.2 Let $\Phi : \mathbf{R}^m \to \mathbf{R}^m$ be a Lipschitz map.

1. Prove that Φ is Lipschitz with respect to the Euclidean metric in $\mathbf{R}^m$.
2. Denote by $\mathrm{Lip}_e(\Phi)$ the Lipschitz constant of Φ with respect to the Euclidean metric in $\mathbf{R}^m$, and verify that

$$\mathrm{Lip}_e(\Phi) \leq \sqrt{m}\,\mathrm{Lip}(\Phi) \quad \text{and} \quad \mathrm{Lip}(\Phi) \leq \sqrt{m}\,\mathrm{Lip}_e(\Phi)\,.$$

   Show by example that these inequalities may be strict.
3. From Remark 12.5.4 recall the definition of the Hausdorff measure $\mathcal{H}$, and prove that

$$\mathcal{H}[\Phi(E)] \leq [\mathrm{Lip}_e(\Phi)]^{m-1}\mathcal{H}(E)$$

   for each $E \subset \mathbf{R}^m$.

We shall need a topological result called *Brouwer's theorem on domain invariance*. Its proof, which is nontrivial, can be found in [47, Chapter 4, Section 8, Theorem 16, p. 199].

Theorem 12.7.3 *If U is an open subset of $\mathbf{R}^m$ and $\Phi : U \to \mathbf{R}^m$ is a homeomorphism, then $\Phi(U)$ is an open set.* ∎

Corollary 12.7.4 *If K is a compact subset of $\mathbf{R}^m$ and $\Phi : K \to \mathbf{R}^m$ is a homeomorphism, then $\partial\Phi(K) = \Phi(\partial K)$.*

PROOF. Since $\Phi(K^\circ) \subset [\Phi(K)]^\circ$ by Theorem 12.7.3, we have

$$\partial\Phi(K) = \Phi(K) - [\Phi(K)]^\circ \subset \Phi(K) - \Phi(K^\circ) = \Phi(\partial K)\,.$$

Applying this result to the map Φ^{-1} produces the desired equality. ∎

The next corollary is a consequence of Corollary 12.7.4 and Propositions 12.7.1 and 12.5.9.

Corollary 12.7.5 *The following statements are true.*

1. *A lipeomorphic image of a solid is a solid; in particular, a lipeomorphic image of a figure is a solid.*
2. *A Lipschitz image of a thin set is a thin set.* ∎

Theorem 12.7.6 *Let $\Phi : A \to B$ be a lipeomorphism from a figure A onto a figure B, and let f be an extended real-valued function defined almost* *everywhere in B. Then* $f \in \overline{\mathcal{FR}}_*(B)$ *if and only if* $(f \circ \Phi)\mathcal{J}_\Phi$ *belongs to* $\overline{\mathcal{FR}}_*(A)$, *in which case*

$$\int_A (f \circ \Phi)\mathcal{J}_\Phi = \int_B f.$$

PROOF. Recall that $\mathcal{J}_\Phi$ has been defined in Section 10.5, and extend it arbitrarily to a function on A, still denoted by $\mathcal{J}_\Phi$. Avoiding a triviality, assume that $A \neq \emptyset$, and observe that in view of symmetry and Proposition 10.3.4, it suffices to prove the theorem for $f \in \mathcal{FR}_*(A)$. Letting $a = 1/\mathrm{Lip}(\Phi^{-1})$ and $b = \mathrm{Lip}(\Phi)$, we have

$$a|x - x'| \leq |\Phi(x) - \Phi(x')| \leq b|x - x'|$$

for all $x, x' \in A$. By Corollary 12.7.4 and Propositions 10.3.4 and 12.7.1, the inequalities

$$a^m |C| \leq |\Phi(C)| \leq b^m |C| \quad \text{and} \quad \boldsymbol{h}[\partial\Phi(C)] = \boldsymbol{h}[\Phi(\partial C)] \leq b^{m-1}\|C\|$$

hold for each figure $C \subset A$. According to Proposition 12.6.5, the indefinite integral of f extends to an additive continuous function F on the family of all subsolids of B. From Corollary 12.7.5, 1 and the above inequalities it is easy to deduce that the map $G : C \mapsto F[\Phi(C)]$ is an additive continuous function in A. We show that G satisfies the conditions of Theorem 12.2.2 with respect to the function $(f \circ \Phi)\mathcal{J}_\Phi$.

Choose an $\varepsilon > 0$, and use Corollary 10.6.10 and Theorem 12.3.2 to find a negligible set $N \subset A$ and a positive function Δ on A so that

$$|f \circ \Phi(x)| \cdot \Big|\mathcal{J}_\Phi(x)|C| - |\Phi(C)|\Big| < \varepsilon|C|$$

for each $x \in A - N$ and each figure $C \subset A \cap U(x, \Delta(x))$ for which $x \in C$ and $r(C) > \varepsilon$. As $\Phi(N)$ is a negligible subset of B, we may assume that $f(y) = 0$ for each $y \in \Phi(N)$, and consequently that the previous inequality holds for all $x \in A$.

By Theorem 12.2.2, there is a gage δ_B on B such that

$$\sum_{i=1}^{p} \Big| f(y_i)|B_i| - F(B_i)\Big| < \varepsilon$$

for each δ_B-fine ε-regular $\mathcal{F}$-partition $\{(B_1, y_1), \ldots, (B_p, y_p)\}$ in B. Without loss of generality, we may suppose that $\delta_B(x) = 0$ for each $x \in \partial B$.

Corollary 12.7.5, 2 implies that $\delta_A = \min\{(\delta_B \circ \Phi)/b, \Delta\}$ is a gage in A. Let $\varepsilon' = (b/a)^m \kappa \varepsilon$, where κ is the constant from Lemma 12.6.4, and select a δ_A-fine ε'-regular $\mathcal{F}$-partition $\{(A_1, x_1), \ldots, (A_p, x_p)\}$ in A. If $B_i = \Phi(A_i)$ and $y_i = \Phi(x_i)$ for $i = 1, \ldots, p$, then $B_i \subset U(y_i, \delta_B(y_i))$ and

$$\frac{|B_i|}{d(B_i)h(\partial B_i)} \geq \left(\frac{a}{b}\right)^m r(A_i) > \kappa \varepsilon.$$

At this point we distinguish two cases.

CASE $m = 1$. As $\kappa \geq 1$, the collection $\{(B_1, y_1), \ldots, (B_p, y_p)\}$ is a δ_B-fine ε-regular $\mathcal{F}$-partition in B. Thus

$$\sum_{i=1}^{p} \Big| [f \circ \Phi(x_i)] \mathcal{J}_\Phi(x_i) |A_i| - G(A_i) \Big|$$

$$\leq \sum_{i=1}^{p} |f \circ \Phi(x_i)| \cdot \Big| \mathcal{J}_\Phi(x_i) |A_i| - |\Phi(A_i)| \Big| + \sum_{i=1}^{p} \Big| f(y_i) |B_i| - F(B_i) \Big|$$

$$< \sum_{i=1}^{p} \varepsilon |A_i| + \varepsilon \leq \varepsilon(|A| + 1)$$

and the theorem follows from Theorem 12.2.2.

CASE $m \geq 2$. Now the solids B_i need not be figures. Nonetheless, it follows from Lemma 12.6.4 that each B_i contains a figure C_i such that

1. $r(C_i) > \varepsilon$,
2. $|f(y_i)| \cdot \Big| |B_i| - |C_i| \Big| < \varepsilon/p$,
3. $|F(C_i) - G(A_i)| < \varepsilon/p$.

Since the figure C_i may not contain the point y_i, an additional adjustment is necessary. Let $z_1, \ldots, z_q$ be all distinct points in the set $\{y_1, \ldots, y_p\}$, and set $I_j = \{i : y_i = z_j\}$ for $j = 1, \ldots, q$. By the choice of δ_B, there are positive numbers $\eta_1, \ldots, \eta_q$ such that $U(z_1, \eta_1), \ldots, U(z_q, \eta_q)$ are disjoint subsets of B. Each set I_j has at most 2^m elements because $\{(A_1, x_1), \ldots, (A_p, x_p)\}$ is an $\mathcal{F}$-partition and Φ is an injective map. Hence there are nonoverlapping cubes $K_{i,j} \subset U(z_j, \eta_j)$ with $z_j \in K_{i,j}$ for each $i \in I_j$ and $j = 1, \ldots, q$. Let $K = \bigcup_{j=1}^{q} \bigcup_{i \in I_j} K_{i,j}$, and for $i \in I_j$ set

$$D_i = (C_i \ominus K) \cup K_{i,j}.$$

As $y_i \in B_i$ for $i = 1, \ldots, p$, the numbers $\eta_1, \ldots, \eta_q$ can be chosen so small that $D_i \subset U(y_i, \delta_B(y_i))$ and that the inequalities 1–3 hold when C_i is replaced by D_i. Thus $\{(D_1, y_1), \ldots, (D_p, y_p)\}$ is a δ_B-fine ε-regular $\mathcal{F}$-partition in B, and we obtain

$$\sum_{i=1}^{p} \left| [f \circ \Phi(x_i)] \mathcal{J}_\Phi(x_i) |A_i| - G(A_i) \right|$$

$$\leq \sum_{i=1}^{p} |f \circ \Phi(x_i)| \cdot \left| \mathcal{J}_\Phi(x_i) |A_i| - |\Phi(A_i)| \right| + \sum_{i=1}^{p} |f(y_i)| \cdot \left| |B_i| - |D_i| \right|$$

$$+ \sum_{i=1}^{p} \left| f(y_i) |D_i| - F(D_i) \right| + \sum_{i=1}^{p} |F(D_i) - G(A_i)|$$

$$< \sum_{i=1}^{p} \varepsilon |A_i| + p\frac{\varepsilon}{p} + \varepsilon + p\frac{\varepsilon}{p} \leq \varepsilon(|A| + 3) .$$

Another application of Theorem 12.2.2 completes the proof. ∎

Note. Z. Buczolich has shown that if $m \geq 2$, then Theorem 12.7.6 is false for the g-integral investigated in Chapter 11 (see [7]). Since figures much like intervals are not rotation invariant, the invariance of the $\mathcal{F}$-integral with respect to lipeomorphisms is somewhat surprising.

12.8 Multipliers

A *multiplier* for a family $\mathcal{E}$ of functions on a set E is a function g on E such that $gf \in \mathcal{E}$ for every $f \in \mathcal{E}$. In this section we show that Lipschitz functions on a figure A are multipliers for the families $\mathcal{R}_*(A)$ and $\mathcal{FR}_*(A)$. In dimension one, this follows from the integration by parts theorems (Theorems 6.2.1 and 12.4.1). In higher dimensions, however, we need a different technique, recently developed by J. W. Mortensen and the author. Before presenting the formal proof, we describe the main idea.

Suppose that a Lipschitz function g on a figure $A \subset \mathbf{R}^m$ maps A into the unit interval $I = [0, 1]$. We observe that for each figure $B \subset A$, the set

$$\Sigma_B = \{(x, t) \in B \times I : t \leq g(x)\}$$

is a solid. If $f \in \mathcal{R}_*(A)$ and $C \subset A \times I$ is a figure, we let

$$H(C) = \int_I \left[\int_{C^t} f(x) \, d\lambda_m(x) \right] d\lambda(t)$$

where $C^t = \{x \in A : (x,t) \in C\}$. Since each C^t is a subfigure of A and $t \mapsto \int_{C^t} f \, d\lambda_m$ is a λ-simple function on I, the map $H : C \mapsto H(C)$ is a well-defined function on the family of all subfigures of $A \times I$. We prove that H is additive and continuous. By Proposition 12.6.5, the function H has a unique extension to an additive continuous function on the family of all subsolids of $A \times I$; this extension is still denoted by H. The argument is completed by showing that the function $G : B \mapsto H(\Sigma_B)$, defined on the family of all subfigures of A, is the indefinite integral of fg in A.

Remark 12.8.1 Although the Fubini theorem does not hold for $\mathcal{R}_*(A)$ (see Example 11.1.2), the above argument is motivated by it. Indeed, assume that $f \in \mathcal{R}(A)$, and that $\mathbf{1}$ denotes the constant function on I equal to 1. By Proposition 8.6.8, the function $f \otimes \mathbf{1}$ belongs to $\mathcal{R}(A \times I)$ and Fubini's theorem (Theorem 10.1.13) implies that

$$H(\Sigma_A) = \int_I \left[\int_{(\Sigma_A)^t} f(x) \, d\lambda_m(x) \right] d\lambda(t) = \int_{\Sigma_A} f \otimes \mathbf{1} \, d\lambda_{m+1}$$

$$= \int_A \left[f(x) \int_0^{g(x)} d\lambda(t) \right] d\lambda_m(x) = \int_A fg \, d\lambda_m .$$

Lemma 12.8.2 *If F is an additive continuous function in a figure A, then*

$$\lim_{\|B\| \to +\infty} \frac{F(B)}{\|B\|} = 0 \qquad and \qquad \sup_{\|B\| < c} |F(B)| < +\infty$$

for each $c > 0$.

PROOF. Clearly, $A \subset U(\mathbf{0}, r)$ for an $r > 0$. Let $a = 2^m r^{m-1}$, and note that the λ_{m-1} measure of each face of $U[\mathbf{0}, r]$ equals $a/2$. Given $c \geq 1$, choose a positive $\varepsilon < 1/[c(1 + 2a)]$ and find an $\eta > 0$ so that $|F(B)| < \varepsilon$ for each figure $B \subset A$ with $\|B\| < 1/\varepsilon$ and $|B| < \eta$. Select an integer $p > (2r)^m/\eta$ and for $i = 1, \ldots, p$, let

$$A_i = \left[-r + (i-1)\frac{2r}{p}, -r + i\frac{2r}{p} \right] \times [-r, r]^{m-1} .$$

If C is a subfigure of A and $\|C\| < c$, then

$$|C \odot A_i| \leq |A_i| < \eta \qquad and \qquad \|C \odot A_i\| \leq \|C\| < c \leq \frac{1}{\varepsilon}$$

for $i = 1, \ldots, p$; the second inequality follows from Exercise 11.2.3. Thus

$$|F(C)| \le \sum_{i=1}^{p} |F(C \odot A_i)| < p\varepsilon \,,$$

and the second claim is proved.

To prove the first claim, let $C \subset A$ be a figure with $\|C\| > \max\{p, 1/\varepsilon\}$, and for each $t \in (-r, r)$, let

$$C_-(t) = C \odot ([-r, t] \times [-r, r]^{m-1}) \,,$$
$$C_+(t) = C \odot ([t, r] \times [-r, r]^{m-1}) \,.$$

Then C is the union of nonoverlapping figures $C_{\pm}(t)$, and it is easy to verify that $\|C_-(t)\| + \|C_+(t)\| \le \|C\| + a$. Observe that $t \mapsto \|C_-(t)\|$ is an increasing function on $(-r, r)$, which raises from 0 to $\|C\|$. Since

$$\lim_{t \to \tau+} \|C_-(t)\| - \lim_{t \to \tau-} \|C_-(t)\| \le a$$

for each $\tau \in (-r, r)$, there is a $\theta \in (-r, r)$ such that

$$\frac{\|C\|}{2} < \|C_-(\theta)\| \le \frac{\|C\|}{2} + a \,.$$

We conclude that $\|C_{\pm}(\theta)\| \le \|C\|/2 + a$. Next find an integer $n \ge 1$ with

$$\frac{\|C\|}{2^n} < \frac{1}{\varepsilon} - 2a \le \frac{\|C\|}{2^{n-1}} \,,$$

and proceeding inductively, construct a division $\{C_1, \ldots, C_{2^n}\}$ of C so that

$$\|C_k\| \le \frac{\|C\|}{2^n} + \sum_{j=0}^{n-1} \frac{a}{2^j} < \frac{\|C\|}{2^n} + 2a < \frac{1}{\varepsilon} \,.$$

Note that the inequality $1 < 1/\varepsilon - 2a$ yields $2^{n-1} < \|C\|$. For $i = 1, \ldots, p$ and $k = 1, \ldots, n$, we have

$$|A_i \odot C_k| \le |A_i| < \eta \quad \text{and} \quad \|A_i \odot C_k\| \le \|C_k\| < \frac{1}{\varepsilon}$$

(see Exercise 11.2.3 for the second inequality). By construction, the collection

$$\{A_i \odot C_k : i = 1, \ldots, p; \ k = 1, \ldots, 2^n\}$$

contains at most $2^n + p - 1$ nonempty figures. Therefore

$$|F(C)| \le \sum_{i=1}^{p} \sum_{k=1}^{2^n} |F(A_i \odot C_k)| < \varepsilon(2^n + p - 1) < \varepsilon(2\|C\| + p),$$

and hence

$$\frac{|F(C)|}{\|C\|} < \varepsilon \left(2 + \frac{p}{\|C\|} \right) < 3\varepsilon. \quad \blacksquare$$

Proposition 12.8.3 *An additive function F in a figure A is continuous if and only if the following condition is satisfied: given $\varepsilon > 0$, there is a $\theta > 0$ such that*

$$|F(B)| < \theta |B| + \varepsilon(\|B\| + 1)$$

for each figure $B \subset A$.

PROOF. As the converse is obvious, assume that F is continuous and choose an $\varepsilon > 0$. According to Lemma 12.8.2, there are positive numbers b and c such that

$$|F(B)| < \varepsilon \|B\| \qquad \text{and} \qquad |F(C)| < b$$

whenever B and C are subfigures of A with $\|B\| \ge c$ and $\|C\| < c$. We can find an $\eta > 0$ so that $|F(C)| < \varepsilon$ for each figure $C \subset A$ for which $\|C\| < c$ and $|C| < \eta$. Now if $\|C\| < c$ and $|C| \ge \eta$, then $|F(C)| < b \le (b/\eta)|C|$. Letting $\theta = b/\eta$, the previous alternatives yield the desired inequality. $\blacksquare$

Exercise 12.8.4 Let F be an additive function in a figure A. Show that F is AC if and only if the following condition is satisfied: given $\varepsilon > 0$, there is a $\theta > 0$ such that $|F(B)| < \theta |B| + \varepsilon$ for each figure $B \subset A$.

Theorem 12.8.5 *Let g be a Lipschitz function on a figure A. If f belongs to $\mathcal{R}_*(A)$ or $\mathcal{FR}_*(A)$, then so does fg, respectively.*

PROOF. Avoiding a triviality, suppose that $A \ne \emptyset$. It suffices to prove the theorem for $\mathcal{FR}_*(A)$; the proof for $\mathcal{R}_*(A)$ is analogous (cf. Remark 12.8.6). Since g is bounded and $\mathcal{FR}_*(A)$ is a linear space containing all constant functions, we may assume that g maps A into the unit interval $I = [0,1]$. Set $c = \max\{1, \mathrm{Lip}(g)\}$, and if $B \subset A$ is a figure, let

$$\Gamma_B = \{(x,t) \in B \times I : t = g(x)\},$$

$$\Sigma_B = \{(x,t) \in B \times I : t \le g(x)\}.$$

The lipeomorphism $(x, 0) \mapsto (x, g(x))$ maps $B \times \{0\}$ onto Γ_B and its Lipschitz constant equals c. If h denotes the m-dimensional quasi-Hausdorff measure in $\mathbf{R}^{m+1}$, then

$$h(\Gamma_B) \leq c^m h(B \times \{0\}) = c^m \lambda_m(B)$$

by Proposition 12.7.1. We conclude that Σ_B is a solid.

Let $f \in \mathcal{FR}_*(A)$, and denote by F the indefinite integral of f in A. Given a figure $C \subset A \times I$ and $t \in I$, the set $C^t = \{x \in A : (x, t) \in C\}$ is a sub-figure of A. Moreover, $t \mapsto \lambda_m(C^t)$, $t \mapsto \|C^t\|$, and $t \mapsto F(C^t)$ are λ-simple functions on I. An easy application of Fubini's theorem (Proposition 8.6.3) and Exercise 11.2.2 reveals that

$$\int_I \lambda_m(C^t)\, d\lambda(t) = \lambda_{m+1}(C) \quad \text{and} \quad \int_I \|C^t\|\, d\lambda(t) = \sum_{i=1}^m \|C\|_i \leq \|C\|.$$

It is also easy to see that $H : C \mapsto \int_I F(C^t)\, d\lambda(t)$ is an additive function in $A \times I$. In view of Propositions 11.4.8 and 12.8.3, for each $\varepsilon > 0$ there is a $\theta > 0$ such that

$$|F(C^t)| < \theta \lambda_m(C^t) + \varepsilon(\|C^t\| + 1)$$

for all $t \in I$. Integrating this inequality over I yields

$$|H(C)| \leq \int_I |F(C^t)|\, d\lambda(t) < \theta \lambda_{m+1}(C) + \varepsilon(\|C\| + 1),$$

which establishes the continuity of H. According to Proposition 12.6.5, the function H has a unique extension to an additive continuous function on the family of all subsolids of $A \times I$; this extension is still denoted by H. The map $G : B \mapsto H(\Sigma_B)$ is an additive function in A. Since $\lambda_{m+1}(\Sigma_B) \leq \lambda_m(B)$ and

$$h(\partial \Sigma_B) \leq \lambda_m(A) + h(\Gamma_A) + \|B\| \leq (1 + c^m)\lambda_m(A) + \|B\|,$$

we see that G is continuous. We complete the argument by showing that G is an indefinite integral of fg.

To this end, choose a positive $\varepsilon \leq c/(1 + c^m)$, and find an $\eta > 0$ so that $|H(C)| < \varepsilon$ for each solid $C \subset A \times I$ with

$$\lambda_{m+1}(C) < [2c\lambda_m(A)]\eta \quad \text{and} \quad h(\partial C) < \frac{2c\lambda_m(A)}{\varepsilon}.$$

By Theorem 12.2.2, there is a gage δ on A such that

$$\sum_{i=1}^p |f(x_i)\lambda_m(A_i) - F(A_i)| < \varepsilon$$

for each δ-fine ε-regular $\mathcal{F}$-partition $\{(A_1, x_1), \ldots, (A_p, x_p)\}$ in A. With no loss of generality, we may assume that $\delta \leq \eta$. Let $\{(A_1, x_1), \ldots, (A_p, x_p)\}$ be a δ-fine ε-regular $\mathcal{F}$-partition in A, and let $J_i = [0, g(x_i)]$ for $i = 1, \ldots, p$. We obtain

$$\sum_{i=1}^{p} |f(x_i)g(x_i)\lambda_m(A_i) - G(A_i)|$$

$$\leq \sum_{i=1}^{p} g(x_i)|f(x_i)\lambda_m(A_i) - F(A_i)| + \sum_{i=1}^{p} |F(A_i)\lambda(J_i) - H(\Sigma_{A_i})|$$

$$< \varepsilon + \sum_{i=1}^{p} |H(A_i \times J_i) - H(\Sigma_{A_i})|.$$

If $S = \sum_{i=1}^{p} |H(A_i \times J_i) - H(\Sigma_{A_i})|$, then after a suitable reordering, we find an integer k with $0 \leq k \leq p$ such that

$$S = \left| \sum_{i=1}^{k} \left[H(A_i \times J_i) - H(\Sigma_{A_i}) \right] \right| + \left| \sum_{i=k+1}^{p} \left[H(A_i \times J_i) - H(\Sigma_{A_i}) \right] \right|$$

$$= \left| H\left[\bigcup_{i=1}^{k} (A_i \times J_i) \right] - H\left[\bigcup_{i=1}^{k} \Sigma_{A_i} \right] \right|$$

$$+ \left| H\left[\bigcup_{i=k+1}^{p} (A_i \times J_i) \right] - H\left[\bigcup_{i=k+1}^{p} \Sigma_{A_i} \right] \right|$$

$$\leq \left| H\left[\bigcup_{i=1}^{k} (A_i \times J_i - \Sigma_{A_i}) \right] \right| + \left| H\left[\bigcup_{i=1}^{k} (\Sigma_{A_i} - A_i \times J_i) \right] \right|$$

$$+ \left| H\left[\bigcup_{i=k+1}^{p} (A_i \times J_i - \Sigma_{A_i}) \right] \right| + \left| H\left[\bigcup_{i=k+1}^{p} (\Sigma_{A_i} - A_i \times J_i) \right] \right|.$$

We let $C = \bigcup_{i=1}^{k} (A_i \times J_i - \Sigma_{A_i})$, and estimate $|H(C)|$ by observing that

$$\lambda_{m+1}(A_i \times J_i - \Sigma_{A_i}) \leq c\,d(A_i)\lambda_m(A_i) < 2c\eta\lambda_m(A_i),$$

$$h\left(\partial[A_i \times J_i - \Sigma_{A_i}] \right) \leq \lambda_m(A_i) + h(\Gamma_{A_i}) + c\,d(A_i)\|A_i\|$$

$$< \left(1 + c^m + \frac{c}{\varepsilon} \right) \lambda_m(A_i) \leq \frac{2c}{\varepsilon}\lambda_m(A_i)$$

for $i = 1, \ldots, k$. Indeed, these estimates imply that

$$\lambda_{m+1}(C) < [2c\lambda_m(A)]\eta \quad \text{and} \quad h(\partial C) < \frac{2c\lambda_m(A)}{\varepsilon},$$

and consequently $|H(C)| < \varepsilon$. Completely analogous verifications show that $S < 4\varepsilon$, and the theorem follows from Theorem 12.2.2. ∎

Remark 12.8.6 In dimension one, it follows from Theorem 6.2.1 and Corollary 5.1.5 that all functions of bounded variation are multipliers for $\mathcal{R}_*(A)$; it was proved in [45] that $\mathcal{R}_*(A)$ has no other multipliers. Non-Lipschitz multipliers for $\mathcal{R}_*(A)$ in higher dimensions also exist (i.e., $g \otimes 1 \otimes \cdots \otimes 1$ where g is any increasing function). On the other hand, whether Lipschitz functions are the only multipliers for $\mathcal{FR}_*(A)$ is an open problem in all dimensions, including dimension one.

A sequence $\{g_n\}$ of functions defined on a set $E \subset \mathbf{R}^m$ is called *equilipschitz* whenever each g_n is Lipschitz and $\sup_n \text{Lip}(g_n) < +\infty$.

Lemma 12.8.7 *If $\{g_n\}$ is an equilipschitz sequence of functions defined on a nonempty figure A, then the following conditions are equivalent.*

1. $\lim g_n = 0$ *uniformly in A.*
2. $\lim g_n = 0$ *almost everywhere in A.*
3. $\lim \int_A |g_n| = 0$.

PROOF. The implication $(1 \Rightarrow 2)$ is obvious, and since A is bounded, $(2 \Rightarrow 3)$ follows from the dominated convergence theorem.

We prove $(3 \Rightarrow 1)$ by contradiction. Choose a $c > \sup_n \text{Lip}(g_n)$, and suppose that there is an $\varepsilon > 0$ such that for $n = 1, 2, \ldots$, we can find a $z_n \in A$ with $|g_n(z_n)| \geq 3\varepsilon$. Passing to a subsequence of $\{g_n\}$, we may assume that for a $z \in A$ all points z_n lie in $B = A \odot U[z, \varepsilon/c]$. It follows that

$$|g_n(x)| \geq |g_n(z_n)| - c|z_n - x| > 3\varepsilon - c(|z_n - z| + |z - x|) > \varepsilon$$

for each $x \in B$ and $n = 1, 2, \ldots$. Thus

$$\lim \int_A |g_n| \geq \lim \int_B |g_n| \geq \varepsilon |B| > 0,$$

a contradiction. ∎

Theorem 12.8.8   *Let $\{g_n\}$ be an equilipschitz sequence of functions defined on a figure A, and let f belong to $\mathcal{R}_*(A)$ or $\mathcal{F}\mathcal{R}_*(A)$. If $\lim g_n = 0$ almost everywhere in A, then $\lim \int_A fg_n = 0$.*

PROOF.   By Lemma 12.8.7, there is a sequence $\{\varepsilon_n\}$ of positive constants such that $\lim \varepsilon_n = 0$ and $|g_n| \leq \varepsilon_n$ for all n. The sequence $\{g_n + \varepsilon_n\}$ is equilipschitz, and $0 \leq g_n + \varepsilon_n \leq 2\varepsilon_n \leq 1$ for all sufficiently large n. Since

$$\lim \int_A fg_n = \lim \int_A f(g_n + \varepsilon_n)$$

whenever either limit exists, we may assume from the onset that each g_n maps A into the unit interval $I = [0, 1]$. Let $c = \sup_n \mathrm{Lip}(g_n)$. Following the proof of Theorem 12.8.5, it is easy to see that the sets

$$\Sigma_n = \{(x, t) \in A \times I : t \leq g_n(x)\}$$

are solids such that

$$\lambda_{m+1}(\Sigma_n) \leq \varepsilon_n \lambda_m(A) \qquad \text{and} \qquad h(\partial \Sigma_n) \leq (1 + c^m)\lambda_m(A) + \|A\|$$

for $n = 1, 2, \ldots$. Now let H be the additive continuous function on the family of all subsolids of A associated with f according to the proof of Theorem 12.8.5. In view of the above estimates, we obtain

$$\lim \int_A fg_n = \lim H(\Sigma_n) = 0. \quad \blacksquare$$

Chapter 13

Recent developments

We devote this final chapter to a brief and somewhat informal discussion on how and why one may wish to modify the $\mathcal{F}$-integral. The main idea is to enlarge the family $\mathcal{F}$ of all figures to a suitable family $\mathcal{C}$ of bounded subsets of $\mathbf{R}^m$ and use $\mathcal{C}$-partitions to define the $\mathcal{C}$-integral.

13.1 The $\mathcal{S}$-integral

We remind the reader that solids were defined in Section 12.6, and denote by $\mathcal{S}$ the family of all solids in $\mathbf{R}^m$.

Definition 13.1.1 An extended real-valued function f defined almost everywhere in a solid S is called $\mathcal{SF}$-*integrable* over S if $f \in \overline{\mathcal{FR}}_*(A)$ for each figure $A \subset S$ and the additive function $A \mapsto \int_A f$ in S is continuous.

The family of all extended real-valued functions defined almost everywhere in a solid S that are $\mathcal{SF}$-integrable over S is denoted by $\overline{\mathcal{SFR}}_*(S)$ or $\overline{\mathcal{SFR}}_*(S, \lambda_m)$. For the sake of consistency, we denote by $\mathcal{SFR}_*(S) = \mathcal{SFR}_*(S, \lambda_m)$ the family of all finite functions from $\overline{\mathcal{SFR}}_*(S)$ that are defined on S. Let S be a solid and $f \in \overline{\mathcal{SFR}}_*(S)$. By Proposition 12.6.5, the additive continuous function $F : A \mapsto \int_A f$ in S has a unique additive continuous extension G to the family of all subsolids of S. The number $G(S)$ is called the $\mathcal{SF}$-*integral* of f over S, denoted by $\int_S f$ or $\int_S f \, d\lambda_m$. The reader can readily verify that this notation is consistent with that of the previous chapters. Deriving the basic properties of the $\mathcal{SF}$-integral is an easy exercise.

To obtain a divergence theorem, we must first define the flux from a solid S of a continuous vector field v defined on ∂S. Fortunately, this is not difficult. Using Tietze's extension theorem ([41, Chapter 8, Theorem 5, p. 148]), extend v to a continuous vector field w on S^-. By Proposition 11.2.8, the flux F of w in S is an additive continuous function in S, which has a unique additive and continuous extension G to the family of all subsolids of S (Proposition 12.6.5). We call the number

$$\int_{\partial S} v \cdot n \, d\lambda_{m-1} = G(S)$$

the *flux* of v from S. It follows from the claim $\lim \rho_n = 0$ in Lemma 12.6.4 that the flux of v from S depends only on v and not on the choice of w.

Remark 13.1.2   At this point the symbol $\int_{\partial S} v \cdot n \, d\lambda_{m-1}$ must be viewed as mere notation. Nonetheless, with some effort, one can define a vector field n on ∂A that corresponds intuitively to the exterior normal of A, and interpret $\int_{\partial S} v \cdot n$ as a McShane-type integral with respect to the Hausdorff measure $\mathcal{H}$ (see Remark 12.5.4 for the definition of $\mathcal{H}$). The reader who wishes to make this statement precise is referred to [9, Chapter 5] or [49, Chapter 5].

Since the definition of the flux from a solid parallels that of the $\mathcal{SF}$-integral, the following divergence theorem is an immediate consequence of Theorem 12.2.5 and Lemma 12.6.4.

Theorem 13.1.3   *Let S be a solid and T be a thin set. If v is a continuous vector field on S^- that is almost differentiable at each point of $S^\circ - T$, then $\operatorname{div} v$ belongs to $\overline{\mathcal{SFR}}_*(S, \lambda_m)$ and*

$$\int_S \operatorname{div} v \, d\lambda_m = \int_{\partial S} v \cdot n \, d\lambda_{m-1} . \quad \blacksquare$$

While the $\mathcal{SF}$-integral has nice properties, its definition displays an asymmetry similar to that of the g-integral. We shall remove it by passing to a less general but still very viable Riemann-type integral: a process resembling the passage from the g-integral to $\mathcal{F}$-integral.

The *S-regularity* of a solid S is the number

$$r_S(S) = \begin{cases} \dfrac{|S|}{d(S)\boldsymbol{h}(\partial S)} & \text{if } d(S)\boldsymbol{h}(\partial S) > 0, \\ 0 & \text{otherwise.} \end{cases}$$

If A is a figure, then $r_S(A) = r(A)$ by Remark 12.5.3. Given $\varepsilon > 0$ and a caliber $\eta = \{\eta_j\}$, we say that a solid S is $(\varepsilon, \eta; S)$-*small* if S is the union of nonoverlapping solids $S_1, \ldots, S_k$ such that $h(\partial S_j) < 1/\varepsilon$ and $|S_j| < \eta_j$ for $j = 1, \ldots, k$. Obviously, each (ε, η)-small figure is $(\varepsilon, \eta; S)$-small; whether the converse holds is unclear.

An S-*partition* is a collection, possibly empty, $\{(S_1, x_1), \ldots, (S_p, x_p)\}$ such that $S_1, \ldots, S_p$ are nonoverlapping solids and $x_i \in S_i$ for $i = 1, \ldots, p$. For a solid S, we define a δ-fine $(\varepsilon; S)$-regular S-partition of S mod $(\varepsilon, \eta; S)$ in the obvious way.

Definition 13.1.4 A function f on a solid S is called S-*integrable* over S if there is a real number I having the following property: given $\varepsilon > 0$, we can find a gage δ on S and a caliber η so that

$$|\sigma(f, P) - I| < \varepsilon$$

for each δ-fine $(\varepsilon; S)$-regular S-partition P of A mod $(\varepsilon, \eta; S)$.

By $SR_*(S)$ or $SR_*(S, \lambda_m)$ we denote the family of all S-integrable functions over a solid S. Using Lemmas 11.3.6 and 12.6.4, it is not hard to show that Proposition 11.3.7 with "mod (ε, η)" replaced by "mod $(\varepsilon, \eta; S)$" remains true when A is a solid. It follows that the number I of Definition 13.1.4 is determined uniquely by the function $f \in SR_*(S)$; we call it the S-*integral* of f over S, denoted by $\int_S f$ or $\int_S f \, d\lambda_m$. This notation is consistent. Indeed, $SR_*(A) \subset FR_*(A)$ for each figure A, and since a result analogous to Proposition 11.4.8 (proved in the same way as Proposition 11.4.8) holds for the S-integral, we infer that $SR_*(S) \subset SFR_*(S)$ for every solid S. It is unknown whether $SR_*(S)$ is a proper subfamily of $SFR_*(S)$, even in the simplest case when S is a one-dimensional cell (cf. [4]).

Remark 13.1.5 The difficulty in comparing the S-integral with the F-integral seems to center about the number n_P of coalescing points x_i in the partition $P = \{(A_1, x_1), \ldots, (A_p, x_p)\}$. If P is an F-partition, then $n_P \leq 2^m$ (recall that we used this fact in the proof of Theorem 12.7.6); if P is an S-partition, then trivially $n_P \leq p$ but there is no bound independent of P.

For a solid S, we define the family $\overline{SR}_*(S) = \overline{SR}_*(S, \lambda_m)$ in the usual way. The standard properties of the S-integral are established by minor modifications of the arguments used in the previous chapters. We only sketch the proofs of the divergence and change of variables theorems.

Lemma 13.1.6 *Let v be a continuous vector field on a closed set $E \subset \mathbf{R}^m$ that is differentiable at $z \in E^{\circ}$. Given $\varepsilon > 0$, there is a $\delta > 0$ such that*

$$\left| \operatorname{div} v(x)|S| - \int_{\partial S} v \cdot n \, d\lambda_{m-1} \right| \leq \varepsilon d(S) h(\partial S)$$

for each solid $S \subset E \cap U(x, \delta)$ with $z \in S$.

PROOF. Choose an $\varepsilon > 0$, and let κ be the constant from Lemma 12.6.4. By the version of Lemma 8.3.4 that applies to figures, there is a $\delta > 0$ such that $U(z, \delta) \subset E$ and

$$\left| \operatorname{div} v(x)|B| - \int_{\partial B} v \cdot n \right| \leq \frac{\varepsilon}{2\kappa} d(B) \|B\|$$

for each figure $B \subset U(z, \delta)$ with $z \in B$. Proceeding towards a contradiction, suppose that

$$\left| \operatorname{div} v(x)|S| - \int_{\partial S} v \cdot n \right| > \varepsilon d(S) h(\partial S)$$

for a solid $S \subset U(z, \delta)$ with $z \in S$. Lemma 12.6.4 implies the existence of a figure $A \subset S$ such that

$$\left| \operatorname{div} v(x)|A| - \int_{\partial A} v \cdot n \right| > \frac{\varepsilon}{\kappa} d(S) \|A\|.$$

Choose a positive $\eta < \delta$ and let $B(\eta) = A \cup U[z, \eta]$. As η approaches zero, $|B(\eta)|$ and $\int_{\partial B(\eta)} v \cdot n$ approach $|A|$ and $\int_{\partial A} v \cdot n$, respectively. Moreover,

$$\lim_{\eta \to 0} d[B(\eta)] \leq d(S) \qquad \text{and} \qquad \lim_{\eta \to 0} \|B(\eta)\| \leq 2\|A\|;$$

the first inequality holds because $z \in S$, and we note that the second cannot be improved if $m = 1$. As $B(\eta) \subset U(z, \delta)$ is a figure containing z, a contradiction follows. ∎

The proof of the following theorem is left to the reader. In view of Lemma 13.1.6, it is similar to that of Theorem 12.2.5.

Theorem 13.1.7 *Let S be a solid and T be a thin set. If v is a continuous vector field on S^- that is almost differentiable at each point of $S^{\circ} - T$, then $\operatorname{div} v$ belongs to $\overline{SR}_*(S, \lambda_m)$ and*

$$\int_S \operatorname{div} v \, d\lambda_m = \int_{\partial S} v \cdot n \, d\lambda_{m-1} . \quad \blacksquare$$

Theorem 13.1.8 *Let A be a solid, let $\Phi : A \to \mathbf{R}^m$ be a lipeomorphism, and let f be an extended real-valued function defined almost everywhere in $\Phi(A)$. Then $f \in \overline{\mathcal{SR}}_*(\Phi(A))$ if and only if $(f \circ \Phi)\mathcal{J}_\Phi$ belongs to $\overline{\mathcal{SR}}_*(A)$, in which case*

$$\int_A (f \circ \Phi)\mathcal{J}_\Phi = \int_{\Phi(A)} f .$$

PROOF. While similar in spirit, the proof is substantially simpler than that of Theorem 12.7.6. Avoiding a triviality, assume that $A^\circ \neq \emptyset$. Extend $\mathcal{J}_\Phi$ arbitrarily to a function on A, and applying Exercise 10.3.5, extend Φ to a lipeomorphism defined on A^-, still denoted by Φ. By Corollary 12.7.5, 1, the set $B = \Phi(A)$ is a solid, and in view of symmetry and Proposition 10.3.4, it suffices to prove the theorem for $f \in \mathcal{SR}_*(B)$. Note that the indefinite integral F of f is an additive continuous function on the family of all subsolids of B. Letting $a = 1/\mathrm{Lip}(\Phi^{-1})$ and $b = \mathrm{Lip}(\Phi)$, we have

$$a|x - x'| \leq |\Phi(x) - \Phi(x')| \leq b|x - x'|$$

for all $x, x' \in A$. By Corollary 12.7.4 and Propositions 10.3.4 and 12.7.1, the inequalities

$$a^m|S| \leq |\Phi(S)| \leq b^m|S| \quad \text{and} \quad \boldsymbol{h}[\partial\Phi(S)] = \boldsymbol{h}[\Phi(\partial S)] \leq b^{m-1}\boldsymbol{h}(\partial S)$$

hold for each solid $S \subset A$. It follows that the map $G : S \mapsto F[\Phi(S)]$ is an additive continuous function on the family of all subsolids of A. Now choose an $\varepsilon > 0$; in doing so one can never go wrong.

We can define the *derivate relative* to $\mathcal{S}$, and prove that each $\mathcal{S}$-integrable function is equal almost everywhere to the derivate relative to $\mathcal{S}$ of its indefinite integral (a result analogous to Theorem 12.3.2). Applying this to Corollary 10.6.10 and changing the values of f to zero on a negligible set, we can find a positive function Δ on A so that

$$|f \circ \Phi(x)| \cdot \left| \mathcal{J}_\Phi(x)|S| - |\Phi(S)| \right| < \varepsilon|S|$$

for each $x \in A$ and every solid $S \subset A$ for which $x \in S$, $r_S(S) > \varepsilon$, and $d(S) < \Delta(x)$. By a result similar to Theorem 12.2.2, there is a gage δ_B on B such that

$$\sum_{i=1}^p \left| f(y_i)|B_i| - F(B_i) \right| < \varepsilon$$

for each δ_B-fine $(\varepsilon, \mathcal{S})$-regular $\mathcal{S}$-partition $\{(B_1, y_1), \ldots, (B_p, y_p)\}$ in B. Corollary 12.7.5, 2 implies that $\delta_A = \min\{(\delta_B \circ \Phi)/b, \Delta\}$ is a gage in A. Let

$\varepsilon' = (b/a)^m \varepsilon$, and select a δ_A-fine $(\varepsilon', \mathcal{S})$-regular $\mathcal{S}$-partition $\{(A_1, x_1), \ldots,$ $(A_p, x_p)\}$ in A. If $B_i = \Phi(A_i)$ and $y_i = \Phi(x_i)$ for $i = 1, \ldots, p$, then it is easy to verify that $\{(B_1, y_1), \ldots, (B_p, y_p)\}$ is a δ_B-fine $(\varepsilon, \mathcal{S})$-regular $\mathcal{S}$-partition in B. Therefore

$$\sum_{i=1}^{p} \left| [f \circ \Phi(x_i)] \mathcal{J}_\Phi(x_i) |A_i| - G(A_i) \right|$$

$$\leq \sum_{i=1}^{p} |f \circ \Phi(x_i)| \cdot \left| \mathcal{J}_\Phi(x_i) |A_i| - |\Phi(A_i)| \right| + \sum_{i=1}^{p} \left| f(y_i) |B_i| - F(B_i) \right|$$

$$< \sum_{i=1}^{p} \varepsilon |A_i| + \varepsilon \leq \varepsilon(|A| + 1),$$

and another application of a result similar to Theorem 12.2.2 completes the proof. ∎

Note. It is unclear whether a change of variables theorem similar to Theorem 13.1.8 can be established for the $\mathcal{SF}$-integral.

13.2 The perimeter

Before beginning this section, recall from Remark 12.5.4 the definition of the $(m-1)$-dimensional Hausdorff measure $\mathcal{H}$.

The solids are simple to define, and they served us well in showing that an operational integral can be obtained directly from a family of sets invariant with respect to lipeomorphisms. This is, however, about the extent of their utility. From the point of view of measure theory, the definition of a solid is ill conceived. Indeed, there is little connection between the topological boundary of a set $E \subset \mathbf{R}^m$ and the measure of E; in particular, neither $h(\partial E)$ nor $\mathcal{H}(\partial E)$ is the counterpart of the perimeter of a figure. While this can be easily seen by adding "whiskers" to a cube, the example below is more instructive. It indicates that the correct definition of perimeter must be based on a measure-theoretic analog of topological boundary.

Example 13.2.1 We modify the Cantor ternary set C of Example 5.3.11 by removing the open middle segments $\Delta_{r,s}$ of length 5^{-r} rather than 3^{-r}. Letting $D = \bigcup_{r=1}^{\infty} \bigcup_{s=1}^{2^{r-1}} \Delta_{r,s}^{-}$, an easy calculation shows that $\lambda(D) = 1/5$ and hence $\lambda(C) = 4/5$. Now the set

$$S = \bigcup_{r=1}^{\infty} \bigcup_{s=1}^{2^{r-1}} (\Delta_{r,s}^{-} \times [0, 5^{-r}])$$

is a solid in $\mathbf{R}^2$, and we have

$$\sum_{r=1}^{\infty} \sum_{s=1}^{2^{r-1}} \| \Delta_{r,s}^- \times [0, 5^{-r}] \| = 4\lambda(D) = h(\partial S) - \lambda(C)$$

$$= h(\partial S - C \times \{0\}) = \mathcal{H}(\partial S - C \times \{0\}).$$

Since $(S^-)^\circ = (S^\circ)^-$, it is difficult to define the set $\partial S - C \times \{0\}$ by topological means. On the other hand, λ-almost all points of C are dispersion points of D by Proposition 6.5.2, and it is not hard to show that these points are also dispersion points of S.

Let $E \subset \mathbf{R}^m$. The set of all nondispersion points of E (see the paragraph preceding Proposition 10.5.3) is called the *essential closure* of E, denoted by E^{-*}. The sets

$$E^{\circ*} = \mathbf{R}^m - (\mathbf{R}^m - E)^{-*} \qquad \text{and} \qquad \partial^* E = E^{-*} - E^{\circ*}$$

are called, respectively, the *essential interior* and *essential boundary* of E. Since $E^\circ \subset E^{\circ*} \subset E^{-*} \subset E^-$, we have $\partial^* E \subset \partial E$. Obviously, $A^{-*} = A$, $A^{\circ*} = A^\circ$, and $\partial^* A = \partial A$ for each figure A. If S is the solid of Example 13.2.1, it is easy to see that

$$\mathcal{H}(\partial^* S) = h(\partial^* S) = \sum_{r=1}^{\infty} \sum_{s=1}^{2^{r-1}} \| \Delta_{r,s}^- \times [0, 5^{-r}] \| .$$

Determining exactly the sets S^{-*}, $S^{\circ*}$, and $\partial^* S$ is harder, and we shall not attempt it.

Warning. The notation for essential closure, essential interior, and essential boundary is not yet standardized. For instance, $\partial^* E$ is often used to denote a subset of the essential boundary of E, called the *reduced boundary* of E, which differs from the essential boundary of E by a set of Hausdorff measure zero (see [9, Section 5.7.1, p. 194, and Section 5.8, Lemma 1, p. 208]).

Exercise 13.2.2 Given a set $E \subset \mathbf{R}^m$, prove the following facts.

1. Each $x \in E^{\circ*}$ is a density point of E, and the converse is true whenever E is measurable.
2. $\partial^* E = E^{-*} \cap (\mathbf{R}^m - E)^{-*}$; in particular, $\partial^* E = \partial^* (\mathbf{R}^m - E)$.

3. Changing E by a negligible set leaves the sets E^{-*}, $E^{\circ*}$, and $\partial^* E$ intact.

Exercise 13.2.3 If A and B are subsets of $\mathbf{R}^m$, show that the following claims are true.

1. $(A \cup B)^{-*} = A^{-*} \cup B^{-*}$ and $(A \cap B)^{\circ*} = A^{\circ*} \cap B^{\circ*}$.
2. The essential boundaries $\partial^*(A \cup B)$, $\partial^*(A \cap B)$, and $\partial^*(A - B)$ are subsets of $(\partial^* A) \cup (\partial^* B)$.

Proposition 13.2.4 *If E is any subset of $\mathbf{R}^m$, then*

$$x \mapsto \overline{\Theta}(E, x) \qquad and \qquad x \mapsto \underline{\Theta}(E, x)$$

are measurable functions on $\mathbf{R}^m$, and $\underline{\Theta}(E, x) = 1$ for almost all $x \in E$. In particular, the sets E^{-*}, $E^{\circ*}$, and $\partial^* E$, as well as the set of all density points of E, are measurable.

PROOF. By Lemma 3.3.14, there is a measurable set A such that $E \subset A$ and $|E| = |A|$. If B is a measurable set, the Carathéodory test for measurability implies that

$$|E| = |E \cap B| + |E - B| \le |A \cap B| + |A - B| = |A|.$$

Thus $|E \cap B| = |A \cap B|$ for each measurable set B, and the upper and lower densities of the sets E and A coincide. An application of Propositions 10.5.3 and 9.3.3 completes the proof. ∎

Corollary 13.2.5 *A set $E \subset \mathbf{R}^m$ is measurable if and only if its essential boundary $\partial^* E$ is negligible.*

PROOF. Let E be measurable. According to Exercise 13.2.2, 1, the essential interior of E consists of all density points of E. Hence by Proposition 10.5.3, the sets E and $E^{\circ*}$ differ by a negligible set. Applying the same argument to the measurable set $\mathbf{R}^m - E$ shows that E^{-*} differs from E by a negligible set. It follows that $\partial^* E$ is negligible.

Conversely, assume that $\partial^* E$ is negligible. By Proposition 13.2.4, the set $E - E^{-*}$ is negligible, and by the assumption, so is the set $E - E^{\circ*}$. In view of this and Exercise 13.2.2, 2, the set

$$E^{-*} - E = [\mathbf{R}^m - (\mathbf{R}^m - E)^{\circ*}] - E = (\mathbf{R}^m - E) - (\mathbf{R}^m - E)^{\circ*}$$

is also negligible. The measurability of E follows from Proposition 13.2.4. ∎

Exercise 13.2.6 Show that for a set $E \subset \mathbf{R}^m$, the following conditions are equivalent:

1. E is measurable;
2. $E^{\circ*}$ is the set of all density points of E;
3. $E^{\circ*}$ and E differ by a negligible set;
4. E^{-*} and E differ by a negligible set;
5. $(E^{\circ*})^{-*} = E^{-*}$.

Exercise 13.2.7 If $E \subset \mathbf{R}^m$ show that $|E| = |E^{-*}|$, and deduce that

$$(E^{-*})^{-*} = E^{-*} \quad \text{and} \quad (E^{-*})^{\circ*} = (E^{\circ*})^{\circ*} = E^{\circ*}.$$

Exercise 13.2.8 Let A and B be measurable sets. Using Lemma 6.5.4, prove that $A^{-*} \cap B^{\circ*} \subset (A \cap B)^{-*}$.

Definition 13.2.9 Let E be a bounded subset of $\mathbf{R}^m$. The *perimeter* of E is the extended real number $\|E\| = \mathcal{H}(\partial^* E)$. If $\|E\| < +\infty$, we say that E is a set of *bounded variation*, abbreviated as BV set.

For figures, the above definition of perimeter agrees with that given in Section 11.2 (see Remark 12.5.4).

Remark 13.2.10 Clearly, a bounded set $E \subset \mathbf{R}^m$ is a BV set if and only if $h(\partial^* E) < +\infty$. On the other hand, the correct perimeter of a BV set E is given by $\mathcal{H}(\partial^* E)$ and not by $h(\partial^* E)$. Indeed, if $m = 2$ and

$$E = \{(x, y) \in \mathbf{R}^2 : |x| + |y| < 1\},$$

then $\|E\| = 4\sqrt{2}$ as anticipated, while $h(\partial^* E) = 4$ (cf. the *Note* following Remark 12.5.4).

Each solid is a BV set, but a cube with infinitely many "whiskers" of the same length shows that the converse is false. According to Exercise 12.5.5 and Corollary 13.2.5, each BV set is measurable. The set

$$D = \{0\} \cup \bigcup_{n=1}^{\infty} \left(\left[\frac{1}{2n}, \frac{1}{2n-1} \right] \times \left[0, \frac{1}{n} \right] \right)$$

is an example of a compact subset of $\mathbf{R}^2$ that is not BV.

Exercise 13.2.11 If A and B are bounded sets, show that

$$\max\{\|A \cup B\|, \|A \cap B\|, \|A - B\|\} \leq \|A\| + \|B\|.$$

Deduce that the family of all BV sets is closed with respect to unions, intersections, and set differences.

If $m = 1$, it is not difficult to see that a bounded set $E \subset \mathbf{R}$ is BV if and only if it differs from a figure A by a negligible set, in which case $\|E\| = \|A\|$. Although, the BV subsets of $\mathbf{R}^m$ for $m \geq 2$ are quite complicated (see Example 13.2.17 below), they admit a Fubini-type characterization.

Let E be a bounded subset of $\mathbf{R}^m$, and let i be an integer with $0 \leq i \leq m$. Following Exercise 11.2.2, denote by E^ξ the section of the set E determined by $\xi \in \mathbf{R}^{m-1}$. If the extended real-valued function $\xi \mapsto \|E^\xi\|$ belongs to $\overline{\mathcal{R}}(\mathbf{R}^{m-1}, \lambda_{m-1})$, let

$$\|E\|_i = \int_{\mathbf{R}^{m-1}} \|E^\xi\| \, d\lambda_{m-1}(\xi);$$

otherwise, let $\|E\|_i = +\infty$.

Theorem 13.2.12 *If E is a bounded measurable set, then*

$$\max_{i=1,\dots,m} \|E\|_i \leq \|E\| \leq \sum_{i=1}^{m} \|E\|_i. \quad \blacksquare$$

The proof of Theorem 13.2.12 is not easy, and we shall comment on it after Theorem 13.2.14 below, where some references are also given.

Lemma 13.2.13 *Let A be a BV set, and let $B_1, \dots, B_n$ be nonoverlapping cubes whose diameters are equal to d. Then*

$$\frac{1}{2m} \sum_{j=1}^{n} \|A \cap B_j\| \leq \frac{|A|}{d} + \|A\|.$$

PROOF. Let K be a one-dimensional figure whose connected components are $K_1, \dots, K_p$, and let $L_1, \dots, L_n$ be nonoverlapping one-cells of the same diameters equal to $d > 0$. Then for each integer r with $1 \leq r \leq p$, the family $\{K_r \cap L_1, \dots, K_r \cap L_n\}$ contains at most $(|K_r|/d) + 2$ one-cells. Thus

$$\sum_{j=1}^{n} \|K \cap L_j\| \leq \sum_{r=1}^{p} \sum_{j=1}^{n} \|K_r \cap L_j\| \leq 2 \sum_{r=1}^{p} \left(\frac{|K_r|}{d} + 2 \right) = 2 \left(\frac{|K|}{d} + \|K\| \right).$$

In view of Theorem 13.2.12, applying Fubini's Theorem 10.1.13 to the above inequality yields

$$\sum_{j=1}^{n} \|A \cap B_j\| \leq \sum_{j=1}^{n} \sum_{i=1}^{m} \|A \cap B_j\|_i$$

$$= \sum_{i=1}^{m} \int_{\mathbf{R}^{m-1}} \left(\sum_{j=1}^{n} \|A^\xi \cap (B_j)^\xi\| \right) d\lambda_{m-1}(\xi)$$

$$\leq 2 \sum_{i=1}^{m} \left[\frac{1}{d} \int_{\mathbf{R}^{m-1}} |A^\xi| \, d\lambda_{m-1}(\xi) + \int_{\mathbf{R}^{m-1}} \|A^\xi\| \, d\lambda_{m-1}(\xi) \right]$$

$$= 2 \left(\frac{m}{d} |A| + \sum_{i=1}^{m} \|A\|_i \right) \leq 2m \left(\frac{|A|}{d} + \|A\| \right). \quad \blacksquare$$

Theorem 13.2.14 *If E is a bounded measurable set, then*

$$\|E\| = \sup \int_E \operatorname{div} v \, d\lambda_m$$

where the supremum is taken over all continuously differentiable vector fields v on $\mathbf{R}^m$ such that $\|v(x)\| \leq 1$ for all $x \in \mathbf{R}^m$. $\blacksquare$

Theorem 13.2.14 states, perhaps, the most essential characteristic of BV sets, which is usually presented as their defining property (cf. [9, Section 5.1, p. 167] or [49, Definition 5.4.1, p. 229]). Its proof is difficult and can be found in [9, Section 5.11, Theorem 1, p. 222]. The equivalence between Theorems 13.2.14 and 13.2.12 has been established in [26].

Corollary 13.2.15 *If $\{E_n\}$ is a sequence of measurable sets whose union is a bounded set E, then $\|E\| \leq \sum_{n=1}^{\infty} \|E_n\|$.*

PROOF. We may assume that $\sum_{n=1}^{\infty} \|E_n\| < +\infty$. Let $C_k = \bigcup_{n=1}^{k} E_n$ for $k = 1, 2, \ldots$, and choose a continuously differentiable vector field v on $\mathbf{R}^m$ such that $\|v(x)\| \leq 1$ for each $x \in \mathbf{R}^m$. Fatou's lemma and Exercise 13.2.11 imply that

$$\int_E \operatorname{div} v = \int_{\mathbf{R}^m} \chi_E \operatorname{div} v = \int_{\mathbf{R}^m} \lim(\chi_{C_k} \operatorname{div} v) \leq \liminf \int_{\mathbf{R}^m} \chi_{C_k} \operatorname{div} v$$

$$\leq \liminf \int_{C_k} \operatorname{div} v \leq \liminf \|C_k\| \leq \lim \sum_{n=1}^{k} \|E_n\| = \sum_{n=1}^{\infty} \|E_n\|$$

and the desired inequality follows. $\blacksquare$

Exercise 13.2.16   Deduce from Corollary 13.2.5 and Theorem 13.2.14 that a set $E \subset \mathbf{R}^m$ is measurable if and only if $\partial^*(\partial^* E) = \emptyset$.

Example 13.2.17   By Corollary 13.2.15 and Exercise 13.2.11, the sets E and $A - E$ of Example 11.7.9 are BV sets, and yet

$$|\partial E| = |\partial(A - E)| = |A - E| > 0.$$

For obvious reasons, we call the sets E "caviar" and the set $A - E$ "Swiss cheese." The complexity of BV sets can be appreciated by the following construction: take a piece of "Swiss cheese" and fill each hole with "caviar," then take each grain of the "caviar" and turn it into "Swiss cheese," and so on, finitely many times.

Another consequence of Theorem 13.2.14 is the following compactness result (see [9, Section 5.2, Theorem 4, and Section 1.3, Theorem 5, pp. 176 and 21]).

Theorem 13.2.18   *Let $\{E_n\}$ be a sequence of BV sets that are contained in a bounded subset of $\mathbf{R}^m$. If $\sup \|E_n\| < +\infty$, then there is a subsequence of $\{E_n\}$ converging to a BV set.*

It follows from [13, Theorem 1.24, p. 22] and Lemma 12.6.4 that for each BV set E there is a sequence $\{A_n\}$ of figures contained in a bounded subset of $\mathbf{R}^m$ such that $A_n \to E$. Combining this with Theorem 13.2.18 yields another characterization of BV sets.

Proposition 13.2.19   *A bounded set $E \subset \mathbf{R}^m$ is BV if and only if there is a sequence $\{A_n\}$ of figures such that $A_n \to E$.* ∎

In comparing Proposition 13.2.19 and Lemma 12.6.4, two points should be noted.

1. If a sequence of subfigures of a bounded set E converges to E, then E need not be a solid.
2. There may be no sequence $\{A_n\}$ of figures converging to a BV set E such that $A_n \subset E$ or $E \subset A_n$ for all n.

Remark 13.2.20   It is not difficult to find a nonmetrizable uniformity τ on the set $\mathcal{F}$ of all figures such that each additive function in a figure is continuous according to Definition 11.2.5 if and only if it is uniformly τ-continuous. Moreover, it follows from Theorem 13.2.18 and Proposition 13.2.19 that the family of all BV sets is the completion of the uniform space $(\mathcal{F}, \tau)$ (see [8, Theorem 3.11.12 and Problem 8.5.13, pp. 275 and 568]).

13.3 The flux

We say that BV sets B and C *overlap* whenever $|B \cap C| > 0$. The reason for this definition, which for solids coincides with that of Section 12.6, is the existence of nonnegligible BV sets whose interiors are empty (e.g., the set of all irrational numbers in $[0,1]$ or the set $A - E$ of Example 13.2.17 are such BV sets in $\mathbf{R}$ and $\mathbf{R}^2$, respectively). A function F defined on the family of all BV subsets of a set $E \subset \mathbf{R}^m$ is called

1. *additive* if $F(B \cup C) = F(B) + F(C)$ for all BV sets $B, C \subset E$ that do not overlap;
2. *continuous* if, given $\varepsilon > 0$, there is an $\eta > 0$ such that $|F(B)| < \varepsilon$ for each BV set $B \subset E$ with $\|B\| < 1/\varepsilon$ and $|B| < \eta$.

If F is additive or continuous according to the above definitions, then its restriction to the family of all subsolids of E is, respectively, additive or continuous in the sense of Section 12.6. In view of Proposition 13.2.19, a proof similar to that of Proposition 12.6.5 yields the following result.

Proposition 13.3.1 *Each additive continuous function F on the family $\mathcal{F}$ of all figures has a unique extension to an additive continuous function on the family of all BV sets.* ∎

Lemma 13.3.2 *Let F be an additive function defined on the family of all BV subsets of a set $E \subset \mathbf{R}^m$, and suppose that, given $\varepsilon > 0$, there is a $\delta > 0$ such that $|F(C)| \leq \varepsilon \|C\|$ for each BV set $C \subset E$ with $d(C) < \delta$. Then F is continuous.*

PROOF. Choose a positive $\varepsilon \leq 1$, and find a $\delta > 0$ so that

$$|F(B)| < \frac{\varepsilon^2}{4m} \|B\|$$

for each BV set $B \subset E$ with $d(B) \leq \delta$. Select a BV set $A \subset E$ with $\|A\| < 1/\varepsilon$ and $|A| < \delta$, and cover it by nonoverlapping cubes $B_1, \ldots, B_n$ of the same diameters equal to δ. Applying Lemma 13.2.13, we obtain

$$|F(A)| \leq \sum_{j=1}^{n} |F(A \cap B_j)| \leq \frac{\varepsilon^2}{4m} \sum_{j=1}^{n} \|A \cap B_j\| \leq \frac{\varepsilon^2}{2\delta} |A| + \frac{\varepsilon^2}{2} \|A\| < \varepsilon. \quad ∎$$

Let E be a BV set and let $v = (f_1, \ldots, f_m)$ be a continuous vector field on ∂E. We model the definition of the flux of v from E according to

Exercise 11.2.4. Choose an integer i with $1 \leq i \leq m$, and for each $\xi \in \mathbf{R}^{m-1}$ denote by E^ξ the section of E determined by ξ (see Exercise 11.2.2). It follows from Theorem 13.2.12 that for λ_{m-1}-almost all $\xi \in \mathbf{R}^{m-1}$, the section E^ξ differs from a one-dimensional figure $A(\xi)$ by a λ-negligible set. We choose such a point $\xi = (\xi_1, \ldots, \xi_{i-1}, \xi_{i+1}, \ldots, \xi_m)$ in $\mathbf{R}^{m-1}$ and let

$$f_i^\xi(t) = (f_i)^\xi(t) = f_i(\xi_1, \ldots, \xi_{i-1}, t, \xi_{i+1}, \ldots, \xi_m)$$

for each t in the ξ-section $(\partial E)^\xi$ of ∂E. Observe that $A(\xi) = \bigcup_{j=1}^{k(\xi)} [a_j^\xi, b_j^\xi]$ where $k(\xi) \geq 0$ is an integer and $a_1^\xi < b_1^\xi < \cdots < a_{k(\xi)}^\xi < b_{k(\xi)}^\xi$ are points of $\partial^*(E^\xi)$. Since $\partial^*(E^\xi) \subset \partial(E^\xi) \subset (\partial E)^\xi$, the number

$$f_i^\xi(E^\xi) = \sum_{j=1}^{k(\xi)} \left[f_i^\xi\left(b_j^\xi\right) - f_i^\xi\left(a_j^\xi\right) \right]$$

is well-defined, and we show that the function $\xi \mapsto f_i^\xi(E^\xi)$ belongs to $\mathcal{R}(\mathbf{R}^{m-1}, \lambda_{m-1})$. Indeed, if f_i is a polynomial, then

$$f_i^\xi(E^\xi) = \sum_{j=1}^{k(\xi)} \int_{a_j^\xi}^{b_j^\xi} \frac{d}{dt} f_i^\xi(t) \, d\lambda(t) = \int_{E^\xi} \left(\frac{\partial f_i}{\partial \xi_i} \right)^\xi (t) \, d\lambda(t)$$

by the fundamental theorem of calculus (Theorem 2.4.8), and the claim follows from Fubini's Theorem 10.1.13. In the general case, f_i is the uniform limit of a sequence $\{\varphi_n\}$ of polynomials by the Stone–Weierstrass approximation theorem ([42, Theorem 7.32, p. 162]). Thus

$$f_i^\xi(E^\xi) = \lim_{n \to \infty} \varphi_n^\xi(E^\xi) \quad \text{and} \quad |\varphi_n^\xi(E^\xi)| \leq \|E^\xi\| \left(1 + \sup_{x \in \partial E} |f(x)| \right)$$

for all $\xi \in \mathbf{R}^{m-1}$ and all sufficiently large n. In view of Theorem 13.2.12, the claim follows from the dominated convergence theorem.

Definition 13.3.3 In the above notation, the number

$$\int_{\partial E} v \cdot n \, d\lambda_{m-1} = \sum_{i=1}^m \int_{\mathbf{R}^{m-1}} f_i^\xi(E^\xi) \, d\lambda_{m-1}(\xi)$$

is called the *flux* of v from E.

We claim that Definition 13.3.3 extends the previous definitions of the flux of a vector field from a set. Indeed, the claim is true for figures by Exercise 11.2.4; this and Proposition 13.3.4 below imply that it is also true for solids. It follows directly from Definition 13.3.3 that Proposition 8.3.1 remains valid if A is a BV set.

Proposition 13.3.4 *Let w be a uniformly continuous vector field on $\mathbf{R}^m$, and let $F(E) = \int_{\partial E} w \cdot n \, d\lambda_{m-1}$ for each BV set E. Then F is an additive continuous function on the family of all BV sets.*

PROOF. The additivity is established by a direct verification left to the reader. Let $w = (f_1, \ldots, f_m)$ and $\varepsilon > 0$. There is a $\delta > 0$ such that $|f_i(x) - f_i(y)| \leq \varepsilon/m$ for all points $x, y \in \mathbf{R}^m$ with $|x - y| < \delta$ and $i = 1, \ldots, m$. If E is a BV set with $d(A) < \delta$, then

$$|F(E)| \leq \sum_{i=1}^{m} \int_{\mathbf{R}^{m-1}} |f_i^\xi(E^\xi)| \, d\lambda_{m-1}(\xi) \leq \frac{\varepsilon}{m} \sum_{i=1}^{m} \int_{\mathbf{R}^{m-1}} \|E^\xi\| \, d\lambda_{m-1}(\xi)$$

$$= \frac{\varepsilon}{m} \sum_{i=1}^{m} \|E\|_i \leq \varepsilon \|E\|.$$

An application of Lemma 13.3.2 completes the argument. ∎

Now if E is a BV set and v is a continuous vector field on ∂E, we can give an alternative definition of the flux of v from E. Indeed, using Tietze's extension theorem ([8, Theorems 2.1.8, p. 97]) and Lemma 10.4.1, v can be extended to a uniformly continuous vector field w on $\mathbf{R}^m$. According to Proposition 13.2.19, there is a sequence $\{A_n\}$ of figures converging to E, and it follows from Proposition 13.3.4 that

$$\int_{\partial E} v \cdot n \, d\lambda_{m-1} = \lim \int_{\partial A_n} w \cdot n \, d\lambda_{m-1} .$$

Remark 13.3.5 We note without proof that the flux of a vector field v from a BV set E depends only on v restricted to $\partial^* E$. The proof of this nontrivial fact is given in [9, Chapter 5] or [49, Chapter 5] (cf. Remark 13.1.2).

13.4 The BV-integral

At the beginning of Section 13.2 we suggested that defining the integral by means of solids may not be appropriate. Indeed, this is true: the discussion of the $\mathcal{SF}$-integral and $\mathcal{S}$-integral was included for motivational purposes only. Considering the facts presented in Sections 13.2 and 13.3, there is little doubt that the correct definition of the integral must be based on a family of BV sets whose topological boundaries are not restricted. We sketch the main ideas but omit the details, which are often formidable.

Let $\mathcal{BV}$ denote the family of all BV sets $E \subset \mathbf{R}^m$ such that $E^{-*} - E$ is a thin set. In view of Exercise 13.2.6, this is not a severe restriction. Note that $\mathcal{S} \subset \mathcal{BV}$, and that $E^{-*} \in \mathcal{BV}$ for each BV set E (see Exercise 13.2.7). Also note that the family $\mathcal{BV}$ is closed with respect to finite unions and intersections. Given $\varepsilon > 0$ and a caliber η, a set $E \in \mathcal{BV}$ is called $(\varepsilon, \eta; \mathcal{BV})$-*small* if E is the union of nonoverlapping sets $E_1, \ldots, E_k$ from $\mathcal{BV}$ such that $\|E_j\| < 1/\varepsilon$ and $|E_j| < \eta_j$ for $j = 1, \ldots, k$. Obviously, each $(\varepsilon, \eta; \mathcal{S})$-small solid is $(\varepsilon, \eta; \mathcal{BV})$-small, but we do not know whether the converse is true. The *regularity* of a set $E \in \mathcal{BV}$ is the number

$$r(E) = \begin{cases} \dfrac{|E|}{d(E)\|E\|} & \text{if } d(E)\|E\| > 0, \\ 0 & \text{otherwise.} \end{cases}$$

For a figure A, the number $r(A)$ defined above coincides with that introduced in Section 11.2. On the other hand, if S is a solid, then $\|S\| \leq \mathcal{H}(\partial S)$ and Remark 12.5.4 implies that $r_S(S) \leq \omega_{m-1} m^{(m-1)/2} r(S)$; these inequalities are generally strict.

A $\mathcal{BV}$-*partition* is a collection, possibly empty, $\{(E_1, x_1), \ldots, (E_p, x_p)\}$ such that $E_1, \ldots, E_p$ are nonoverlapping elements of $\mathcal{BV}$ and $x_i \in E_i$ for $i = 1, \ldots, p$. For an $E \in \mathcal{BV}$, we define a δ-fine ε-regular $\mathcal{BV}$-partition of E mod $(\varepsilon, \eta; \mathcal{BV})$ in the obvious way. A proof that such a partition actually exists is given in [38, Section 2] by a nontrivial elaboration of Proposition 11.3.7.

Denote by BV the family of all BV subsets of $\mathbf{R}^m$. It is clear that the above concepts defined in terms of $\mathcal{BV}$ have their counterparts defined in terms of BV. The following example shows, however, that for an $E \in BV$, a δ-fine BV-partition of E mod $(\varepsilon, \eta; BV)$ need not exist, even when δ is a positive function and ε-regularity is not required.

Example 13.4.1   Let $m = 1$, and let C and α be, respectively, the Cantor ternary set and function (see Remark 5.3.12). For each BV set $B \subset [0,1]$ there is a unique figure $\hat{B} \subset [0,1]$ such that $\lambda[(B - \hat{B}) \cup (\hat{B} - B)] = 0$. The map $\hat{\alpha} : B \mapsto \alpha(\hat{B})$ is an additive continuous function on the family of all BV subsets of $[0,1]$. Choose a positive $\varepsilon < 1$ and use Proposition 6.7.3 to find a caliber η so that $\alpha(B) < \varepsilon$ for each figure $B \subset [0,1]$ that is (ε, η)-small. On a BV set $E = [0,1] - C$ define a positive function $\delta : x \mapsto \text{dist}(x, C)$, and assume that a δ-fine BV-partition P of E mod $(\varepsilon, \eta; BV)$ exists. Without difficulty we obtain that $\hat{\alpha}(\bigcup P) = 0$ and $\hat{\alpha}(E - \bigcup P) < \varepsilon$. Since α is additive and $\hat{\alpha}(E) = 1$, we have arrived at a contradiction.

Definition 13.4.2 A function f on a set $E \in \mathcal{BV}$ is called $\mathcal{BV}$-*integrable* over E if there is a real number I having the following property: given $\varepsilon > 0$, we can find a gage δ on E and a caliber η so that

$$|\sigma(f, P) - I| < \varepsilon$$

for each δ-fine ε-regular $\mathcal{BV}$-partition P of E mod $(\varepsilon, \eta; \mathcal{BV})$.

By $\mathcal{BVR}_*(E) = \mathcal{BVR}_*(E, \lambda_m)$ we denote the family of all $\mathcal{BV}$-integrable functions over a set $E \in \mathcal{BV}$. The number I of Definition 13.4.2 is determined uniquely by the function $f \in \mathcal{BVR}_*(E)$; we call it the $\mathcal{BV}$-*integral* of f over E, denoted by $\int_E f$ or $\int_E f \, d\lambda_m$. The reader can verify that this is a consistent notation. It is easy to see that $\mathcal{BVR}_*(S) \subset \mathcal{SR}_*(S)$ for every solid S and that equality holds if $m = 1$.

For a set $E \in \mathcal{BV}$, we define the family $\overline{\mathcal{BVR}}_*(E) = \overline{\mathcal{BVR}}_*(E, \lambda_m)$ in the usual way. While most of the standard properties can be established for the $\mathcal{BV}$-integral, we note that only a weaker version of Proposition 11.4.4 is true. The interested reader is referred to [37, Example 5.21]. We restrict our attention to outlining the proofs of the divergence and change of variables theorems.

Proposition 13.4.3 *Let E be a BV set, and let w be a differentiable vector field defined on an open set U containing E^-. If $\operatorname{div} w$ belongs to $\mathcal{R}(E, \lambda_m)$, then*

$$\int_E \operatorname{div} w \, d\lambda_m = \int_{\partial E} w \cdot n \, d\lambda_{m-1}.$$

PROOF. Let $w = (g_1, \dots, g_m)$. The fundamental theorem of calculus implies that

$$g_i^\xi(E^\xi) = \int_{E^\xi} \frac{d}{dt} g_i^\xi(t) \, d\lambda(t) = \int_{E^\xi} \left(\frac{\partial g_i}{\partial \xi_i} \right)^\xi (t) \, d\lambda(t)$$

for each $\xi \in \mathbf{R}^{m-1}$ and $i = 1, \dots, m$. Since $\operatorname{div} w$ belongs to $\mathcal{R}(E, \lambda_m)$, by Fubini's Theorem 10.1.13, we obtain

$$\int_{\partial E} v \cdot n \, d\lambda_{m-1} = \sum_{i=1}^{m} \int_{\mathbf{R}^{m-1}} g_i^\xi(E^\xi) \, d\lambda_{m-1}(\xi)$$

$$= \sum_{i=1}^{m} \int_{\mathbf{R}^{m-1}} \left[\int_{E^\xi} \left(\frac{\partial g_i}{\partial \xi_i} \right)^\xi (t) \, d\lambda(t) \right] d\lambda_{m-1}(\xi)$$

$$= \sum_{i=1}^{m} \int_E \frac{\partial g_i}{\partial \xi_i} \, d\lambda_m = \int_E \operatorname{div} w \, d\lambda_m. \quad \blacksquare$$

Using Proposition 13.4.3, the reader can verify that Lemma 8.3.4 and its proof hold for any BV set B. Thus, together with Lemma 12.2.3, we have the ingredients necessary for proving the divergence theorem.

If v is a vector field defined on a set $E \subset \mathbf{R}^m$ that is almost differentiable at each $x \in E$, then v is σ-Lipschitz by Proposition 10.6.8. Following the pattern described at the beginning of Section 10.6, we can define $\operatorname{div} v$ almost everywhere in E. To connect $\operatorname{div} v$ with the flux of v, we need the following proposition, which is a version of Whitney's extension theorem (cf. [48, Chapter VI, Section 2]). It is established by scrutinizing the proof of [48, Theorem 3, Chapter VI, p. 174].

Proposition 13.4.4 *A continuous function f on a closed set $C \subset \mathbf{R}^m$ can be extend to a continuous function g on $\mathbf{R}^m$ so that g is almost differentiable at each $x \in C$ at which f is almost differentiable.* ∎

Theorem 13.4.5 *Let $A \in \mathcal{BV}$ and let T be a thin set. If v is a continuous vector field on A^- that is almost differentiable at each point of $A^{\circ *} - T$, then $\operatorname{div} v$ belongs to $\overline{\mathcal{BVR}}_*(A, \lambda_m)$ and*

$$\int_A \operatorname{div} v \, d\lambda_m = \int_{\partial A} v \cdot n \, d\lambda_{m-1}.$$

PROOF. Note that v is continuous or almost differentiable if and only if its components are continuous or almost differentiable, respectively. Thus, by Proposition 13.4.4, the vector field v has a continuous extension w defined on $\mathbf{R}^m$ that is almost differentiable at each point of $A^{\circ *} - T$. In view of Lemma 10.4.1, we may asume that w is uniformly continuous. It follows from Stepanoff's Theorem 10.6.7 and Lemma 10.5.5 that almost everywhere in $A^{\circ *} - T$, the vector field w is differentiable and $\operatorname{div} v = \operatorname{div} w$. Proposition 13.3.4 implies that the function $F : B \mapsto \int_{\partial B} w \cdot n \, d\lambda_{m-1}$ is additive and continuous on the family BV of all BV sets. Since $F(A) = \int_{\partial A} v \cdot n \, d\lambda_{m-1}$, it suffices to show that $\operatorname{div} w$ belongs to $\overline{\mathcal{BVR}}_*(A)$ and $\int_A \operatorname{div} w \, d\lambda_m = F(A)$. But this is completely analogous to the proof of Theorem 12.2.5. ∎

Lemma 13.4.6 *Let $\Phi : E \to \mathbf{R}^m$ be a lipeomorphism of a set $E \subset \mathbf{R}^m$. Then Φ has a unique extension to a lipeomorphism $\Psi : E^- \to \mathbf{R}^m$ and $\Psi(E^{-*}) = [\Phi(E)]^{-*}$.*

PROOF. Avoiding a triviality, assume that the set E is not empty. By Exercise 10.3.5, the lipeomorphism Φ has a unique extension to a lipeomorphism

$\Psi : E^- \to \mathbf{R}^m$. If $a = 1/\mathrm{Lip}(\Phi^{-1})$ and $b = \mathrm{Lip}(\Phi)$, then

$$a|x - x'| \leq |\Phi(x) - \Phi(x')| \leq b|x - x'|$$

for each $x, x' \in E^-$. Select a point $x \in E^- - E^{-*}$ and let $y = \Psi(x)$. Since

$$\Phi(E) \cap U[y, \varepsilon] \subset \Phi(E \cap U[x, \varepsilon/a]),$$

$$|\Phi(E \cap U[x, \varepsilon/a])| \leq b^m |E \cap U[x, \varepsilon/a]|$$

for each $\varepsilon > 0$, we obtain

$$\limsup_{\varepsilon \to 0+} \frac{|\Phi(E) \cap U[y, \varepsilon]|}{|U[y, \varepsilon]|} \leq \left(\frac{b}{a}\right)^m \limsup_{\varepsilon \to 0+} \frac{|E \cap U[x, \varepsilon/a]|}{|U[x, \varepsilon/a]|} = 0.$$

Consequently y is a dispersion point of $\Phi(E)$ and the lemma follows by symmetry. ∎

Under the assumptions of Lemma 13.4.6, it is also true that $\Psi(E^{\circ *}) = [\Phi(E)]^{\circ *}$ whenever the set E is measurable; a proof, which is nontrivial, can be found in [5]. In combination with Lemma 13.4.6, this implies that $\Psi(\partial^* E) = \partial^*[\Phi(E)]$. Using Exercise 12.7.2 and Corollary 12.7.5, 2, we obtain the following proposition needed in the proof of the change of variables theorem.

Proposition 13.4.7 *Let $E \in \mathcal{BV}$ and let $\Phi : E \to \mathbf{R}^m$ be a lipeomorphism. Then $\Phi(E) \in \mathcal{BV}$ and $\|\Phi(E)\| \leq [\mathrm{Lip}_e(\Phi)]^{m-1}\|E\|$.* ∎

Now the proof of the next theorem is similar to that of Theorem 13.1.8, and we leave it to the reader.

Theorem 13.4.8 *Let $A \in \mathcal{BV}$, let $\Phi : A \to \mathbf{R}^m$ be a lipeomorphism, and let f be an extended real-valued function defined almost everywhere in $\Phi(A)$. Then $f \in \overline{\mathcal{BVR}}_*(\Phi(A))$ if and only if $(f \circ \Phi)\mathcal{J}_\Phi$ belongs to $\overline{\mathcal{BVR}}_*(A)$, in which case*

$$\int_A (f \circ \Phi)\mathcal{J}_\Phi = \int_{\Phi(A)} f.$$ ∎

Remark 13.4.9 For every figure A, we have the inclusions

$$\mathcal{R}(A) \underset{\neq}{\subset} \mathcal{BVR}_*(A) \subset \mathcal{SR}_*(A) \subset \mathcal{FR}_*(A) \underset{\neq}{\subset} \mathcal{R}_*(A).$$

If $m = 1$, then each g-integrable function over a figure A is integrable over a subcell of A ([44, Chapter VII, Theorem (1.4), p. 244]). On the other hand, Z. Buczolich constructed a function $f \in \mathcal{BVR}_*([0,1]^2)$ that is integrable over *no* two-dimensional subcell of $[0,1]^2$ (see [6]). This shows that for $m \geq 2$, the $\mathcal{BV}$-integral over figures cannot be obtained from the McShane integral by a process similar to Denjoy's constructive definition (see [44, Chapter VIII, Section 5]). In view of the above inclusions, the same is true for the $\mathcal{SF}$-integral, $\mathcal{F}$-integral, and g-integral.

Bibliography

[1] M. Artin. *Algebra*. Prentice-Hall, Englewood Cliffs, 1991.

[2] A.S. Besicovitch. On sufficient conditions for a function to be analytic, and behaviour of analytic functions in the neighbourhood of non-isolated singular points. *Proc. London Math. Soc.*, 32:1–9, 1931.

[3] P. Billingsley. *Probability and Measure*. John Wiley, New York, 1979.

[4] B. Bongiorno, M. Giertz, and W.F. Pfeffer. Some conditionally convergent integrals in the real line. *Boll. Un. Mat. Italiana*, 7:371–402, 1992.

[5] Z. Buczolich. Density points and bi-Lipschitz functions in $\mathbf{R}^m$. *Proc. Am. Math. Soc.*, 116:53–56, 1992.

[6] Z. Buczolich. A v-integrable function which is not Lebesgue integrable on any portion of the unit square. *Acta Math. Hung.*, 59:383–393, 1992.

[7] Z. Buczolich. The g-integral is not rotation invariant. *Real Analysis Exchange*, 18:2:437–447, 1992–93.

[8] R. Engelking. *General Topology*. PWP, Warsaw, 1977.

[9] L.C. Evans and R.F. Gariepy. *Measure Theory and Fine Properties of Functions*. CRC Press, London, 1992.

[10] K.J. Falconer. *The Geometry of Fractal Sets*. Cambridge Univ. Press, Cambridge, 1985.

[11] H. Federer. *Geometric Measure Theory*. Springer-Verlag, New York, 1969.

[12] J. Foran and S. Meinershagen. Some answers to a question of P. Bullen. *Real Analysis Exchange*, 13:1:265–277, 1987–88.

[13] E. Giusti. *Minimal Surfaces and Functions of Bounded Variation*. Birkhäuser, Boston, 1984.

[14] R. Henstock. *Theory of Integration*. Butterworth, London, 1963.

[15] R. Henstock. *Lectures on the Theory of Integration*. World Scientific, Singapore, 1988.

[16] R. Henstock. *The General Theory of Integration*. Clarendon Press, Oxford, 1991.

[17] M.W. Hirsch. A proof of nonretractability of a cell onto its boundary. *Proc. American Math. Soc.*, 14:364–365, 1963.

[18] E.J. Howard. Analyticity of almost everywhere differentiable functions. *Proc. American Math. Soc.*, 110:745–753, 1990.

[19] J. Jarník and J. Kurzweil. A nonabsolutely convergent integral which admits C^1-transformations. *Časopis Pěst. Mat.*, 109:157–167, 1984.

[20] J. Jarník and J. Kurzweil. A nonabsolutely convergent integral which admits transformation and can be used for integration on manifolds. *Czechoslovak Math. J.*, 35:116–139, 1986.

[21] J. Jarník and J. Kurzweil. A new and more powerful concept of the PU-integral. *Czechoslovak Math. J.*, 38:8–48, 1988.

[22] L. Kuipers and H. Niederreiter. *Uniform Distribution of Sequences*. John Wiley, New York, 1974.

[23] J. Kurzweil. *Nichtabsolut konvergente Integrale*. Teubner, Leibzig, 1980.

[24] J. Kurzweil and J. Jarník. The PU-integral: its definition and some basic properties. In *New Integrals*, pages 66–81. P.S. Bullen, P.Y. Lee, J.L. Mawhin, P. Muldowney, and W.F. Pfeffer, eds., Lecture Notes in Math. 1419, Springer-Verlag, New York, 1990.

[25] J. Kurzweil, J. Mawhin, and W.F. Pfeffer. An integral defined by approximating BV partitions of unity. *Czechoslovak Math. J.*, 41:695–712, 1991.

[26] J. Mařík. The surface integral. *Czechoslovak Math. J.*, 6:522–558, 1956.

[27] R.M. McLeod. *The Generalized Riemann Integral*. Mathematical Association of America, Washington, D.C., 1980.

[28] E.J. McShane. A Riemann-type integral that includes Lebesgue–Stieltjes, Bochner and stochastic integrals. *Mem. American Math. Soc.*, 88, 1969.

[29] E.J. McShane. A unified theory of integration. *American Math. Monthly*, 80:349–359, 1973.

[30] E.J. McShane. *Unified Integration*. Academic Press, New York, 1983.

[31] J.W. Mortensen and W.F. Pfeffer. Multipliers for the generalized Riemann integral. *J. Math. Anal. Appl.*, in press.

[32] A. Novikov and W.F. Pfeffer. An invariant Riemann type integral defined by figures. *Proc. American Math. Soc.*, in press.

[33] W.F. Pfeffer. *Integrals and Measures*. Marcel Dekker, New York, 1977.

[34] W.F. Pfeffer. The divergence theorem. *Trans. American Math. Soc.*, 295:665–685, 1986.

[35] W.F. Pfeffer. A note on the generalized Riemann integral. *Proc. American Math. Soc.*, 103:1161–1166, 1988.

[36] W.F. Pfeffer. A descriptive definition of a variational integral and applications. *Indiana Univ. Math. J.*, 40:259–270, 1991.

[37] W.F. Pfeffer. The Gauss–Green theorem. *Advances Math.*, 87:93–147, 1991.

[38] W.F. Pfeffer. A Riemann type definition of a variational integral. *Proc. American Math. Soc.*, 114:99–106, 1992.

[39] W.F. Pfeffer. The Riemann–Stieltjes approach to integration. *NRIMS-CSIR*, TWISK 187:1980, Pretoria.

[40] W.F. Pfeffer and B.S. Thomson. Measures defined by gages. *Canadian J. Math.*, 44:1303–1316, 1992.

[41] H.L. Royden. *Real Analysis*. Macmillan, New York, 1968.

[42] W. Rudin. *Principles of Mathematical Analysis*. McGraw-Hill, New York, 1976.

[43] W. Rudin. *Real and Complex Analysis*. McGraw-Hill, New York, 1987.

[44] S. Saks. *Theory of the Integral*. Dover, New York, 1964.

[45] W.L.C. Sargent. On the integrability of a product. *J. London Math. Soc.*, 23:28–34, 1948.

[46] V.L. Shapiro. The divergence theorem for discontinuous vector fields. *Ann. Math.*, 68:604–624, 1958.

[47] E.H. Spanier. *Algebraic Topology*. McGraw-Hill, New York, 1966.

[48] E.M. Stein. *Singular Integrals and Differentiability Properties of Functions*. Princeton Univ. Press, Princeton, 1970.

[49] W.P. Ziemer. *Weakly Differentiable Functions*. Springer-Verlag, New York, 1989.

List of symbols

■   Denotes the end of a proof, or the end of a statement for which no proof has been given.

Euclidean spaces

$\mathbf{R}$	3		$\overline{\mathbf{R}}$	37
$\mathbf{R}^m$	133		$n^i_+,\ n^i_-$	148
$U(x,r)$	4, 133		$U[x,r]$	4, 133
$\operatorname{dist}(A,B)$	133		$x \cdot y$	133
$|x|$	133		$\|x\|$	133
$d(A)$	133		$d_e(A)$	189

Partitions and integrals

$\bigcup P$	125		$\sigma(f,P;\alpha)$	7, 141
$\int_A f\,d\alpha$	9, 144		$\int_a^b f\,d\alpha$	9
$\int_A^* f$	219		$\int_E f d\lambda_m$	182
$U(f,A;\alpha)$ $L(f,A;\alpha)$	72, 159		$\int_{\partial A} v \cdot n\,d\lambda_{m-1}$	149, 212 274, 286
$\operatorname{mod}(\varepsilon,\eta;\mathcal{S})$	275		$\operatorname{mod}(\varepsilon,\eta)$	216
$\operatorname{mod}(\varepsilon,\eta;\mathcal{BV})$	288		$\operatorname{mod}(\varepsilon,\eta;BV)$	288

Sets

E^-, E°			E^{-*}, $E^{\circ*}$	
∂E	4, 133		$\partial^* E$	279
A^i_+, A^i_-	148		$A_{(i)}$	148, 211
E_x, E^y	164		E^ξ	211, 286
$A_n \to X$	225		$|E|$	168
$A^\bullet$	86, 170		$A^\star$	171
$s(A)$	139, 246		$r(A)$	239, 288
$r_S(S)$	274		$\|A\|_i$	211, 282
$\|A\|$	125, 148 210, 281		$A \odot B$ $A \ominus B$	125, 210

Functions and measures

λ	5		λ_m	139
λ_0	210		f^+, f^-	14
$f_n \nearrow f$ $f_n \searrow f$	18		$\alpha(c+)$ $\alpha(c-)$	37
α^δ	38, 153		α^*	38, 153
α_A	60, 157		χ_E	59, 156
$F \otimes G$	137		$f \otimes g$	166
f_x, f^y	161		f^ξ	213, 286
f^E	180		h_η, h	253, 254
$\mathcal{H}_\eta$, $\mathcal{H}$	255		ω_m	189
$V(F, B)$	81, 168		VF	82, 169
$\mathrm{Lip}(F)$	121, 190			

Families of sets and functions

$\mathcal{S}$	273		$\mathcal{F}$	242
BV	288		$\mathcal{BV}$	288
$\mathcal{R}(A, \alpha)$	9, 144		$\overline{\mathcal{R}}(A, \alpha)$	68, 158
$\mathcal{R}_*(A, \alpha)$	103		$\overline{\mathcal{R}}_*(A, \alpha)$	112
$\mathcal{R}(E)$	182		$\overline{\mathcal{R}}(E)$	183
$\mathcal{R}_*(A)$	219		$\overline{\mathcal{R}}_*(A)$	222
AC	83, 169		AC_*	115, 229
$\mathcal{FR}_*(A)$ $\overline{\mathcal{FR}}_*(A)$	243		$\mathcal{SFR}_*(S)$ $\overline{\mathcal{FSR}}_*(S)$	273
$\mathcal{SR}_*(S)$ $\overline{\mathcal{SR}}_*(S)$	275		$\mathcal{BVR}_*(E)$ $\overline{\mathcal{BVR}}_*(E)$	289

Differentiation

F'	25, 175		$\operatorname{grad} f$	150
$D\Phi$	150		$\det \Phi$	193
J_Φ	191, 199		$\mathcal{J}_\Phi$	202, 203
$\operatorname{div} v$	151, 290		$\operatorname{DIV} v$	233
$d_v f, \overline{d}_v f$	195		F'_α	91, 175
$\mathcal{D}^\alpha F(x)$ $\mathcal{D}_\alpha F(x)$	71, 158		$D^\alpha F(x)$ $D_\alpha F(x)$	91, 174
$F_\alpha^{sym}(x)$	100, 179		$\mathcal{F}D_{\lambda_m}F(x)$	247
$\mathcal{F}F'(x)$	248		$|D\eta|F(x)$	230
$\overline{\Theta}(E, x)$ $\underline{\Theta}(E, x)$ $\Theta(E, x)$	118, 200			

Index

almost
  differentiable, 120, 205
  derivable, 230
    relative to $\mathcal{F}$, 251
  everywhere, 56
  all, 56

bi-Lipschitz map, 190
boundary, 4, 133
  essential, 279
  reduced, 279
Brouwer theorem
  on fixed point, 199
  on domain invariance, 262

caliber, 126, 216
Cantor
  function, 96
  set, 96
cell, 3, 134
  dyadic, 53
  face of, 148, 211
  vertex of, 211
change of variables, 28, 187, 202, 204,
    262, 277, 291
concassage, 203
convergence
  dominated, 20, 113
  of figures, 225
  of functions, 18
  monotone, 18, 68, 113
countable
  additivity, 48
  set, 3
  subadditivity, 40
cube, 134
  dyadic, 134
    fundamental property of, 134

density, 118, 200
  point, 118, 200

derivate, 91, 174
  Lebesgue, 71, 158
  relative to $\mathcal{F}$, 248
  relative to $\mathcal{S}$, 277
  symmetric, 100
derivative, 25
  directional, 195
  partial, 150
diameter, 133
  Euclidean, 189
differentiable
  function, 25
  map, 150
  vector field, 151
Dirac
  length, 146
  volume, 146
distance, 133
divergence, 151, 290
  mean, 233
  theorem, 152, 224, 232, 233, 237, 244,
      274, 276, 290
division, 81, 135, 211
  cellular, 211
dyadic
  cell, 53
  cube, 134
    fundamental property of, 134

enumeration, 3
equilipschitz sequence, 271
essential
  boundary, 279
  closure, 279
  interior, 279
exterior normal, 149, 274

figure, 125, 210
  (ε, η)-small, 126, 216
  regularity of, 239
  shape of, 239

flux, 149
  from
    BV set, 286
    cell, 149
    figure, 212, 213
    solid, 274
function, 4
  AC, 83, 169
  AC$_*$, 115, 229
  additive
    associated, 4
    of BV sets, 285
    continuous, 213, 258, 285
    of figures, 126, 212
    of intervals, 4, 136
    of solids, 258
  almost derivable, 230
    relative to $\mathcal{F}$, 251
  of bounded variation, 82, 168, 228
  Cantor, 96
  characteristic, 59, 156
  decreasing, 4
  derivable, 91, 175
    relative to $\mathcal{F}$, 248
    relative to $\mathcal{S}$, 277
  differentiable, 25
    almost, 120
  $\mathcal{F}$AC$_*$, 249
  gap, 30, 50, 107
  extended real-valued, 38
    continuous, 69
  increasing, 4
  integrable
    $\mathcal{BV}$-, 289
    $\mathcal{F}$-, 243
    g-(gage), 219, 222
    Henstock–Kurzweil, 102, 112
    McShane(–Stieltjes), 8, 68, 143,
      158, 180, 182
    P-, 102, 112
    $\mathcal{S}$-, 275
    $\mathcal{SF}$-, 273
  Lipschitz, 15, 121
  measurable, 56, 156
  monotone, 4
  semicontinuous, 69

    simple, 59
    singular, 98
    variation of 81, 168

gage, 117, 215
  null set of, 117, 215
gradient, 151

Hausdorff measure, 255
homeomorphism, 190, 262
hyperplane, 155

integral, 7, 9, 143
  $\mathcal{BV}$-, 289
  conditionally convergent, 103
  depending on parameter, 23
  descriptive definition of
    full, 94, 97, 177
    partial, 93, 115, 229, 249
  $\mathcal{F}$-, 243
  double, 161
  g-(gage), 219, 222
  generalized Riemann, 9
  Henstock–Kurzweil, 102, 112
  improper, 105, 225
  indefinite, 81, 112, 222, 243
  iterated, 161
  Lebesgue, 80
  McShane(–Stieltjes), 9, 68, 143, 158,
    180, 182
  P-, 102, 112
  $\mathcal{S}$-, 275
  $\mathcal{SF}$-, 273
integration by parts, 32, 108, 251
interval, 3, 134
  degenerate, 3, 134

lemma
  Cousin, 6, 140
  Fatou, 20
  Henstock, 17
Lebesgue
  decomposition, 99, 178
  integral, 80
  length, 5
  measure, 81, 168

partition, 5, 139
volume, 138
length, 5
  Dirac, 146
  Lebesgue, 5
lipeomorphism, 190
Lipschitz
  constant, 121, 190
  function, 15, 121
  map, 190
Luzin
  condition, 185
  map, 185

majorant, 72, 159
map, 3
  bi-Lipschitz, 190
  determinant of, 193, 199, 202, 203
  diagonal, 192
  differentiable, 150
    almost, 205
  differential of, 150
  empty, 3
  Lipschitz, 190
  Luzin, 185
  σ-Lipschitz, 203
measurable
  function, 56, 156
  set, 44, 154
measure
  countable additivity of, 48
  countable subadditivity of, 40
  Hausdorff, 255
  in $\mathbf{R}$, 40
  in $\mathbf{R}^m$, 153
  Lebesgue, 81, 168
    Hausdorff evaluation of, 188
  metric, 40
  monotonicity of, 40
  normalization of, 40
  of set, 38, 153
  on σ-algebra, 48
  outer, 40
  quasi-Hausdorff, 254
metric, 133

Euclidean, 133
minorant, 72, 159

negligible set, 55, 156
normal, 149, 274
null set, 117, 215

overlapping
  BV sets, 285
  figures, 125, 210
  intervals, 3, 134
  solids, 258

partition, 5, 139
  anchored in, 38, 139
  body of, 125, 139
  $\mathcal{BV}$-, 288
  δ-fine, 5, 117, 139, 216
  empty, 6
  ε-shapely, 139
  ε-regular, 243
  $\mathcal{F}$-, 242
  in, 6, 139
  Lebesgue, 5, 139
  of, 6, 139
    A mod (ε, η), 127, 216
    E mod $(\varepsilon, \eta; \mathcal{BV})$, 288
    S mod $(\varepsilon, \eta; \mathcal{S})$, 275
  P-(Perron), 5, 139
  $\mathcal{S}$-, 275
perimeter of
  bounded set, 281
  cell, 148
  figure, 125, 210, 211
point of
  density, 118, 200
  dispersion, 118, 200

quasi-Hausdorff measure, 254

reduced boundary, 279
regularity of
  BV set, 288
  figure, 239
rotation, 192

segment, 3
segmentation, 41, 136
shape, 139, 239, 246
set
  of bounded variation (BV), 281
    $(\varepsilon, \eta; \mathcal{BV})$-small, 288
  Cantor, 96
  countable, 3
  enumeration of, 3
  F_σ, G_δ, 50
  measurable, 44, 154
  negligible, 55, 156
  null, 117, 215
  projection of, 211
  relatively open, 187
  section of, 211
  shape of, 246
  thin, 214
σ-algebra, 48
σ-Lipschitz map, 203
singular
  function, 98
  volume, 178
solid, 258
  $(\varepsilon, \eta, \mathcal{S})$-small, 275
  $\mathcal{S}$-regularity of, 274
Stieltjes sum, 7, 142, 218

test
  Cauchy, 10
  Carathéodory, 52
  Lebesgue, 79
  Perron, 76, 159

theorem
  Brower
    on fixed point, 199
    on domain invariance, 262
  change of variables, 28, 187, 202, 204,
    262, 277, 291
  covering, 86, 170, 246
  divergence, 152, 224, 232, 233, 237,
    244, 274, 276, 290
  dominated convergence, 20, 113
  Fubini, 165, 184
  fundamental of calculus, 28, 103, 124
  integration by parts, 30, 108, 251
  Kirszbraun, 190
  Lebesgue decomposition, 99, 178
  mean value, 34, 109
  monotone convergence, 18, 68, 113
  Rademacher, 198
  Stepanoff, 123, 205
  Tonelli, 165
  Vitali, 87, 171, 246
  Vitali–Carathéodory, 78

vector field, 149
  divergence of, 151, 233
  fluxing, 237
vertex, 211
Vitali
  cover, 87, 171, 246
    centered, 90
  covering theorem, 87, 171, 246
volume, 138
  Dirac, 146
  Lebesgue, 138
  singular, 178